Perceived Exertion Laboratory Manual

Luke Haile • Michael Gallagher, Jr.
Robert J. Robertson

Perceived Exertion Laboratory Manual

From Standard Practice to Contemporary Application

Springer

Luke Haile
Lock Haven University of Pennsylvania
Lock Haven, PA, USA

Michael Gallagher, Jr.
University of Central Arkansas
Conway, AR, USA

Robert J. Robertson
University of Pittsburgh
Pittsburgh, PA, USA

ISBN 978-1-4939-1916-1 ISBN 978-1-4939-1917-8 (eBook)
DOI 10.1007/978-1-4939-1917-8
Springer New York Heidelberg Dordrecht London

Library of Congress Control Number: 2014948551

Printed on acid-free paper

Springer is part of Springer Science+Business Media (www.springer.com)

I dedicate this book to a few very important people in my life. First, to my parents, for always pushing me to do my best while pursuing my dreams every step of the way. Second, to the teachers who had the most profound influence on me throughout my education: Greg Laubach, Joseph Andreacci, Robert Robertson, and Fredric Goss. Each of them showed me how to be an educator, researcher, and advisor by being an all-around good man. Third, to my wife, Amanda, and son, Samuel, for being my unending motivation to live with purpose and passion.

LH

I dedicate this book to my family, friends, teachers, and students. Each of them continually challenges me to be the best version of myself; as a son, a brother, a colleague, and as a mentor. I would not be where I am today without their love and support.

MG

To my wife Margaret, whose sage advice and sound judgment have supported me throughout my professional career.

RJR

Preface

This unique laboratory manual contains a series of conceptually linked research paradigms that are based on an empirical framework for tailoring individualized exercise programs to client perceptions and behaviors. The various laboratory experiments are teaching tools that analyze perceptual and psychosocial variables that influence participation in physical activities, describe methods for assessing these factors in clients, examine guidelines for exercise prescription and program evaluation, and feature practical applications of current research. The rationale underlying each of the experiments is based on a combination of scientific findings and psychological insights that will teach students and practitioners how to create effective strategies for increasing physical activity in clients at various stages of health and illness. The manual is formatted such that the experiments can be performed individually or sequentially in groups of two or more depending on the extent of the knowledge base that informs the learning experience.

In the late 1950s and early 1960s, research was conducted by Swedish psychologist Gunnar Borg, together with medical doctor and clinical physiologist Hans Dahlstrom, that explored psychophysics and perception in the context of sports and occupational endeavors. Borg's 1962 monograph of his doctoral dissertation research, *Physical Performance and Perceived Exertion*, introduced the field of perceived exertion to the world. Bruce Noble invited Borg to visit the University of Pittsburgh in the fall of 1967, marking the beginning of perceived exertion research in the United States. The following spring of 1968, Borg continued his trip with a short drive Northeast to the Pennsylvania State University, where he visited Ellsworth Buskirk, James Skinner, and Oded Bar-Or. The visit culminated with the American College of Sports Medicine annual meeting, which was held in State College, Pennsylvania. The presentations made at this conference, most notably by Borg, Noble, and Michael Sherman, were the first broad exposure of perceived exertion research to an audience of United States exercise scientists.

In 1973, Borg made a second visit to the University of Pittsburgh to continue his work with Noble and his doctoral students, Robert Robertson, Kent Pandolf, and Enzo Cafarelli. During this trip to the United States, Borg also visited the University of Wisconsin to collaborate with William Morgan. These five scientists Noble,

Robertson, Pandolf, Cafarelli, and Morgan began the proliferation of perceived exertion research throughout the United States and other parts of the world. Most exercise scientists conducting perceived exertion research today can be linked to one of these researchers through their scientific genealogy, connecting student to mentor back through the generations.

In 1996, almost 30 years after Borg's first visit to the University of Pittsburgh, Noble and Robertson published a book entitled *Perceived Exertion*. This was the first all-encompassing synthesis of the burgeoning field of perceived exertion. It provided both an empirical and theoretical resource for researchers and clinicians. The text included a historical review, from the roots of modern psychology to the advancement of psychophysical scaling and the development of the field by Borg. It described the development, administration, and experimental use of Borg's rating of perceived exertion (RPE) scale and discussed the theoretical models developed to explain how psychophysiological signals mediate the intensity of exertional ratings. Great detail is devoted to both specific and broad explanations for the involvement of the physiological (i.e., respiratory-metabolic and peripheral mediators) and psychological inputs to the effort sense, whether conscious or unconsciously monitored. In addition, the text includes a summary of research involving the use of RPE for exercise testing and prescription by both the exercise leader and client.

In 2004, Robertson published a book entitled *Perceived Exertion for Practitioners: Rating Effort with the OMNI Picture System*. In contrast to Noble and Robertson's *Perceived Exertion*, this book was written as a practical guide for health-fitness, clinical and therapeutic professionals that can also be applied to sports medicine, physical education, and coaching. It explains how to use RPE scales to assess physical fitness and to prescribe and regulate exercise intensity for individuals of varied fitness level and clinical status. In addition, the text introduces and explains the rationale behind the OMNI Scale of Perceived Exertion, developed as a more easily understood alternative to the Borg Scale, especially when used with children. The book details the use of the OMNI Scale for exercise testing and programming for a wide variety of exercise types and settings.

This book, entitled *Perceived Exertion Laboratory Manual: From Standard Practice to Contemporary Application*, serves as the third installment of works by Robert Robertson. It is authored along with two of his doctoral students, Luke Haile and Michael Gallagher. As this book comes to press, we are approaching the 50th year of perceived exertion research in the United States.

This book serves a number of purposes that are shared with Robertson's previous books. *Perceived Exertion Laboratory Manual* includes updated summaries of the research for multiple areas within the field of perceived exertion. These varied content areas pertain to exercise assessment, prescription, and program monitoring that are linked to underlying psychophysical and physiological rationale. However, these reviews are written not only for researchers and clinicians but also for educators and exercise science students alike. *Perceived Exertion Laboratory Manual* includes structured experiments that yield practical explanations for the use of RPE scales in both teaching laboratories and field-based physical fitness assessments, as well as their use for exercise prescription and intensity self-regulation. A unique feature of the manual is that these explanations are provided in the format of detailed exercise experiments with both a literature review and step-by-step methods for administration in a research and/or educational setting.

Unique to *Perceived Exertion Laboratory Manual* is the inclusion of full chapters devoted to the constructs of *exercise-induced muscle pain* and the *affective response to exercise*. Although there are many perceptual and psychosocial variables that may play a role in mediating the perceived exertion response to exercise, research has paid special attention to these two, studied both individually and together with perceived exertion. Therefore, following an introductory chapter, Part I of this book (Chaps. 2–4) is devoted to three principal variables: perceived exertion, exercise-induced muscle pain, and the affective response to exercise. All three of these constructs can be easily used to develop effective and cost-efficient strategies to promote the adoption and maintenance of physical activity by healthy and clinical populations. Each chapter in Part I includes a brief historical review and conceptual framework for the constructs and the scales used to measure them during exercise.

Parts II and III present the conceptual framework and methodology for a series of laboratory exercise experiments to study perceived exertion, the principle variable used for the development of each experimental design. The authors chose this presentation style to align with the literature. The field of perceived exertion was conceived by Borg nearly 30 years prior to studies of pain and affective responses to exercise were seen in the literature. Consequently, many of the research models used to study pain and affect during exercise were developed through research conducted on the perceived exertion response. Each chapter in these two parts includes detailed methodology to examine exertional perceptions during aerobic exercise (treadmill exercise and/or cycle ergometry) and resistance exercise (where appropriate).

Part II (Chaps. 5–8) is devoted to the use of perceived exertion during exercise assessment. Chapter 5 presents laboratory experiments that examine perceived exertion scaling procedures, the mastery of which is a prerequisite to understanding the use of perceived exertion during exercise for individuals of varying characteristics. The experiment in Chap. 6 covers the use of perceived exertion scales during graded exercise testing, i.e., perceptual estimation protocols. This includes procedures for both maximal oxygen consumption (VO_2max) and one-repetition maximum (1RM) assessments, which are the primary methods for the determination of RPE scale validity. Chapter 7 presents an experiment that focuses on the determination of a target RPE for use during exercise intensity self-regulation. The experiment in Chap. 8 examines the use of RPE to predict both VO_2max and 1RM.

The information derived from the experiments presented in Chaps. 5 and 6 is a necessary prerequisite to the use of RPE in any type of exercise test or prescription scenario. However, Chap. 7 begins the presentation of content and laboratory methods that are only appropriate in specific situations. Therefore, Chaps. 7 and 8, as well as each chapter in Part III of the book, contain case studies providing an example of an individual for which the methods presented in the laboratory experiment would be appropriate.

Part III (Chaps. 9–13) presents experiments that use perceived exertion for exercise prescription and program evaluation. The experiment in Chap. 9 employs a perceptual estimation-production paradigm for exercise intensity self-regulation. This paradigm is a staple for most perceived exertion research involving exercise intensity regulation pacing strategies for sport performance. Chapter 10 builds upon this through an experiment that examines aerobic interval exercise, a form of exercise intensity modulation that is growing in popularity. The experiment in Chap. 11 takes the paradigm one step further by determining the just noticeable difference (JND) in perceived exertion. This experiment presents the JND as a measure of perceptual acuity with application towards the study of pacing strategy during endurance exercise performance. The experiment in Chap. 12 compares self-selected and imposed exercise intensities, an important consideration prior to exercise intensity prescription especially for novice exercisers. Chapter 13 is the final segment of Part III and includes an experiment that evaluates "off-stimulus" measurements of perceived exertion, predicted RPE and session RPE, which are assessed prior to exercise or following exercise, respectively.

As noted above, the laboratory experiments that are described in Parts II and III focus solely on the measurement and prescriptive application of perceived exertion. Part IV brings focuses on the research variables pain and affect; both of which can be examined independently or in concert with perceived exertion. Chapter 14 presents a series of brief literature summaries (power reviews) that explain how research models initially intended to examine perceived exertion can also be applied to the study of pain and affective responses to exercise. Then, Chaps. 15–17 highlight topics that are of growing interest in the current literature, with a focus on the interplay between exertion, pain, and affect. These applied perceptual and psychosocial research topics include the effects of caffeine, carbohydrate ingestion, and music on the perceived exertion, pain, and affective responses to exercise performance.

This fourth and final section of the book illustrates the direction in which perceived exertion research has been traveling, especially over the past 15 years. Investigations involving exercise performance and adherence to physical activity programs have taken a multidisciplinary approach, with theoretical and empirical roots embedded in the disciplines of physiology and psychology. Recent research has examined many psychoperceptual variables in addition to perceived exertion, pain, and affect. The conceptual models and research methodology necessary to study these three variables, along with mediating physiological variables, provide the health-fitness professional with a solid foundation for exercise assessment and prescription with the goal of promoting the adoption of long-term physical activity participation.

Lock Haven, PA, USA — Luke Haile
Conway, AR, USA — Michael Gallagher Jr
Pittsburgh, PA, USA — Robert J. Robertson

Acknowledgements

First, we would like to acknowledge the work of Gunnar Borg, the founding father of perceived exertion, whose models and methods will be used by current and future generations of exercise scientists and behavioral psychologists.

We would also like to acknowledge Bruce Noble, who first invited Borg to the United States. He began a lineage of exercise science educators and researchers with academic diaspora across the continent and beyond. His knowledge and mentoring has been passed through each of the authors of this book, whether directly or indirectly, and will continue for generations to come.

We would like to acknowledge the work of retired educators and researchers William Morgan, Kent Pandolf, and Enzo Cafarelli, who are responsible for so many great scientists and teachers continuing their work today. We gratefully acknowledge all those who are important contributors to the more recent literature cited throughout this book, especially Roger Eston, Rod Dishman, Panteleimon Ekkekakis, Steven Petruzzello, Dane Cook, and Patrick O'Connor.

We owe a huge debt of gratitude to our colleagues, advisors, and friends from the University of Pittsburgh Center for Exercise and Health-Fitness Research, especially Fredric Goss, Kenneth Metz, Deborah Aaron, and Elizabeth Nagle. They are responsible for much of the research presented in this book, but also the support and guidance that made the book possible.

We extend special appreciation to Donna Farrell, whose support (both moral and clerical) has been a lifeline for the writing of this book. Finally, we must thank a few exceptional students; Haley Frerichs, Caroline Proctor, and Curtiss Scroggins; whose laboratory assistance was important for the exercise experiments presented in the book.

Contents

Part IV Applied Perceptual and Psychosocial Research

List of Abbreviations

1RM	One-repetition maximum
AR	Affective response
AT	Anaerobic threshold
ATP	Adenosine triphosphate
BP	Blood pressure
CERT	Children's effort rating table
EDI	Exercise discomfort index
FS	Feeling Scale
GXT	Graded exercise test
HR	Heart rate
HR-VT	Heart rate corresponding to the ventilatory threshold
JND	Just noticeable difference
LT	Lactate threshold
PA	Physical activity
PACES	Physical Activity Enjoyment Scale
PAE	Physical activity enjoyment
PO	Power output
POmax	Maximal power output
POpeak	Peak power output
RPE	Rating(s) of perceived exertion
RPE-A	Differentiated rating of perceived exertion for the arms
RPE-AM	Differentiated rating of perceived exertion for the active muscle
RPE-C	Differentiated rating of perceived exertion for the chest/breathing
RPE-L	Differentiated rating of perceived exertion for the legs
RPE-O	Undifferentiated rating of perceived exertion for the overall/total body
RPE-VT	Rating of perceived exertion corresponding to the ventilatory threshold
VCO_2	Volume of carbon dioxide production
V_E	Pulmonary (minute) ventilation

VO_2	Volume of oxygen consumption/uptake
VO_2max	Maximal oxygen consumption/uptake, maximal aerobic power
VO_2peak	Peak oxygen consumption/uptake, peak aerobic power
VT	Ventilatory threshold

Chapter 1
Introduction

Regular participation in physical activity (PA) has many physiological and psychological benefits, whether undertaken as part of a professionally prescribed exercise program or a leisure pursuit. PA includes structured aerobic and resistance exercise, competitive and recreational sport, occupational activity, as well as activities of daily living. Increasing one's level of PA is gaining wide recognition as one of the essential elements of behavioral change necessary to promote overall health-fitness. From young to old, from those with chronic illnesses to elite athletes, regular PA can play an important role in optimizing performance and enhancing functional capacity, even preventing and treating a wide variety of diseases and disorders. This laboratory manual presents data based on experiments that are intended to provide advanced understanding of those perceptual and psychosocial factors that influence promotion, adoption, and maintenance of PA of the type that provides positive health-fitness benefits.

1.1 Benefits of Physical Activity

Regular participation in PA has shown moderate to high associations with the prevention of obesity, cardiovascular disease (especially coronary heart disease), stroke, type 2 diabetes, osteoporosis, certain cancers (including colon, rectal, breast, and prostate), as well as the decline of mental function (England Department of Health 2004). PA has also shown moderate to high associations with the successful treatment of obesity, coronary heart disease, peripheral vascular disease, osteoarthritis, and low back pain (England Department of Health 2004). Increasing daily levels of PA has shown beneficial effects by attenuating lung deterioration in smokers and decreasing their risk for chronic obstructive pulmonary disease (Garcia-Aymerich et al. 2007), increasing high-density lipoprotein (HDL) cholesterol while decreasing total cholesterol and triglycerides (Codina et al. 1999), and preventing falls in the elderly (Gillespie et al. 2012). Given these salutary effects, increasing PA

L. Haile et al., *Perceived Exertion Laboratory Manual*,
DOI 10.1007/978-1-4939-1917-8_1

level leads to a significantly lower mortality rate and longer life expectancy (Kujala et al. 1998; Manini et al. 2006).

The psychological effects of increasing PA participation are extensive as well. PA can prevent age-related decline in mental function, treat depression and low levels of mental well-being, improve mood, reduce anxiety, and enhance self-esteem (England Department of Health 2004; Fox 1999; Fox et al. 2007). Increasing the habitual PA level can improve overall cognitive function by alleviating stress and improving sleep quality (England Department of Health 2004). Overall, PA is closely tied to health-related quality of life (Elley et al. 2003; Pedersen and Saltin 2006). Of importance is the opinion of many scientists and health professionals that quality of life, as influenced by regular PA participation, is closely related to mortality rate and life expectancy.

1.2 Physical Activity Guidelines

The American College of Sports Medicine (ACSM) and the Centers for Disease Control and Prevention (CDC) have developed guidelines specifying the minimum amount of aerobic PA in which most adults should participate to promote and maintain overall health. The initial guidelines, published in 1995 (Pate et al. 1995), were updated in collaboration with the American Heart Association in 2007 (Haskell et al. 2007) and are consistent with the recommendations put forth by the U.S. Department of Health and Human Services (USDHHS) in its Dietary Guidelines for Americans (USDHHS 2005). The primary recommendations for aerobic exercise are that an adult should participate in at least 30 min of moderate intensity activity at least five times per week. The two key words in these recommendations as written here are *at least*. The guidelines indicate the minimum amount of aerobic activity in which every adult should engage. Additional health benefits can be achieved, including further reduction of chronic disease risk, by increasing the duration, frequency, and intensity of aerobic exercise above minimum recommendations. The progressively greater salutary effects of increased levels of PA reflect the dose–response relation that has been shown in large-scale, prospective studies over the past decade. In addition, increasing the amount of weight-bearing or high-impact aerobic exercise, such as brisk-walking, jogging or running, enhances skeletal health. The updated guidelines also include recommendations for resistance training and calisthenic exercises to increase and subsequently maintain muscular strength and endurance, ultimately improving functional capacity and promoting physical independence (Haskell et al. 2007).

1.3 The Physical Inactivity Epidemic

Regardless of the detailed guidelines set forth through the cooperative work of multiple professional and governmental organizations, the adoption and maintenance of regular PA to improve health-fitness, although significantly important in decreasing one's risk for premature death, seems to be a goal that many people fail to accomplish. In 2012, 23.1 % of United States (US) adults reported performing no leisure-time PA (CDC 2014). Data from 2011 indicate that only 51.6 % of US adults met the ACSM guidelines for aerobic PA and only 20.9 % of US adults met the guidelines for both aerobic PA and muscle strengthening exercise (CDC 2014).

The physical inactivity epidemic is not limited to the US. According to the World Health Organization (WHO), physical inactivity is a principle risk factor for chronic diseases and premature death in many industrialized nations (WHO 2014). Data from 2008 indicate that only 18 % of Europeans reported engaging in moderate PA on a regular basis (Allender et al. 2008). The Spanish National Health Survey of 2001 reported that 46.6 % of those over age 15 years did not exercise in their free time, with only 8.5 % reporting exercising on a regular basis (Gine-Garriga et al. 2009). In addition, according to 1998 data from the England Department of Health, PA, Health Improvement and Prevention, only 31 % of those 16 years of age or older achieved the ACSM recommended levels of moderate PA (England Department of Health 2004).

The low prevalence rates of PA participation throughout the industrialized world may reflect the quality of an individual's available clinical care. However, patients seldom receive medical recommendations for PA from their physicians and nurses. Recent research has shown that hospital staff frequently report not having enough time or knowledge to prescribe appropriate exercise to those seeking care (Puig Ribera et al. 2005, 2006). This makes it clear that addressing physical inactivity is not seen as effective health care advice warranting significant use of time spent on patient care in the hospital setting. Regardless, given recent data from both the United States and Europe, it would appear most adults do not follow PA recommendations to achieve positive health outcomes and improve overall fitness.

1.4 Economic Cost of Physical Inactivity

Low levels of PA not only affect the health of nations, but the wealth of nations. In a report by Chenoweth and Leutzinger (2006), the economic cost of physical inactivity alone was estimated at $93.32 billion per year, and that only accounted for seven states in the US. The yearly cost of physical inactivity and overweight combined, when projected to 2008 for the entire US population, was estimated at $708 billion. These cost estimates include direct medical care, worker's compensation, and loss of occupational productivity (Chenoweth and Leutzinger 2006). Data from England in 2004 estimate the economic cost of physical inactivity at £8.2 billion (approximately $13 billion), with obesity adding another £2.5 billion (England Department of Health 2004). Therefore, attaining optimal levels of overall health-fitness by increasing regular PA participation is one of the foremost public health initiatives of our time, in both the US and Europe. Controlling medical costs through PA interventions is especially important in these stringent economic times.

1.5 Exercise Is Medicine®

Recent initiatives by ACSM in partnership with the US Surgeon General have culminated in the *Exercise is Medicine®* initiative (Exercise is Medicine 2014). This initiative promotes the assessment of "exercise vitals" at every visit to the doctor's office. The campaign stresses that self-report of the frequency and intensity of PA participation should be recorded and discussed just as are standard clinical assessments of height, weight, heart rate (HR), and blood pressure (BP). The goal of the *Exercise is Medicine®* initiative is to promote awareness of the importance of PA in advancing overall health to both patient and doctor. The measurement of "exercise vitals" is intended to begin the discussion of the importance of exercise and PA in both the prevention and treatment of disease. It follows that exercise and PA can be prescribed like any medication for those patients who must increase their habitual PA level and/or decrease excess body weight. Such an exercise prescription is particularly important in treating a wide range of chronic diseases and clinical disorders for which PA has shown significant physiological and health benefit. However, this awareness of the health implications of regular exercise, whether made known at the doctor's office or hospital, at school, or through a public service announcement, is only the beginning step in a systematic process to increase the individual's level of PA participation.

1.6 Physical Activity Adherence

Although the increasing prevalence of physical inactivity is a health care burden around the world, there is little evidence concerning how to increase participation in PA programs (Gine-Garriga et al. 2009). Substantial efforts to develop PA intervention programs have been undertaken by professional and governmental public health organizations throughout the world. The guidelines set forth by these organizations are designed to have the greatest impact on the overall public health of nations and do not necessarily describe the exercise or PA programs that address the needs of each individual.

The process of individual behavior change is complex and involves numerous constructs including personal, programmatic, social and environmental factors. Health care and fitness professionals must move beyond the traditional prescriptive exercise program that is based on factors such as medical clearance, supervision (Haskell et al. 2007), and structured guidelines to identify overload training intensity using target HR, percent of maximal oxygen consumption (VO_2max), or specific resistance settings on various types of exercise equipment. An effective PA program must take into consideration an individual's interests, needs, daily schedule, environment, family structure, work, social environments, travel, and even the possibility of inclement weather (Haskell et al. 2007). The promotion of long-term adherence to PA programs must use strategies to overcome participation barriers, through individual behavioral change and by employing an ecological perspective at multiple community levels. In addition, exercise programs must be sensitive to the financial status of individuals of all socioeconomic levels and cannot exclusively depend on the availability of specific equipment and/or enrollment in exercise classes. Morgan (2005) reviewed approaches to increase PA through exercise-referral schemes. The review concluded that additional research is needed to develop intervention strategies that could increase long-term adherence in a wide range of populations for differing activities. It was observed that there must be a move to prescribe an exercise program that is physiologically and psychologically effective, is cost-effective and can be performed in a wide range of indoor and outdoor settings. In addition, studies have found that most people will remember PA advice only if it is linked to a serious health condition for which they have been diagnosed. We must work to change the fact that many people are not concerned about their own PA habits until they require therapeutic exercise for medical care. In this context they receive an exercise prescription that is part of the treatment for a chronic disease or clinical disorder rather than a preventive recommendation to promote health-fitness (Flocke and Stange 2004).

Most individuals have their own unique characteristics regarding the decision to participate in PA. These individual characteristics include perceptions of exertion and pain associated with PA intensity, affective responses associated with PA, attitudes toward various types of PA, and self-efficacy for performance of PA. These perceptual and psychosocial variables help to define the PA needs and preferences of each individual beyond their living environment and financial abilities. They must be used to shape each individual's PA program. Such individualized programs should use strategies for behavior change that meets perceptual and psychosocial needs as they pertain to PA adoption and adherence while simultaneously attaining optimal physiological stimulus levels to enhance overall health-fitness. It should be the goal of the health professional to help people engage in a new way of thinking about an active lifestyle and see PA as an opportunity to promote individual health. A physically active lifestyle can bring with it enjoyment, improved vitality, a sense of achievement, physical fitness, optimal body weight, and lasting health (England Department of Health 2004).

1.7 Laboratory Manual of Perceptual and Psychosocial Exercise Experiments

Numerous population-based studies have demonstrated that many individuals who reside in industrialized nations are confronted with perceptual and psychosocial barriers that prevent regular participation in PA and exercise conditioning programs. It is the goal of this manual to present laboratory experiments and accompanying literature reviews that help the student learn about programmatic innovations to improve PA adoption and maintenance with a unique emphasis on understanding the mitigating role of perceptual and psychosocial dynamics. This task can be accomplished by teaching the next generation of fitness professionals, exercise scientists and public health researchers how to study, test, and use perceptual and psychosocial variables to help their clients and patients achieve overall health-fitness. The PA-related variables that are the principal focus of this laboratory manual are *perceived exertion*, *exercise-induced muscle pain*, and the *affective response (AR) to exercise*. These perceptual and psychosocial constructs can be easily used to develop effective and cost-efficient strategies to promote and improve the adoption and maintenance of PA by healthy and clinical populations. The conceptual framework and methodology for each laboratory experiment is presented in terms of perceived exertion, the principle variable used for the development of each experimental design. Chapter 14, which introduces Part 4, Applied Perceptual and Psychosocial Research, details the use of each conceptual model with exercise-induced muscle pain and AR during exercise. Each of these variables has been an important consideration in the recent literature concerning exercise assessment, prescription, and program monitoring. Their continued application in practice and further exploration in research are critical to the development of future health-fitness professionals.

References

Allender S, Scarborough P, Peto V, Rayner M, Leal J, Luengo-Fernandez R, Gray A (2008) European cardiovascular disease statistics: 2008 edition. European Heart Network, Brussels, Belgium www.ehnheart.org. Accessed 28 Mar 2014.

Centers for Disease Control and Prevention (2014) Behavioral risk factor surveillance system, Prevalence and trends data. http://apps.nccd.cdc.gov/brfss/. Accessed 28 Mar 2014.

Chenoweth D, Leutzinger J. The economic cost of physical inactivity and excess weight in American adults. J Phys Act Health. 2006;3:148–63.

Codina O, Elosua R, Marrugat J. Actividad fisica y arteriosclerosis: efectos de la actividad fisica sobre la oxidacion lipidica, la hemostasia y la function endotelial. Med Clin (Barc). 1999;112: 508–15.

Elley CR, Kerse N, Arroll B, Robinson E. Effectiveness of counselling patients on physical activity in general practice: cluster randomised controlled trial. BMJ. 2003;326:793.

England Department of Health, PA, Health Improvement and Prevention. At least five a week: evidence on the impact of physical activity and its relationship to health. London, UK: Department of Health; 2004.

Exercise is Medicine (2014) About exercise is medicine. http://www.exerciseismedicine.org/about.htm. Accessed 3 July 2014.

Flocke SA, Stange KC. Direct observation and patient recall of health behavior advice. Prev Med. 2004;38:343–9.

Fox KR. The influence of physical activity on mental well-being. Public Health Nutr. 1999;2: 411–8.

Fox KR, Stathi A, McKenna J, Davis MG. Physical activity and mental well-being in older people participating in the better ageing project. Eur J Appl Physiol. 2007;100:591–602.

Garcia-Aymerich J, Lange P, Benet M, Schnohr P, Anto JM. Regular physical activity modifies smoking-related lung function decline and reduces risk of chronic obstructive pulmonary disease: a population-based cohort study. Am J Respir Crit Care Med. 2007;175:458–63.

Gillespie LD, Robertson MC, Gillespie WJ, Sherrington C, Gates S, Clemson LM, Lamb SE. Interventions for preventing falls in older people living in the community. Cochrane Database Syst Rev. 2012;9, CD007146.

Gine-Garriga M, Martin C, Martin C, Puig-Ribera A, Anton JJ, Guiu A, Cascos A, Ramos R. Referral from primary care to a physical activity programme: establishing long term adherence? A randomized controlled trial. Rationale and study design. BMC Public Health. 2009; 9:1–9.

Haskell WL, Lee I, Pate RR, Powell KE, Blair SN, Franklin BA, Macera CA, Heath GW, Thompson PD, Bauman A. Physical activity and public health: updated recommendations for adults from the American College of Sports Medicine and the American Heart Association. Med Sci Sports Exerc. 2007;39:1423–34.

Kujala UM, Kaprio J, Sarna S, Koskenvuo M. Relationship of leisure-time physical activity and mortality: the Finnish twin cohort. JAMA. 1998;279:440–4.

Manini TM, Everhart JE, Patel KV, Schoeller DA, Colbert LH, Visser M, Tylavsky F, Bauer DC, Goodpaster BH, Harris TB. Daily activity energy expenditure and mortality among older adults. JAMA. 2006;296:171–9.

Morgan O. Approaches to increase physical activity: reviewing the evidence for exercise-referral schemes. Public Health. 2005;119:361–70.

Pate RR, Pratt M, Blair SN, Haskell WL, Macera CA, Bouchard C, Buchner D, Ettinger W, Heath GW, King AC, Kriska A, Leon AS, Marcus BH, Morris J, Paffenbarger RS, Patrick K, Pollock ML, Rippe JM, Sallis J, Wilmore JH. Physical activity and public health: a recommendation from the Centers for Disease Control and Prevention and the American College of Sports Medicine. JAMA. 1995;273:402–7.

Pedersen BK, Saltin B. Evidence for prescribing exercise therapy in chronic disease. Scand J Med Sci Sports. 2006;16:3–63.

Puig Ribera A, McKenna J, Riddoch C. Attitudes and practices of physicians and nurses regarding physical activity promotion in the Catalan primary health-care system. Eur J Public Health. 2005;15:569–75.

Puig Ribera A, McKenna J, Riddoch C. Physical activity promotion in general practices of Barcelona: a case study. Health Educ Res. 2006;21:538–48.

United States Department of Health and Human Services (2005) Dietary guidelines for Americans. http://www.health.gov/dietaryguidelines/dga2005/document/. Accessed 31 Mar 2014.

World Health Organization (2014) Physical inactivity: a global public health problem. http://www.who.int/dietphysicalactivity/factsheet_inactivity/en/. Accessed 28 Mar 2014.

Part I
Perceptual and Psychosocial Variables

Chapter 2
Perceived Exertion

Perceived exertion can be defined as *the subjective intensity of effort, strain, discomfort and/or fatigue that is felt during exercise* (Robertson and Noble 1997). The exertional experience involves detecting and interpreting sensations arising from the body during any type of PA (Noble and Robertson 1996). The underlying processes that are subjectively monitored during PA, referred to as exertional mediators, are classified as physiological, psychosocial, performance-related and symptomatic in nature. Perceived exertion can be assessed during aerobic and resistance exercise, leisure time or daily living activities, occupational physical activity, or a wide variety of recreational and competitive sport activities. Individuals can rate their level of perceived exertion by selecting a number, or rating of perceived exertion (RPE), from a range of numerical categories displayed on a perceived exertion scale. These RPE scales may include verbal and pictorial descriptors that are placed in juxtaposition to numerical categories representing the range of perceptual responsiveness from very low to very high intensity. The Borg RPE Scale and OMNI RPE Scales have been used in perceptual paradigms designed to quantify and predict physiological responses to acute exercise and adaptations to exercise training. RPE is an important variable used to monitor exercise programming, ensuring the attainment of optimal exercise intensity for the achievement of health-fitness benefits and to promote PA adherence.

2.1 Mediators of Exertional Perceptions

Over the past 50 years, the perceived exertion knowledge base has grown exponentially. Research has studied many aspects of this gestalt-like perceptual response to exercise, described so because it is a complex pattern of physical, biological, and psychological phenomena. The Global Explanatory Model for Perceived exertion (Fig. 2.1) illustrates the mechanisms, both internal and external, by which an exercise stimulus results in an individual's unique perceptual response (Noble et al. 1986;

L. Haile et al., *Perceived Exertion Laboratory Manual*,
DOI 10.1007/978-1-4939-1917-8_2

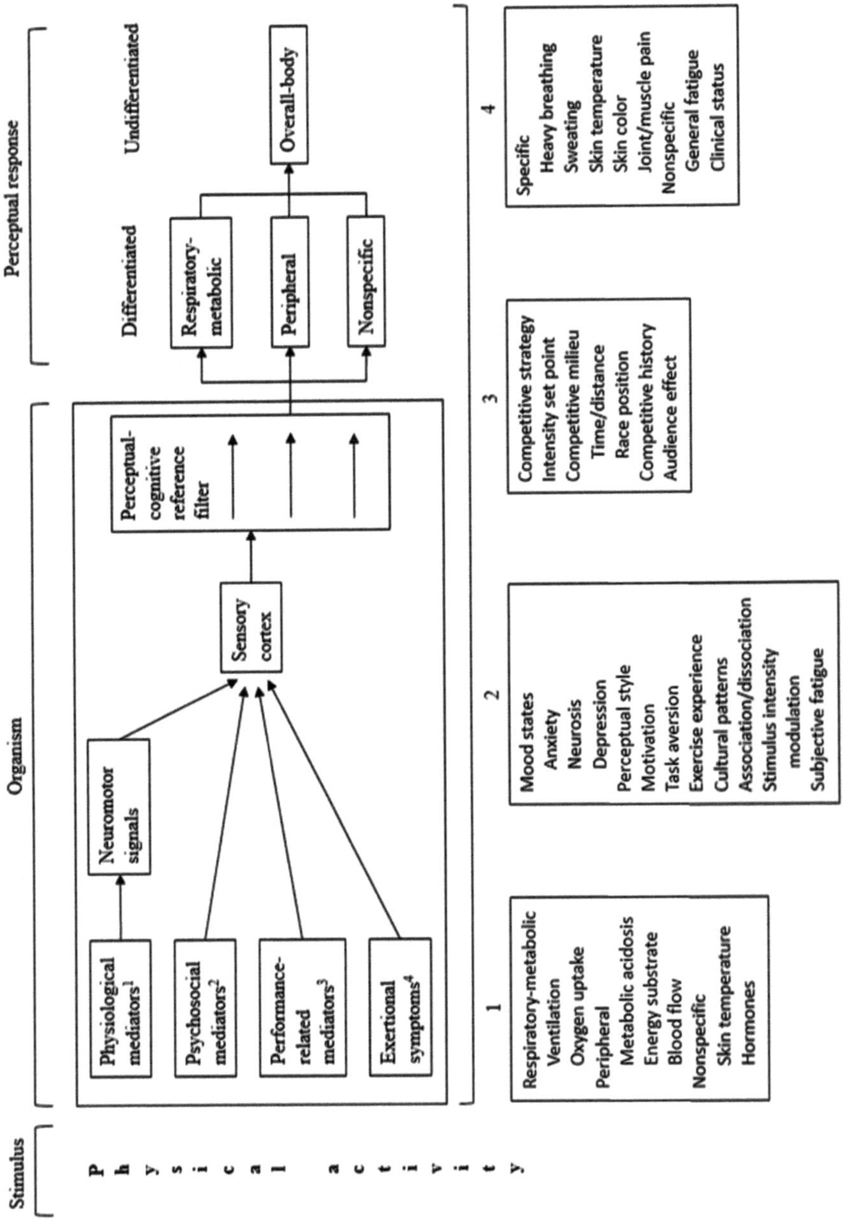

Fig 2.1 Global explanatory model of perceived exertion (adapted from Noble and Robertson 1996)

Noble and Robertson 1996; Robertson et al. 1986). Exertional mediators are the underlying physiological, psychosocial, performance-related and symptomatic processes that an individual subjectively monitors during PA. These mediators function collectively and interactively to ultimately shape the RPE response. Gaining knowledge about how these exertional mediators act to influence adoption and maintenance of PA has been a primary focus of perceived exertion research and its application in both health-fitness and clinical settings.

2.1.1 Physiological Mediators

Physiological mediators of exertional perceptions can be subdivided into *respiratory-metabolic, peripheral*, and *nonspecific* categories. Respiratory-metabolic physiological mediators include those processes that are influenced by aerobic metabolic demand during PA. These include pulmonary ventilation (V_E), oxygen uptake (VO_2), carbon dioxide production (VCO_2), HR and BP. Peripheral physiological mediators include factors such as metabolic acidosis (pH, lactic acid), muscle blood flow, muscle fiber type and glycogen content, as well as plasma glucose and free fatty acid concentrations. Nonspecific physiological mediators include systemic events that occur during exercise, such as hormonal regulation and increases in both skin and body core temperature.

Physiological mediators of exertional perceptions play a primary role in shaping the effort sense due to their effect of altering tension-producing properties of skeletal muscle. Muscle contractions are monitored through a neurophysiological pathway between the motor and sensory cortex (Robertson 2001). Developed tension in both peripheral and respiratory muscle is monitored and ultimately interpreted as effort sensation. As PA intensity increases, a feed-forward mechanism rooted in the motor cortex increases skeletal muscle motor unit recruitment and firing frequency. Corollary signals branching from the motor efferents and terminating in the sensory cortex also increase in frequency, intensifying perception of effort. These efferent signals are integrated with afferent proprioceptive feedback from muscles and joints that help fine-tune the RPE response.

2.1.2 Psychosocial Mediators

Physiological mediators of exertion generally function similarly for most individuals and have been a major focus of RPE research since conception of the discipline by Borg. However, experimental research focusing on the identification of psychosocial mediators that may account for inter-individual differences in the perceived exertion response is gaining substantial interest. Morgan (2001) separated the psychosocial mediators of exertional perceptions into four distinct classifications: (1) affective mediators that are linked to emotions and mood states, including anxiety, depression, introversion, and extroversion; (2) cognitive mediators that include association/dissociation, self-efficacy, and personality type; (3) perceptual process mediators that include pain tolerance, somatic perception, perceptual augmentation, and perceptual reduction; (4) social/situational mediators that include music, sex of the counselor or test administrator, and social setting. Recently, there have been substantial increases in the knowledge base concerning the AR to exercise and its relation to RPE, largely due to the work of Ekkakakis (2003) and his colleagues. This work has identified affect as a potential factor mediating the adoption and maintenance of PA. As such, AR is one of the primary variables employed in this laboratory manual.

2.1.3 Performance-Related Mediators

Performance-related exertional mediators can be defined as variables that describe and provide feedback regarding the intensity of an acute exercise bout and the prediction of exercise performance outcomes. Measures of these variables may be provided to the individual by a coach or teammate or may be monitored by the individual using a watch or digital display of an exercise machine. Variables such as time/distance traveled, time/distance remaining, speed/pace or even characteristics of a competitor's exercise performance may affect the RPE response.

2.1.4 Exertional Symptoms

Exertional symptoms, ultimately, are the final outward expression of the internal physiological and psychological processes that are experienced by an individual during PA and exercise. Physiological, psychological and/or performance-related exertional mediators are uniquely integrated such that exertional symptoms are linked to the individual's conscious perceptual report. Exertional symptoms can be divided into two separate classifications: somatic and psychological. The most pronounced of the somatic exertional symptoms and, arguably, all symptoms, is fatigue. Thus, fatigue is a primary term in the definition of perceived exertion. Aches, cramps, muscle and joint pain, feelings of heaviness and dyspnea (breathlessness) are somatic symptoms felt in varying degrees when performing many different exercise modalities. Psychological symptoms that may directly affect the perceptual response include task aversion and low motivation.

2.2 Perceptual-Cognitive Reference Filter

The final step in the formation of the perceptual response is the overall integration of the various signals generated by exertional mediators that pass through the perceptual-cognitive reference filter. It is proposed that this filter is located in the sensory cortex and provides a sensory weighting to past and present PA experiences and environments. These are ultimately expressed as an individual's perceptual style with the weighting often dominated by specific physiological or psychosocial exertional mediators. In this final step, the exertional signals that arise from the physiological responses to an acute exercise performance are mediated by the array of stored information in the perceptual-cognitive reference filter. This mediating process ultimately shapes the intensity of perceived exertion that is rated by the individual using a category metric (Robertson 2001).

2.3 Rating Perceived Exertion

The origin of perceived exertion is rooted in psychophysics. This science studies human sensation by establishing a mathematical relation between physical stimuli and sensory responses (Noble and Robertson 1996). More specifically, psychophysics has been defined as the study of the relation between sensation and stimulus when both are measured as quantities (Marks 1974). Gunnar Borg, a Swedish psychologist, sought to do just that when he pioneered the measurement of perceived exertion and developed the first RPE scale. His initial work in the late 1950s and throughout the 1960s sought to define perceived exertion as it applied to individuals' subjective adaptation to various types of exercise and occupational activities (Borg 1962a, 1962b, 1970, 1971; Borg et al. 1971). Borg introduced his 15-category (i.e. 6–20) RPE scale (Fig. 2.2) in the mid 1960s while on a sabbatical visit to the

6	No exertion at all	
7	Extremely light	
8		
9	Very light	
10		
11	Light	
12		
13	Somewhat hard	
14		
15	Hard	(heavy)
16		
17	Very hard	
18		
19	Extremely hard	
20	Maximal exertion	

Fig 2.2 Borg RPE scale (Borg 1998) © Gunnar Borg 1998

University of Pittsburgh and the Pennsylvania State University. The Borg Scale is used to measure RPE and determine the relation between exertional perceptions and a wide array of physical, physiological and psychosocial factors that are linked to exercise performance. RPE, measured using Borg's scale as well as other, newer category scales such as Robertson's OMNI Scale, is one of the most commonly employed variables assessed in exercise science research.

Borg designed the numerical format of the first perceived exertion scales to align closely with HR responses, which were taken as a good general indicator of physical strain during PA. This physiological response is easily measured and, therefore, often used as a primary variable for exercise intensity prescription. Borg's initial category metric did not yield perceptual responses that met the expected linear relation between HR and RPE as measured using a 21-category scale. The scale included numerical categories that ranged from 0 to 20 with verbal descriptors linked to every odd integer from 3 to 19, such as "Extremely light" at 3 and "Extremely laborious" at 19 (Borg 1962a, 1962b, 1970). Borg then fine-tuned the original scale, shortening it to 15 categories that ranged from 6 to 20 with the goal of predicting exercise HR from RPE responses of a normal, healthy, middle-aged man performing cycle ergometry. This prediction was accomplished by multiplying the RPE response by 10 (e.g., an RPE of $13 \times 10 =$ exercise HR of 130 $b{\cdot}min^{-1}$) (Borg 1985; Borg and Lindblad 1976). This 6–20 scale, commonly known as the Borg Scale (Fig. 2.2), is used in clinical and health-fitness settings where cycle ergometry is employed worldwide. However, the validity of predicting HR from RPE responses using a simple multiplication factor of 10 was never truly realized due to the great inter-individual variability in HR responses under varying types of exercise, clinical and environmental conditions. This fact notwithstanding, many studies have shown a linear relation between workload, HR and RPE, establishing validity of the Borg (6–20) RPE Scale.

2.4 Perceived Exertion Scales for Children

Prior to 2000, relatively few investigations studied RPE responses of children. Oded Bar-Or pioneered the study of children's perception of effort in a 1977 study involving cycle ergometer exercise where RPE was measured using Borg's 6–20 category metric validated for adults (Bar-Or 1977). However, subsequent research demonstrated that the Borg (6–20) RPE Scale may be unsuitable for children (Lamb and Eston 1997a). Not until 1994 was consideration given to the design of a child-specific RPE scale. Roger Eston and colleagues (Eston et al. 1994; Williams et al. 1994) designed the Children's Effort Rating Table (CERT) to address the semantic limitations of children when they attempted to use RPE scales that were formatted using adult vocabulary. Verbal descriptors for CERT were chosen by children because they were commonly used expressions understood as descriptions of exertion during PA. Ten descriptors were placed along a numerical rating range from 1 to 10. This resulted in a more familiar rating scale format as opposed to the adult oriented 6–20 Borg Scale. However, a follow-on investigation found a nonlinear relation between perceptual and physiological responses. CERT had a diminished sensitivity across the upper HR range during dynamic exercise (Lamb and Eston 1997b; Robertson et al. 2000).

2.5 The OMNI Scale

In response to growing clinical and experimental interest in investigating children's perceptions of effort and in recognition of the potential methodological and semantic limitations of available RPE scales, Robertson and colleagues developed the OMNI picture system for rating effort in children (Robertson 2004; Robertson et al. 2000). For the 0–10 OMNI RPE scale, both verbal and pictorial descriptors were chosen by children to aid in linking exertional symptoms with the perceptual rating. The name OMNI is a contemporary contraction of the word omnibus, meaning "of, relating to, or providing for many things at once" (Merriam-Webster Online 2014). By extension, when used in the context of exertion scaling the word OMNI refers to a category metric having broadly generalizable properties. This was of practical importance because the OMNI scales were designed for use by individuals of varying ages participating in a wide range of PA modalities. The first OMNI scale (Fig. 2.3), developed for cycle ergometer exercise, demonstrated a high level of validity for use by male and female children of mixed race (Robertson et al. 2000). Later, the adult format of the OMNI-Cycle RPE scale (Fig. 2.4) was developed using age appropriate verbal and pictorial descriptors (Robertson 2004; Robertson et al. 2004). The Adult OMNI-Cycle RPE Scale demonstrated high concurrent and construct validity for use by both men and women (Robertson 2004; Robertson et al. 2004). Subsequently, different OMNI Scale formats were developed and

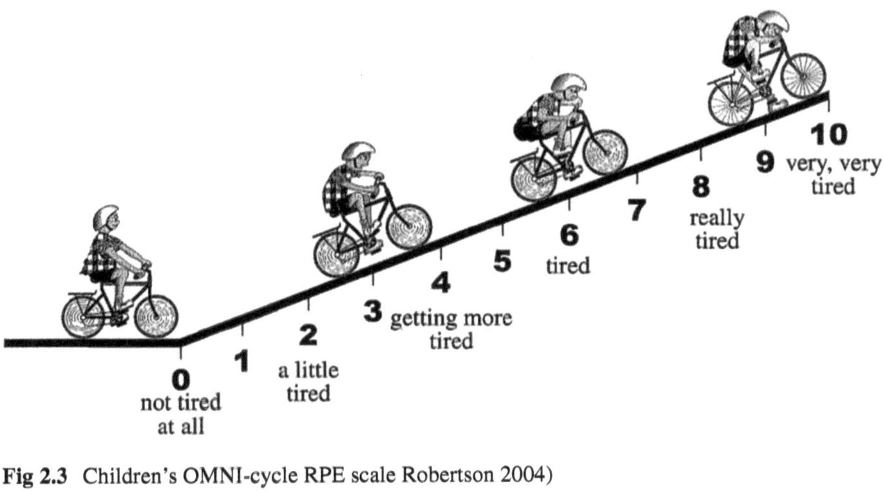

Fig 2.3 Children's OMNI-cycle RPE scale Robertson 2004)

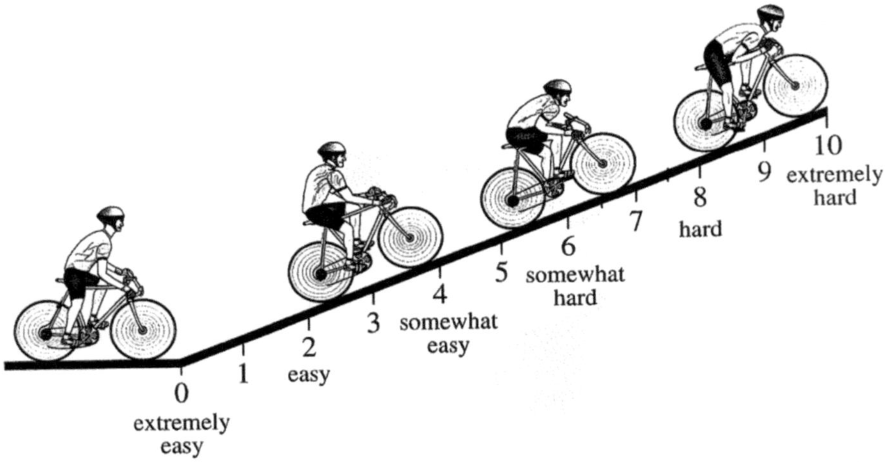

Fig 2.4 Adult OMNI-cycle RPE scale (Robertson 2004)

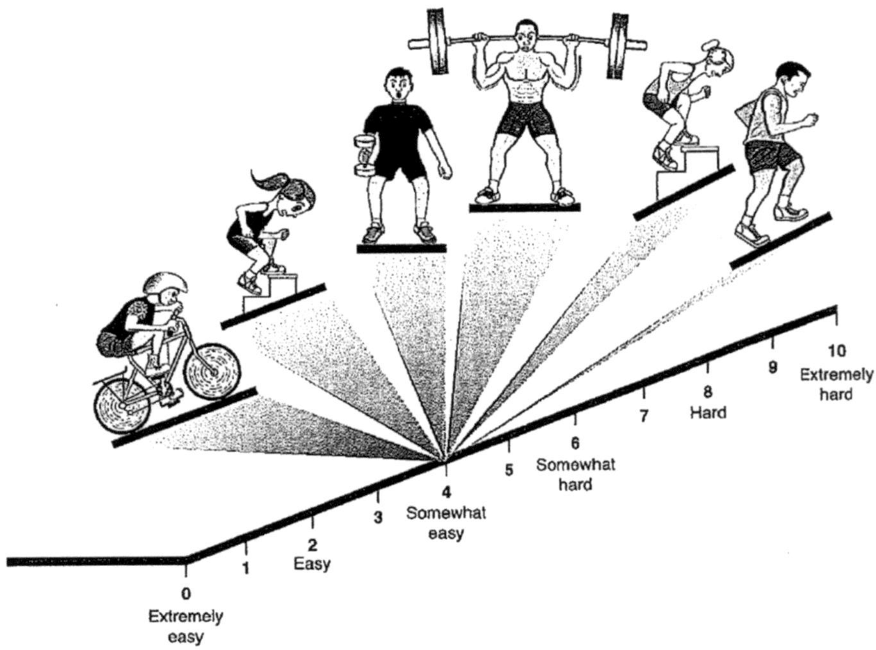

Fig 2.5 Montage of pictorial descriptors for the OMNI RPE scales (Robertson 2004)

validated for use by children and/or adults performing such PA modalities as walking and running (Utter et al. 2002, 2004; Robertson 2004), resistance exercise (Lagally and Robertson 2006; Robertson et al. 2003, 2004, 2005), stepping (Krause et al. 2012; Robertson et al. 2004, 2005), and elliptical ergometry (Mays et al. 2010) (Fig. 2.5; see Appendix A for figures of additional OMNI Scale formats).

2.6 Use of RPE Scales

The field of perceived exertion was originated by Gunnar Borg as he sought a new way to describe adaptations to exercise, initially in an occupational setting and later fitness and sport settings. Borg hypothesized that an individual's perceptual responses to exercise testing could provide information to both quantify and predict physiological and performance adaptations. In addition, RPE is a measurable construct that most individuals can understand and use after proper instructions from an exercise scientist or coach. This fact is largely responsible for the growth of the perceived exertion knowledge base in the published literature and its continued popularity in clinical and sport settings. The uses of RPE presented in this laboratory manual include, but are not limited to, the following: (1) the determination of maximal work capacity during fitness testing, (2) the indication of impending exercise test termination, (3) the prediction of maximal aerobic power or one-repetition maximum from submaximal exercise responses, (4) the identification of physiological responses such as the ventilatory threshold, and subsequent use of these measures to prescribe a "target" RPE for exercise conditioning, (5) exercise intensity self-regulation and monitoring of exercise intensity self-regulation error, and (6) monitoring exercise programs to determine if the exercise intensity is appropriate to achieve an overload training stimulus.

References

Bar-Or O. Age-related changes in exercise prescription. In: Borg G, editor. Physical work and effort. New York, NY: Pergamon; 1977. p. 255–66.

Borg G (1962a) Physical performance and perceived exertion. In Studia psychologica et paedogogica, series altera, investigations XI, Gleerup, Lund, Sweden. p. 1–63.

Borg G. A simple rating scale for use in physical work tests. K Fysiogr Sallsk Lund Forh. 1962b;2:7–15.

Borg G. Perceived exertion as an indicator of somatic stress. Scand J Rehab Med. 1970;2:92–8.

Borg G. The perception of physical performance. In: Shepherd R, editor. Frontiers of fitness. Springfield, IL: Charles C. Thomas; 1971. p. 280–94.

Borg G. An introduction to Borg's RPE-scale. Ithaca, NY: Mouvement; 1985.

Borg G, Edgren B, Marklund G. A simple walk test of physical work capacity. In reports from the institute of applied psychology, vol. 18. Stockholm, Sweden: University of Stockholm; 1971.

Borg G, Lindblad I. The determination of subjective intensities in verbal descriptors of symptoms. In Reports from the institute of applied psychology, vol. 75. Stockholm, Sweden: University of Stockholm; 1976.

Borg G. Borg's perceived exertion and pain scales. Champaign, IL: Human Kinetics; 1998.

Ekkakakis P. Pleasure and displeasure from the body: perspectives from exercise. Cogn Emot. 2003;17:213–39.

Eston RG, Lamb KL, Bain A, Williams AM, Williams JG. Validity of a perceived exertion scale for children: a pilot study. Percept Mot Skills. 1994;78:691–7.

Krause MP, Goss FL, Robertson RJ, Kim K, Elsangedy HM, Krinski K, da Silva S. Concurrent validity of an OMNI rating of perceived exertion scale for bench stepping exercise. J Strength Cond Res. 2012;26:506–12.

Lagally KM, Robertson RJ. Construct validity of the OMNI resistance exercise scale. J Strength Cond Res. 2006;20:252–6.

Lamb KL, Eston RG. Effort perception in children. Sports Med. 1997a;23:139–48.
Lamb KL, Eston RG. Measurement of effort perception: time for a new approach. In: Welsman N, Armstrong N, Kirby B, editors. Children and exercise XIX Vol II J. Exeter, UK: Washington Singer; 1997b. p. 11–23.
Marks LE. Sensory processes: the new psychophysics. New York, NY: Academic; 1974.
Mays RJ, Goss FL, Schafer MA, Kim KH, Nagle-Stilley EF, Robertson RJ. Validation of adult OMNI perceived exertion scales for elliptical ergometry. Percept Mot Skills. 2010;111: 848–62.
Merriam-Webster Online. Dictionary and thesauraus. http://www.merriam-webster.com/netdict.html (2014). Accessed 28 Mar 2014.
Morgan WP. Utility of exertional perception with special reference to underwater exercise. Int J Sport Psychol. 2001;32:137–61.
Noble BJ, Kraemer WJ, Allen JG, Plank JS, Woodard LA. The integration of physiological cues in effort perception: stimulus strength vs. relative contribution. In: Borg G, Ottoson D, editors. The perception of exertion in physical work. London: Macmillan; 1986. p. 83–96.
Noble BJ, Robertson RJ. Perceived exertion. Champaign, IL: Human Kinetics; 1996.
Robertson RJ. Development of the perceived exertion knowledge base: an interdisciplinary process. Int J Psychol. 2001;12:189–96.
Robertson RJ. Perceived exertion for practitioners: rating effort with the OMNI picture system. Champaign, IL: Human Kinetics; 2004.
Robertson RJ, Falkel JE, Drash AL, Swank AM, Metz KF, Spungen SA, LeBoeuf JR. Effect of blood pH on peripheral and central signals of perceived exertion. Med Sci Sports Exerc. 1986;18:114–22.
Robertson RJ, Noble BJ. Perception of physical exertion: methods, mediators and applications. Exerc Sport Sci Rev. 1997;25:407–52.
Robertson RJ, Goss FL, Boer NF, People JA, Foreman AJ, Dabayebeh IM, Millich NB, Balasekaran G, Riechman SE, Gallagher JD, Thompkins T. Children's OMNI scale of perceived exertion: mixed gender and race validation. Med Sci Sports Exerc. 2000;32:452–8.
Robertson RJ, Goss FL, Rutkowski J, Lenz B, Dixon C, Timmer J, Frazee K, Dube J, Andreacci J. Concurrent validation of the OMNI perceived exertion scale for resistance exercise. Med Sci Sports Exerc. 2003;35:333–41.
Robertson RJ, Goss FL, Dube J, Rutkowski J, Dupain M, Brennan C, Andreacci J. Validation of the adult OMNI scale of perceived exertion for cycle ergometer exercise. Med Sci Sports Exerc. 2004;36:102–8.
Robertson RJ, Goss FL, Andreacci J, Dube JJ, Rutkowski JJ, Snee BM, Kowallis RA, Crawford K, Aaron DJ, Metz KF. Validation of the children's OMNI RPE scale for stepping exercise. Med Sci Sports Exerc. 2005;37:290–8.
Utter AC, Robertson RJ, Nieman DC, Kang J. Children's OMNI scale of perceived exertion: walking/running evaluation. Med Sci Sports Exerc. 2002;34:139–44.
Utter AC, Robertson RJ, Green JM, Suminski RR, Mcanulty SR, Nieman DC. Validation of the adult OMNI Scale of perceived exertion for walking/running exercise. Med Sci Sports Exerc. 2004;36:1776–80.
Williams JG, Eston RG, Furlong B. CERT: a perceived exertion scale for young children. Percept Mot Skills. 1994;79:1451–8.

Chapter 3
Exercise-Induced Muscle Pain

Pain has been defined by the International Association for the Study of Pain as *an unpleasant sensory and emotional experience associated with actual or potential tissue damage, or described in terms of such damage* (Merskey and Bogduk 1994). This definition implies that the pain experience is subjective, has an affective component, and may not require actual tissue damage (Borg 1998; O'Connor and Cook 1999). The purpose of this laboratory manual is to focus on the intensity of pain that occurs naturally in active skeletal muscles during exercise, even in individuals who are healthy and injury-free. This perceptual construct is independent of perceived exertion, but the two variables have been measured concurrently during both aerobic and resistance exercise in adults and children (Cook et al. 1997, 1998; Kane et al. 2010; Robertson et al. 2009). Similar to perceived exertion, individuals can rate their intensity of perceived pain by selecting a rating from a range of numerical categories displayed on a scale. Some scales, such as the Pain Intensity Scale and the Children's OMNI Muscle Hurt Scale, include construct-specific verbal and pictorial descriptors placed in juxtaposition to numerical categories representing the range of perceptual responsiveness from no pain at all to unbearable pain. Exercise-induced muscle pain may be an important variable to monitor during exercise testing and prescription because of its potentially powerful implications for the adoption and maintenance of regular PA.

3.1 Pain Threshold During Exercise

In addition to exercise-induced muscle pain intensity, other aspects of pain that have been examined include pain threshold, pain tolerance and the affective components of pain (Cook et al. 1997). The threshold of naturally occurring muscle pain has been assessed during exercise as the time-point at which pain is "just noticeable". This has been measured using a timer that the subject activates to indicate the time-point during an exercise test at which pain sensation is detected (Cook et al. 1997).

L. Haile et al., *Perceived Exertion Laboratory Manual*,
DOI 10.1007/978-1-4939-1917-8_3

In contrast, the construct of perceived exertion does not allow for an onset threshold. It is generally expected that as soon as exercise begins, a certain level of exertion is perceived, the intensity of which can range from very low to very high. However, the onset of muscle pain varies greatly between individuals and in some cases may not occur until 90 % of peak exercise intensity (Cook et al. 1998).

3.2 Mechanisms for Pain During Exercise

Muscle pain during exercise may occur as a result of the stimulation of two separate nociceptive pathways: mechanoreceptors and chemoreceptors. Both of these afferent pathways send information to the central nervous system regarding tissue damage or the potential for tissue damage (O'Connor and Cook 1999). With respect to mechanoreceptor pathways, as physical measures of exercise such as power output (PO), break resistance, or weight lifted increase, there is a corresponding deformation of nociceptive mechanoreceptors. Second, as specific noxious byproducts of metabolism accumulate, such as bradykinin, there may be a greater stimulation of nociceptive chemoreceptors as well as a sensitization of the aforementioned mechanoreceptors. Metabolites such as hydrogen ions sensitize both types of nerve fibers, increase in contracting skeletal muscle as a function of time, and have been shown to accumulate at a faster rate with a greater anaerobic contribution to energy metabolism (Stebbins et al. 1990). In addition, increasing the amount of active musculature during exercise may increase stimulation of both of these nociceptive pathways (Cook et al. 1998).

Noxious sensations detected by the body during exercise are interpreted as a specific level of perceived exertion. As exertional perceptions increase during exercise, they can also be accompanied by increases in muscle-specific pain sensation. The onset of muscle pain sensation usually occurs at higher exercise intensity or at a later point in time during exercise. Muscular sensations of pain change in quality when the noxious stimulus increases in strength, stimulating nociceptors. This neurophysiological sequence results in pain and subsequent voluntary actions or even reflexes to withdraw from the stimulus (Borg 1998). However, it is important to effectively differentiate between perceptions of exertion and pain, especially when measuring them concurrently, as they are not isomorphic constructs. This can be accomplished by using standardized scaling instructions and separate, construct specific scales (Cook et al. 1997; Robertson et al. 2009).

3.3 Clinical Conditions and Pain

For certain clinical conditions, it is important to differentiate between naturally occurring muscle pain during exercise and pain that is symptomatic of a disease or disorder. An individual with cardiovascular disease can experience pain as a result of ischemia (or decreased tissue oxygen supply relative to demand) that is often reproducible during exercise at a specific level of exertion or exercise intensity. This includes angina pectoris, which is chest pain secondary to coronary artery disease, and intermittent claudication, which is pain in the legs secondary to peripheral artery disease. Pain is also a common symptom in diabetic patients with peripheral neuropathies and various types of arthritis. There are also certain disorders that are primarily characterized by pain, such as fibromyalgia and lower back pain. For these conditions, a well-supervised PA program or exercise prescription may be an important aspect of the treatment plan to help improve fitness levels and psychological well-being. Depending on the clinical condition and the degree of functional limitation as a result of a particular disease state, it may be more important to monitor clinically specific pain rather than naturally occurring muscle pain during exercise.

3.4 Rating Exercise-Induced Muscle Pain

Although the concept of pain and related clinical symptoms has been recognized for centuries, the study of naturally occurring muscle pain during exercise is relatively new (Cook et al. 1997). The earliest known investigations include those of Lloyd (1972), who reported pain threshold and pain tolerance during isometric biceps contractions, and Weiser and colleagues (1973), who asked subjects to rate the intensity of "leg aches" and "leg cramps" using a 5-point Likert scale immediately following moderate intensity cycle ergometer exercise. Prior to these studies, research involving the measurement of pain during exercise focused on pain that occurred as a symptom of a disease process, such as claudication secondary to peripheral artery disease. Because research that involved the measurement of muscle and limb pain intensity in healthy individuals employed a mechanical occlusion model to induce claudication pain, the findings of such experiments cannot be considered naturally occurring muscle pain during exercise (O'Connor and Cook 1999).

Borg, best known for his work with perceived exertion, began his study of exercise-induced pain as reported by patients undergoing clinical exercise stress tests. It was found that the quality of patients' rating of perceived exertion was affected by pain often in the form of angina pectoris or induced by musculoskeletal problems (Borg 1998). This led, in part, to the development of a new perceptual scale, the Borg (0–10) Category-Ratio (CR) Scale (CR10 Scale), that could be used for the measurement of perceived exertion and pain. The CR10 Scale was designed with easy-to-use numerical categories and verbal descriptors, similar to the Borg RPE Scale. However, a more complicated format was employed to give the new scale ratio properties similar to the observed changes in the physical and physiological responses often measured along with perceived exertion (Borg 1998). The original CR10 Scale (Borg 1977) had numerical categories ranging from 0 to 10 and contained responses to avoid bottom and ceiling effects which were expanded in a later version (Borg 1998). To avoid bottom effects, the original CR10 Scale contained an additional numerical category, 0.5, and later was expanded to include 0.3, 1.5 and 2.5. To avoid ceiling effects, the scale allowed subjects to rate a number higher than 10 (higher than 11 in the newer version), in case they experience a very high level of perception never encountered before (Borg 1998).

In early studies of exercise-induced pain, Borg et al. (1985) and Ljunggren et al. (1987) used the original version of the CR10 Scale to rate pain sensation, described as "aches and pain in the legs", reported during cycle ergometer testing. The same scale was used to simultaneously rate perceived exertion during cycle ergometry. For both load-incremented and constant PO exercise protocols, moderate to strong correlations were found between pain and perceived exertion (Borg et al. 1985; Ljunggren et al. 1987). Later, it was argued that the measurement of both muscle pain and perceived exertion during exercise should not be so closely linked, and previously reported strong correlations may have been a "*demand artifact*" consequent to using the same perceptual scale to simultaneously measure two independent perceptual constructs (Cook et al. 1997). In response, Cook and colleagues (1997) developed the Pain Intensity Scale (Fig. 3.1), a construct specific scale based on the original numerical category format of the Borg CR10 Scale but containing verbal descriptors specific to exercise-induced muscle pain. The scale included the option to rate a sensory level higher than ten if the pain experienced by the subjects was higher than that ever encountered before. In the first investigation that employed the Pain Intensity Scale, Cook et al. (1997) presented detailed instructions, evidence for reliability and validity, and a study of pain threshold.

0	No pain at all
0.5	Very faint pain (just noticeable)
1	Weak pain
2	Mild pain
3	Moderate pain
4	Somewhat strong pain
5	Strong pain
6	
7	Very strong pain
8	
9	
10	Extremely intense pain (almost unbearable)
•	Unbearable pain

Fig. 3.1 Pain intensity scale (Cook et al. 1998)

3.5 The Children's OMNI-Hurt Scale

Due to the complex construction of the CR10 Scale format (especially the presence of an undesignated upper response category) and the strong validity of the OMNI Scales for rating perceived exertion in children, Robertson and colleagues (2009) developed the Children's OMNI-Hurt Scale. The term hurt was used rather than pain because it is more commonly expressed by children to describe their nociceptive feelings (Hicks et al. 2001). Robertson et al. (2009) extended the observation of Cook et al. (1997) that construct specific perceptual scales were necessary to concurrently and differentially rate perceived exertion and muscle pain when assessed during the same bout of exercise, especially when children were evaluated. Therefore, the Children's OMNI-Hurt Scale was developed based on the principles of the Children's OMNI RPE Scales. The scale contains numerical categories and construct specific verbal and pictorial descriptors appropriate for children (Fig. 3.2). The pictorial descriptors were adapted from the Faces Pain Scale used to assess clinical pain experienced during the treatment of diseases (Hockenberry et al. 2005).

The Children's OMNI-Hurt Scale was used initially to measure naturally occurring muscle hurt (i.e., pain) during resistance exercise in 10–14 year old children (Robertson et al. 2009). The results of this investigation indicated that the children could differentially rate the intensity of both muscle hurt and perceived exertion when measured during isotonic resistance exercise. Perceived exertion was measured by the Children's OMNI-Resistance Exercise RPE Scale. Even though both scales share a common foundational format having the same numerical categories, children were able to identify muscle hurt and perceived exertion as separate perceptual constructs. This conclusion was based on the moderate correlations

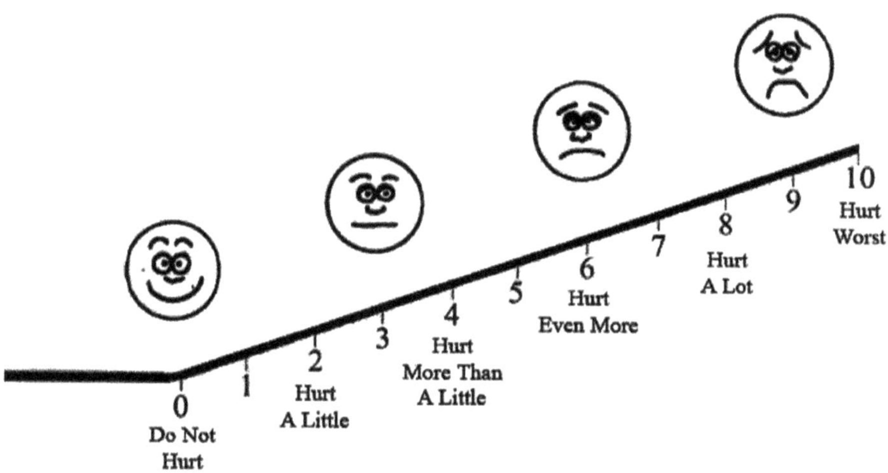

Fig. 3.2 Children's OMNI-hurt scale

(ranging from $r=0.19$–0.82) between the two perceptual ratings during unilateral arm curl and leg extension exercise (Robertson et al. 2009). It was observed that construct specific verbal and pictorial descriptors, as well as standardized instruction with special attention paid to the differentiation between muscle hurt and perceived exertion, may be necessary for this level of perceptual differentiation, especially in children.

3.6 The Use of Scales for Exercise-Induced Pain

In its formative stages, measures of exercise-induced pain were used to model, measure and monitor the pain symptoms of disease, clinical disorders and injury processes (Borg 1998; O'Connor and Cook 1999). Over time, the field progressed to include pain assessment in healthy and injury-free individuals in order to quantify a potential barrier to exercise participation. Using the same measurement principles as applied in the assessment of perceived exertion, exercise-induced pain is easy to assess in individuals ranging widely in demographic characteristics. Whether the individual is a child or adult, special attention should be paid to differentiating between the constructs of perceived exertion and pain, especially when the two are measured concurrently. The applications of exercise-induced pain are presented in this laboratory manual in conjunction with perceptual paradigms developed originally, in most cases, for perceived exertion. However, for many individuals the measurement of exercise-induced pain, such as when used to determine the exercise intensity corresponding to the pain threshold, may be an important factor in optimizing PA programming to achieve health-fitness goals.

References

Borg G, editor. Physical work and effort. Oxford, UK: Pergamon; 1977.

Borg G. Borg's perceived exertion and pain scales. Champaign, IL: Human Kinetics; 1998.

Borg G, Ljunggren G, Ceci R. The increase of perceived exertion, aches and pain in the legs, heart rate and blood lactate during exercise on a bicycle ergometer. Eur J Appl Physiol. 1985;54: 343–9.

Cook DB, O'Connor PJ, Eubanks SA, Smith JC, Lee M. Naturally occurring muscle pain during exercise: assessment and experimental evidence. Med Sci Sports Exerc. 1997;29:999–1012.

Cook DB, O'Connor PJ, Oliver SE, Lee Y. Sex differences in naturally occurring muscle pain and exertion during maximal cycle ergometry. Intern J Neurosci. 1998;95:183–202.

Hicks CL, von Baeyer CL, Spafford P, van Korlaar I, Goodenough B. The faces of pain scale-revised: toward a common metric in pediatric pain measurement. Pain. 2001;93:173–83.

Hockenberry MJ, Wilson D, Winkelstein ML. Wong's essentials of pediatric nursing. 7th ed. St. Louis, MO: Mosby; 2005.

Kane I, Robertson RJ, Fertman CI, McConnaha WR, Nagle EF, Rabin BS, Rubinstein EN. Predicted and actual exercise discomfort in middle school children. Med Sci Sports Exerc. 2010;42: 1013–21.

Ljunggren G, Ceci R, Karlsson J. Prolonged exercise at a constant load on a bicycle ergometer: ratings of perceived exertion and leg aches and pain as well as measurements of blood lactate accumulation and heart rate. Int J Sports Med. 1987;8:109–16.

Lloyd AJ. Auditory EMG feedback during a sustained submaximum isometric contraction. Res Q. 1972;43:39–46.

Merskey H, Bogduk N. Classification of chronic pain: description of chronic pain syndromes and definitions of pain terms. Seattle, WA: IASP; 1994.

O'Connor PJ, Cook DB. Exercise and pain: the neurobiology, measurement, and laboratory study of pain in relation to exercise in humans. Exerc Sport Sci Rev. 1999;29:119–66.

Robertson RJ, Goss FL, Aaron DJ, Nagle EF, Gallagher Jr M, Kane IR, Tessmer KA, Schafer MA, Hunt SE. Concurrent muscle hurt and perceived exertion of children during resistance exercise. Med Sci Sports Exerc. 2009;41:1146–54.

Stebbins CL, Carretero OA, Mindroiu T, Longhurst JC. Bradykinin release from contracting skeletal muscle of the cat. J Appl Physiol. 1990;69:1225–30.

Weiser PC, Kinsman RA, Stamper DA. Task-specific symptomatology changes resulting from prolonged submaximal bicycle riding. Med Sci Sports. 1973;5:79–85.

Chapter 4
The Affective Response to Exercise

An affective response (AR) can be defined as *the general psychological state of an individual, including but not limited to emotions and mood, within a given situation* (Ekkekakis and Petruzzello 2002). In its most basic context, AR is used to describe an individual's subjective experience (i.e., intrapersonal or experiential core) of all valenced responses; i.e., those that involve the potential for both positive and negative dimensions. Similar to RPE and exercise-induced muscle pain responses, individuals can rate the AR to exercise and PA by selecting a number from a range of numerical categories (both positive and negative) displayed on an affect scale. Scales designed to measure AR, such as the Feeling Scale (FS), include verbal descriptors placed in juxtaposition to numerical categories representing the range of affective responsiveness from the most negative feelings to the most positive feelings regarding the exercise situation. The Feeling Scale has been used to assess the range of overall feelings an individual may experience before, during and after PA (Ekkekakis 2008; Haile et al. 2013; Hardy and Rejeski 1989). In addition, scales have been designed to measure specific affective domains, such as enjoyment during PA. The measurement of AR and enjoyment during exercise, along with perceived exertion, may be crucial to identifying the types of exercise programs that provide health-fitness benefits and promote psychological well-being.

4.1 Cognitive Appraisal in the Formation of AR

An individual's overall AR to a given situation, such as a bout of exercise, may be shaped by more complex affective phenomena such as specific domains, emotions and mood states (Ekkekakis and Petruzzello 2000). For example, an individual's overall feelings during a given situation may range from very good or pleasurable to very bad or displeasurable. However, when an individual decides the specific point along a continuum that accurately describes overall AR, a process of cognitive appraisal is engaged. This process, whether subconscious or conscious, integrates

L. Haile et al., *Perceived Exertion Laboratory Manual*,
DOI 10.1007/978-1-4939-1917-8_4

information from the internal and external environment to evaluate the current situation regarding its meaning for the individual's survival and well-being (Ekkekakis 2003; Lazarus 1991).

4.2 Significance of Affective Responses to Exercise

Since the development of the Borg RPE Scale over 50 years ago (Borg 1962), perceived exertion has dominated the literature attempting to explain the subjective response to an exercise stimulus. As demonstrated by the Global Explanatory Model of Perceived Exertion in Chap. 2 (see Fig. 2.1), RPE is a gestalt-like perceptual response that involves a complex pattern of physiological, psychosocial, performance-related and symptomatic process mediators. These mediators, functioning individually or collectively, ultimately shape the RPE response to exercise. Hardy and Rejeski (1989) cited Borg (1962) when they suggested "Because RPE represents a 'gestalt' of various sensations related to the stress and strain of physical work, it may not accurately reflect the affect a person feels during exercise." In their example, two individuals performing exercise may give the same RPE for a given workload (e.g., 15 on the 6–20 Borg Scale). However, one individual may feel "good" while the other feels "bad" at the same level of exertion (Hardy and Rejeski 1989). Research has suggested that one's overall AR during exercise, as well as specific affective states such as exercise enjoyment and social physique anxiety, can play an important role in regular PA participation, the adherence to an exercise program, and the potential of an individual to withdraw from participating in exercise (Ekkekakis and Lind 2006; Parfitt et al. 2006; Wankel 1993). Therefore, it is not only important to determine "what" one is feeling during exercise (i.e., RPE), but also "how" one feels during exercise (i.e., AR) (Hardy and Rejeski 1989).

4.3 The Affect Circumplex Model

In the early factor analyses of terms used to describe various affective states reported by subjects, researchers identified up to a dozen independent affective states (Nowlis and Nowlis 1956), including anger, anxiety, elation, sadness, and tension, to name a few. It was generally agreed that each of these affective states could be treated as separate dimensions, which served as the basis for early monopolar scales used to measure the degrees of these various states (Izard 1972; McNair et al. 1971; Nowlis 1965; Thayer 1967). Another school of thought was that these various self-reported affective states were related to one another and that this relation could be described using two bipolar scales that bisect one another: one ranging from pleasantness/pleasure to unpleasantness/displeasure; the other from attention/arousal to rejection (i.e., degrees of arousal) (Schlosberg 1952). This was termed as the *circumplex* model. Studies involving subject identification of facial expressions (Cliff and Young 1968; Royal and Hays 1959; Schlosberg 1952; Shepard 1962a, 1962b) and vocal expressions of emotions (Green and Cliff 1975) agreed with this circumplex model of affect. Whereas each observed affective state could be placed on the continuum of each of these bipolar scales, certain affective states were found to be better explained as a combination of these two bipolar scales (Russell and Pratt 1980). For example, one could feel excitement, an affective state being pleasant and having a high degree of arousal. Other examples include contentment (pleasant, low arousal), distress (unpleasant, high arousal), and depression (unpleasant, low arousal) (Russel 1980).

4.4 The Challenge of Measuring AR During Exercise

Affect includes numerous dimensions that can each be analyzed separately. In addition, the study of PA can include physiologically and psychologically diverse subjects performing exercise of varying intensity and modalities in diverse environments (Ekkekakis 2008). Therefore, numerous metrics have been used to study the relation between PA and AR. Two such metrics involving affective domains include the State-Trait Anxiety Inventory (Spielberger et al. 1970) and the Profile of Mood States (McNair et al. 1971). However, the use of these metrics has assumed that very specific affective domains can represent the whole affective spectrum as may be represented in a diverse group of individuals (Ekkekakis 2008). Other metrics developed specifically for the study of the PA–AR relation have defined, and therefore have limited, this relation to include specific affective domains. These include the Exercise-Induced Feeling Inventory (Gauvin and Rejeski 1993), the Subjective Exercise Experiences Scale (McAuley and Courneya 1994), and their derivations (Annesi 2006; Lox et al. 2000; Rejeski et al. 1999). However, these metrics were developed in healthy, physically active individuals and may not represent the affective experience of other subject populations (Ekkekakis 2008), especially those who have significant barriers to the adoption and maintenance of regular PA. Ekkekakis (2008) described this effect as domain under-representation, which has been demonstrated in sedentary adults (Gauvin et al. 1997) and older adults with osteoarthritis (Focht et al. 2004).

Other limitations of scales used to measure the PA–AR relation is that they are often difficult for the subject to understand and difficult to use during the actual exercise performance. It has been suggested that the average person's understanding of affective states may be overly simplified if not mistaken (Russel 1980). However, from a public health perspective, it is important to understand an individual's decision making process regarding PA participation. Therefore, PA research should employ an easy-to-use AR metric that can accurately describe the overall feelings of a wide range of individuals. Many AR scales involve questionnaires with numerous test items each targeting specific affective domains. These questionnaires cannot be used effectively to assess a subject's AR during acute PA. This is especially the case during high intensity exercise when one's affective state can change rapidly due to increases in the intensity of interoceptive physiological cues originating from exercising muscle and cardiorespiratory responses (Ekkekakis 2003). Therefore, to measure AR during acute exercise, it is necessary to have a metric that can be used quickly by subjects. Therefore, it would be beneficial to have affect rating scales that assess the overall AR rather than specific affective domains during an acute bout of exercise. This type of rating scale could be similar to an RPE scale. The overall AR is formed by incorporating all of the various affective domains important to that individual. This type of scale allows individuals to interpret their own subjective feelings about the exercise experience using their own unique thought processes. In addition, rating the overall AR may provide a construct that is easier to understand for the average individual, especially a child, rather than measuring an entire set of affective domains.

4.5 Rating AR using The Feeling Scale

Rejeski and colleagues (1987) developed the Feeling Scale (FS, Fig. 4.1) to measure AR during exercise. The scale was intended to be easy to use and easily understandable. The FS is not intended to measure various categories or specific domains of affect because it was believed that the strongest aspect of AR is the initial determination of good or bad, a more global response (Hardy and Rejeski 1989; Weiner et al. 1979). The FS was based on one of the core bipolar constructs of Russell's (1980) circumplex model that could be used to differentiate between feelings along the continuum of core emotions ranging from pleasantness/pleasure to unpleasantness/displeasure (Frijda 1988). The FS is an 11-point bipolar (i.e., valenced) metric with numerical categories ranging from –5 to 5. Verbal descriptors were situated in

+5	Very Good
+4	
+3	Good
+2	
+1	Fairly Good
0	Neutral
- 1	Fairly Bad
- 2	
- 3	Bad
- 4	
- 5	Very Bad

Fig. 4.1 Feeling scale (Rejeski et al. 1987)

juxtaposition to each odd integer, with numerical categories ranging from "very bad" (at −5), representing maximal displeasure/unpleasantness, to "neutral" (at 0), to "very good" (at +5) representing maximal pleasure/pleasantness (Rejeski et al. 1987).

The FS has demonstrated face, content and construct validity (Hardy and Rejeski 1989; Rejeski et al. 1987) and has been widely used to measure AR in various subject populations performing a variety of exercise situations (Ekkekakis et al. 2000; Ekkekakis et al. 2004; Ekkekakis and Petruzzello 2002; Lind et al. 2008; Parfitt et al. 2006). This line of research has indicated that, although affective states are considered as a psychosocial correlate of exertional perceptions, AR measured using the FS and RPE are not isomorphic constructs. When subjects were asked to rate both AR and RPE during cycle ergometer exercise bouts of various workloads (30, 60, and 90 % of peak VO_2), the constructs were only moderately inversely correlated, with resultant coefficients ranging from $r = -0.33$ to -0.55 (Hardy and Rejeski 1989).

The FS was based on the bipolar construct of the circumplex model. This model differentiates between feelings distributed along the affective continuum from pleasantness/pleasure to unpleasantness/displeasure (Russel 1980). Therefore, consideration has been given to the bipolar construct of the circumplex model that bisects pleasure–displeasure during exercise, and analogously differentiates between feelings along the affective continuum from low to high arousal (or activation). The single-item Felt Arousal Scale (FAS), developed by Svebak and Murgatroyd (1985), is a 6-point scale that has been used to measure the degree of perceived activation associated with exercise (Kerr and Vlaswinkel 1993; Kerr and van den Wollenberg 1997; Hall et al. 2002). Previous research has measured the AR to exercise using both the FS and FAS to investigate the application of the circumplex model in explaining exercise behavior (Hall et al. 2002). Evidence suggests that FS ratings of AR that distribute along the pleasure–displeasure continuum have a strong practical application to exercise intensity prescription and PA program adherence. The application of ratings of perceived activation using the FAS requires further investigation. As such, this manual will only focus its discussion of AR as measured using the FS. The assumption here is that the FS provides a global AR measurement indicating good or bad feelings during exercise (Hardy and Rejeski 1989).

4.6 Exercise Intensity and AR: The Dual-Mode Model

Ekkekakis (2003) proposed the "dual-mode model" to explain inter-individual differences in AR during exercise of varying intensities. As described by Parfitt and colleagues (2006), the model "is based upon the interplay of relevant cognitive processes and interoceptive cues prior to and following the transition from aerobic to anaerobic metabolism", namely, the anaerobic threshold (AT) as identified using the ventilatory threshold (VT) or lactate threshold (LT). At exercise intensities below the AT, when the energy metabolism is supported by aerobic pathways, the acute AR is primarily influenced by cognitive processes such as appraisal, self-efficacy, and social context. These cognitive processes are shaped by personal experience, individual personality variables, personal goal achievement, etc. All of these factors are unique to the individual (Ekkekakis 2003). Thus, there may be heterogeneity in AR at intensities below the AT due to the inter-individual differences in interpretation of the exercise (Rose and Parfitt 2007). At exercise intensities above the AT, when energy metabolism is supported by ever-increasing anaerobic sources, lactacidemia in active muscle and decreasing tissue and blood pH become significant mediators of the acute AR. In addition, AR is influenced by interoceptive cues from baroreceptors, thermoreceptors, and visceroreceptors in the heart and lungs (Rose and Parfitt 2007). Thus, there would be less inter-individual variability in AR during exercise at intensities above the AT because the affect experience is shaped less by cognitive processes and more by physiological cues that disrupt metabolic homeostasis (Ekkekakis 2003; Parfitt et al. 2006).

Research has supported the "dual-mode model" in that the acute AR to exercise declined once the intensity exceeded the AT (Ekkekakis et al. 2005; Hall et al. 2002). It was shown that, while 47 % of subjects exhibited a decline in AR during 15 min of treadmill exercise intensities below their VT, 80 % of subjects exhibited a decline in affect during similar exercise at intensities above their VT (Ekkekakis et al. 2005). Similar results were reported by Parfitt and colleagues (2006) when subjects performed 20 min of treadmill exercise. AR was more positive and stable when subjects performed intensities below their LT, with only 25 % of subjects exhibiting a decline in AR during exercise. When subjects performed intensities above their LT, AR became increasingly more negative with 83 % of subjects exhibiting a decline in AR during exercise (Parfitt et al. 2006).

4.7 PA Enjoyment: A Specific Affective Domain

The AR to exercise is an all-encompassing estimation of the overall feelings an individual experiences during exercise. Each individual's AR to PA is an integration of specific affective domains, shaped by emotions and mood states after evaluation of the exercise situation through the cognitive appraisal process (Ekkekakis 2003). Enjoyment can be defined as a positive affective or emotional state that reflects the feelings of fun, liking or pleasure (Wankel 1993). Enjoyment experienced during PA (PAE) may be a critical element in promoting adherence to a PA program and improving psychological well-being (Wankel 1993). In fact, enjoyment has been found to be associated with higher levels of PA participation in studies involving healthy children (DiLorenzo et al. 1998; Sallis et al. 1999; Trost et al. 1997) and adults (Dacey et al. 2008; McArthur and Raedeke 2009; Williams et al. 2006), as well as adults with clinical conditions (Hagberg et al. 2009).

4.7.1 Questionnaire Methodology to Assess PAE

The assessment of PAE has been undertaken primarily using questionnaire methodology that involves multiple test items with questions that are rated on 5- or 7-point Likert scales. The most prominent example in the literature is the Physical Activity Enjoyment Scale (PACES, Kendzierski and DeCarlo 1991) which asks participants to rate how they feel at the present moment about previous PA using 18 different test items. Abbreviated versions of PACES have been developed that have decreased the number of test items (Mullen et al. 2011; Paxton et al. 2008). Investigations using PACES have focused on PAE measurement in youth (Crisp et al. 2012; Grieser et al. 2008; Schneider and Graham 2009; Schneider and Cooper 2011). Grieser and colleagues (2008) found that black and Hispanic girls reported lower PAE, as well as lower levels of support for PA within the school setting, compared to white girls of the same age. Schneider and Graham (2009) found PAE to be significantly correlated with scores from a questionnaire regarding the Behavioral Activation System (BAS) in adolescent males and females. In theory, the BAS is a neurobehavioral mechanism that governs one's drive to undertake goal-oriented behavior and experience positive emotions. In contrast, PAE was not correlated to scores regarding the Behavioral Inhibition System (BIS). The BIS is proposed as a mechanism that drives an individual away from threatening or unknown situations and causes negative emotions (Carver and White 1994; Schneider and Graham 2009). Schneider and Cooper (2011) found that baseline levels of PAE moderated exercise behavior change across a 9-month, school-based PA intervention in adolescent girls. Specifically, those girls reporting low PAE before undertaking the intervention significantly increased their vigorous PA level across the 9-month participation period. In contrast, those girls reporting high baseline PAE levels showed no change in vigorous PA during the intervention (Schneider and Cooper 2011). Crisp and colleagues (2012) measured PAE following two cycle ergometer exercise bouts in normal and overweight boys: a continuous cycle ergometer exercise bout performed at the intensity of maximal fat oxidation, and a continuous cycle ergometer exercise bout performed at the same intensity with added sprint intervals. Adding sprint intervals to the continuous cycle ergometer exercise bout did not change PAE scores compared to the continuous cycle ergometer exercise performed without the sprint intervals, even though the subjects reported to prefer the exercise that included sprints. In addition, energy expenditure during exercise was increased with added sprints and post-exercise energy intake was not different between trials (Crisp et al. 2012). Two recent investigations have used PACES to assess PAE in adults. Bartlett and colleagues (2011) found that high intensity interval running induced higher post-exercise PAE than moderate intensity continuous running in recreationally active young men. In addition, Mullen and colleagues (2011) recently validated an abbreviated version of PACES to measure PAE in older adults.

4.7.2 Limitations of Questionnaire Methodology to Assess PAE

The results of research involving PACES and other PAE questionnaires have provided potentially valuable information regarding individuals' memory of enjoyment during previous PA that may predict future PA participation. A limitation of survey PAE metrics is that they are administered off-stimulus, i.e., not while the subject is actually performing a bout of PA. Completion of these metrics can be time-consuming and, as such, are not administered on-stimulus, i.e., during an acute bout of PA. When using a questionnaire such as the PACES, individuals can be asked to rate PAE for (a) recent PA in general, (b) after participation in a specific PA program or intervention, or (c) immediately after a specific bout of PA. However, these questionnaires are not designed to indicate the change in PAE that may occur across the acute exercise bout (before, during and after) or to differentiate between specific intensities of exercise.

4.7.3 Single-Item Metrics for Assessment of PAE During Exercise

Research has found considerable inter-individual variability in AR during exercise, especially of a moderate intensity (Ekkekakis 2008). In addition, post-exercise estimations of AR to previous exercise may not be indicative of the acute exercise experience (Haile et al. 2013). Therefore, a single-item bipolar scale similar to the FS was developed to measure AR during exercise. It was anticipated that this single-item metric would allow an evaluation of the dynamics of PAE across the exercise stimulus similar to previous research involving AR. The FS integrates all valenced affective states into a continuum of numerical categories. A subject is asked to estimate AR based on the initial, and perhaps strongest, emotional aspect of the exercise experience that places overall feelings on a scale ranging from very bad to very good. Using a similar measurement construct, a scale was developed by Haile and colleagues (2012) specific to PAE. In this context, PAE was considered as an affective domain with important implications regarding PA adoption and maintenance. The scale was modeled after the FS to assess the most basic aspect of enjoyment by asking subjects to rate overall feelings of PAE on a continuum ranging from very unenjoyable to very enjoyable (Fig. 4.2). The investigation by Haile and colleagues (2012) measured acute exercise PAE, physiological variables (VO_2, HR), RPE and acute exercise AR during a load-incremented cycle ergometer exercise test that terminated at peak intensity in college-aged adults. Significant low to moderate negative correlations ($r = -0.33$ to -0.53) were found for the relation between PAE and both physiological variables and RPE. As expected, a significant correlation was found between PAE and AR, $r = 0.92$ (Haile et al. 2012). Stanley and colleagues (2009; Stanley and Cumming 2010) developed a similar 7-point, single-item PAE metric which they used to measure the PAE response to cycle ergometer exercise. Significant correlations were found between PAE and FS ratings of AR before, during, and after exercise ($r = 0.41$–0.55) (Stanley et al. 2009).

Fig. 4.2 Bipolar physical activity enjoyment scale (Haile et al. 2012)

+5	**Very Enjoyable**
+4	
+3	**Enjoyable**
+2	
+1	**Slightly Enjoyable**
0	**Neutral**
- 1	**Slightly Unenjoyable**
- 2	
- 3	**Unenjoyable**
- 4	
- 5	**Very Unenjoyable**

4.8 The Use of Scales for AR and PAE

Although many scales and questionnaire methods have been developed to assess specific psychosocial affective domains, overall AR, and exercise specific AR, few metrics are appropriate for the assessment of these responses across the entire time sequence of an exercise test, i.e., before, during, and after exercise. The FS and single-item PAE scales have this measurement capability. A unique characteristic of these scales compared to survey methods is their simplicity. These scales ask the subject to rate the most basic aspects of the AR or PAE. Do you feel good or bad? Is the experience enjoyable or unenjoyable? Subjects are asked to choose the degree to which they feel positive (good/enjoyable) or negative (bad/unenjoyable) about the experience, but can also choose neutral if they feel neither good/enjoyable nor bad/unenjoyable? This characteristic allows the scales to be easily understood by adults and children because it asks them to rate their PA experience using terms they are familiar with and use to rate behavior and experiences of everyday life. The simplicity of these scales does not take away from the significance of the information they can provide about the constructs they measure. AR and PAE have long been proposed as important links in the chain between appropriate exercise intensity and the eventual adoption and maintenance of regular PA. However, it must be understood that the exercise intensity that elicits the optimal level of AR and PAE may vary greatly between individuals.

The applications of AR and PAE measurement during exercise that are presented in this laboratory manual are in conjunction with perceived exertion paradigms designed for purposes of effective exercise prescription. The simultaneous measurement of perceived exertion and affective variables during an acute exercise bout provides the health-fitness professional with information that can determine both physiological and psychological benefits to PA programming. At times, especially for previously sedentary individuals at the onset of an exercise program, physiological benefits may have to be considered as secondary in order to promote adherence. However, continued PA participation is not likely if physiological benefits, such as improved cardiorespiratory fitness and weight loss, are not achieved. Therefore, knowledge regarding the measurement of both perceptual and affective variables associated with exercise prepares health-fitness professionals to weigh the costs and benefits of the various characteristics of an effective exercise prescription.

References

Annesi JJ. Preliminary testing of a brief inventory for assessing changes in exercise-induced feeling states. Percept Mot Skills. 2006;102:776–80.

Bartlett JD, Close GL, Maclaren DPM, Gregson W, Drust B, Morton JP. High-intensity interval running is perceived to be more enjoyable than moderate-intensity continuous exercise: implications for exercise adherence. J Sports Sci. 2011;29:547–53.

Borg G. Physical performance and perceived exertion. Lund, Sweden: Gleerup; 1962.

Carver CS, White TL. Behavioral inhibition, behavioral activation, and affective responses to impending reward and punishment: the BIS/BAS scales. J Pers Soc Psychol. 1994;67:319–33.

Cliff N, Young FW. On the relation between unidimensional judgements and multidimensional scaling. Organ Behav Hum Perform. 1968;3:269–85.

Crisp NA, Fournier PA, Licari MK, Braham R, Guelfi KJ. Optimising sprint interval exercise to maximise energy expenditure and enjoyment in overweight boys. Appl Physiol Nutr Metab. 2012;37:1222–31.

Dacey M, Baltzell A, Zaichowsky L. Older adults intrinsic and extrinsic motivation toward physical activity. Am J Health Behav. 2008;32:570–82.

DiLorenzo TM, Stuckey-Ropp RC, Vander Wal JS, Gotham HJ. Determinants of exercise among children. II. A longitudinal analysis. Prev Med. 1998;27:470–7.

Ekkekakis P. Pleasure and displeasure from the body: perspectives from exercise. Cogn Emot. 2003;17:213–39.

Ekkekakis P. Affect circumplex redux: the discussion on its utility as a measurement framework in exercise psychology continues. Int Rev Sport Exerc Psychol. 2008;1:139–59.

Ekkekakis P, Hall EE, Van Lunduyt LM, Petruzzello SJ. Walking in (affective) circles. Can short walks enhance affect? J Behav Med. 2000;23:245–75.

Ekkekakis P, Petruzzello SJ. Analysis of the affect measurement conundrum in exercise psychology: I. Fundamental issues. Psychol Sport Exerc. 2000;2:71–88.

Ekkekakis P, Petruzzello SJ. Analysis of the affect measurement conundrum in exercise psychology: IV. A conceptual case for the affect circumplex. Psychol Sport Exerc. 2002;3:35–63.

Ekkekakis P, Hall EE, Petruzzello SJ. Practical markers of the transition from aerobic to anaerobic metabolism during exercise: rationale and a case for affect-based exercise prescription. Prev Med. 2004;38:149–59.

Ekkekakis P, Hall EE, Petruzzello SJ. Variations and homogeneity in affective responses to physical activity of varying intensities: an alternative perspective on dose–response based on evolutionary considerations. J Sport Sci. 2005;23:477–500.

Ekkekakis P, Lind E. Exercise does not feel the same when you are overweight: the impact of self-selected and imposed intensity on affect and exertion. Int J Obes. 2006;30:652–60.

Focht BC, Knapp DJ, Gavin TP, Raedeke TD, Hickner RC. Affective and self-efficacy responses to acute aerobic exercise in sedentary older and younger adults. J Aging Phys Act. 2004;15:123–38.

Frijda NH. The laws of emotion. Am Psychol. 1988;43:349–58.

Gauvin L, Rejeski WJ, Norris JL, Lutes L. The curse of inactivity: failure of acute exercise to enhance feeling states in a community sample of sedentary adults. J Health Psychol. 1997;2:509–23.

Gauvin L, Rejeski WJ. The exercise-induce feeling inventory: development and initial validation. J Sport Exerc Psychol. 1993;15:403–23.

Green RS, Cliff N. Multidimensional comparisons of structures of vocally and facially expressed emotions. Percept Psychophys. 1975;17:429–38.

Grieser M, Neumark-Sztainer D, Saksvig BI, Lee J, Felton GM, Kubik MY. Black, Hispanic, and white girls' perceptions of environmental and social support and enjoyment of physical activity. J School Health. 2008;78:314–20.

Hagberg LA, Lindahl B, Nyberg L, Hellenius M. Importance of enjoyment when promoting physical exercise. Scand J Med Sci Sports. 2009;19:740–7.

Haile AM, Haile L, Taylor M, Shafer A, Wisniewski K, Deldin A, Panzak G, Goss FL, Nagle E, Robertson RJ. Concurrent validity of an exercise enjoyment scale using physiological and psychological criteria. Med Sci Sports Exerc. 2012;44:S645.

Haile L, Goss FL, Robertson RJ, Andreacci JL, Gallagher M, Nagle EF. Session perceived exertion and affective responses to self-selected and imposed cycle exercise of the same intensity in young men. Eur J Appl Physiol. 2013;116:1755–65.

Hall EE, Ekkekakis P, Petruzzello SP. The affective beneficence of vigorous exercise revisited. Br J Health Psychol. 2002;7:47–66.

Hardy CJ, Rejeski WJ. Not what but how one feels: the measurement of affect during exercise. J Sport Exerc Psychol. 1989;11:304–17.

Izard CE. Patterns of emotions. New York, NY: Academic; 1972.

Kendzierski D, DeCarlo K. Physical activity enjoyment scale: two validation studies. J Sport Exerc Psychol. 1991;13:50–64.

Kerr JH, Vlaswinkel EH. Self-reported mood and running under natural conditions. Work Stress. 1993;7:161–77.

Kerr JH, van den Wollenberg A. High and low intensity exercise and psychological mood states. Psychol Health. 1997;12:603–18.

Lazarus RS. Emotion and adaptation. New York, NY: Oxford University Press; 1991.

Lind E, Ekkekakis P, Vazou S. The affective impact of exercise intensity that slightly exceeds the preferred level: 'pain' for no additional 'gain.'. J Health Psychol. 2008;13:464–8.

Lox CL, Jackson S, Tuholski SW, Wasley D, Treasure DC. Revisiting the measurement of exercise-induced feeling states: the physical activity affect scale (PAAS). Meas Phys Educ Exerc Sci. 2000;4:79–95.

McAuley E, Courneya KS. The subjective exercise experiences scale (SEES): development and preliminary validation. J Sport Exerc Psychol. 1994;16:163–77.

McArthur L, Raedeke T. Race and sex differences in college student physical activity correlates. Am J Health Behav. 2009;33:80–90.

McNair DM, Lorr M, Droppleman LF. Manual: profile of mood states. San Diego, CA: Educational and Industrial Testing Services; 1971.

Mullen SP, Olson EA, Phillips SM, Szabo AN, Wojcicki TR, Mailey EL, Gothe NP, Fanning JT, Kramer AF, McAuley E (2011) Measuring enjoyment of physical activity in older adults: invariance of the physical activity enjoyment scale (paces) across groups and time. Int J Behav Nutr Phys Act. doi:10.1186/1479-5868-8-103.

Nowlis V. Research with the mood adjective check list. In: Tomkins SS, Izard CE, editors. Affect, cognition, and personality. New York, NY: Springer; 1965.

Nowlis V, Nowlis HH. The description and analysis of mood. Ann NY Acad Sci. 1956;65:345–55.

Parfitt G, Rose EA, Burgess WM. The psychological and physiological responses of sedentary individuals to prescribed and preferred intensity exercise. Br J Health Psychol. 2006;11:39–53.

Paxton RJ, Nigg C, Motl RW, Yamashita M, Chung R, Battista J, Change J. Physical activity enjoyment scale short form: does it fit for children? Res Q Exerc Sport. 2008;79:423–7.

Rejeski WJ, Best D, Griffith P, Kenney E. Sex-role orientation and the responses of men to exercise stress. Res Q. 1987;58:260–4.

Rejeski WJ, Reboussin BA, Dunn AL, King AC, Sallis JF. A modified exercise-induced feeling inventory for chronic training and baseline profiles of participants in the activity counseling trial. J Health Psychol. 1999;4:97–108.

Rose EA, Parfitt G. A quantitative analysis and qualitative explanation of the individual differences in affective responses to prescribed and self-selected exercise intensities. J Sport Exerc Psychol. 2007;29:335–54.

Royal DC, Hays WL. Empirical dimensions of emotional behavior. Acta Psychol. 1959;15:419.
Russel JA. A circumplex model of affect. J Pers Soc Psychol. 1980;39:1161–78.
Russell JA, Pratt GA. A description of the affective quality attributed to environments. J Pers Soc Psychol. 1980;38:311–22.
Sallis JF, Prochaska J, Taylor W, Hill J, Geraci J. Correlates of physical activity in national sample of girls and boys in grades 4 through 12. Health Psychol. 1999;18:410–5.
Schlosberg H. The description of facial expressions in terms of two dimensions. J Exp Psychol. 1952;44:229–37.
Schneider M, Graham D. Personality, physical fitness, and affective responses to exercise among adolescents. Med Sci Sports Exerc. 2009;41:947–55.
Schneider M, Cooper DM (2011) Enjoyment of exercise moderates the impact of a school-based physical activity intervention. Int J Behav Nutr Phys Act. doi:10.1186/1479-5868-8-64.
Shepard RN. The analysis of proximities: multidimensional scaling with an unknown distance function. I Psychometrika. 1962a;27:125–39.
Shepard RN. The analysis of proximities: multidimensional scaling with an unknown distance function. II Psychometrika. 1962b;27:219–46.
Spielberger CD, Gorsuch RL, Lushene RE. Manual for the state-trait anxiety inventory. Palo Alto, CA: Consulting Psychologists; 1970.
Stanley DM, Williams SE, Cumming J. Preliminary validation of a single-item measure of exercise enjoyment: the exercise enjoyment scale. J Sport Exerc Psychol. 2009;31:S138–9.
Stanley DM, Cumming J. Are we having fun yet? Testing the effects of imagery use on the affective and enjoyment responses to acute moderate exercise. Psychol Sport Exerc. 2010;11:582–90.
Svebak S, Murgatroyd S. Metamotivational dominance: a multimethod validation of reversal theory constructs. J Pers Soc Psychol. 1985;48:107–16.
Thayer RE. Measurement of activation through self-report. Psychol Rep. 1967;20:663–78.
Trost SG, Pate RR, Saunders R, Ward DS, Dowda M, Felton G. A prospective study of determinants of physical activity in rural fifth-grade children. Prev Med. 1997;26:257–63.
Wankel LM. The importance of enjoyment to adherence and psychological benefits from physical activity. Int J Sports Psychol. 1993;24:151–69.
Weiner B, Russell D, Lerman D. The cognitive-emotion process in achievement related contexts. J Pers Soc Psychol. 1979;37:1211–20.
Williams D, Papandonatos G, Napolitano M, Lewis B, Whiteley J. Perceived enjoyment moderats the efficacy of an individually tailored physical activity intervention. J Sport Exerc Psychol. 2006;28:300–9.

Part II
Assessment

Chapter 5
Perceived Exertion Scaling Procedures

Borg has developed and validated two empirical models that explain: (a) psycho-physiological interdependence during exercise (i.e., Effort Continua Model) and (b) provide the psychophysical justification for inter-individual comparisons of effort ratings. Borg's Effort Continua Model describes the functional interdependence of perceptual and physiological responses during exercise. The model provides valuable information regarding the corresponding and interdependent responses of exertional perceptions and underlying physiological mediators as exercise performance intensity increases. Borg's Range Model predicts that for all clinically normal individuals, there exists corresponding and equal perceptual and physiological/physical response ranges during exercise. This model provides the psychophysical rationale for perceived exertion scaling procedures. There are two types of category scale anchoring: (a) memory procedures and (b) exercise procedures. Memory procedures involve asking the individual to think about the level of exertion perceived during previous PA that they have performed and use this exertional memory to establish their feelings that correspond to the low and high response categories. Exercise procedures involve the individual actually experiencing levels of exertion from a very low to a very high or maximal level and cognitively assigning corresponding low and high scale categories to the intensity of these sensations. The use of both procedures depends on an individual's previous experience with rating exertional perceptions that varied widely in intensity and mode. The rationale underlying the experimental purpose of the investigation is embedded in the basic tenet of Borg's Effort Continua Model and Range Model. The primary purpose of this laboratory experiment is to orient an individual to the use of a perceived exertion category metric during aerobic and/or resistance exercise using both memory and exercise scale anchoring procedures.

L. Haile et al., *Perceived Exertion Laboratory Manual*,
DOI 10.1007/978-1-4939-1917-8_5

5.1 Background

5.1.1 Borg's Effort Continua and Range Models

The rationale underlying Borg's development of metrics to measure perceived exertion during exercise was based on the concept of the three effort continua: performance, physiological, and perceptual (Robertson 2001). Each continuum represents the individual's range of possible responses within that specific domain, yet the three continua are closely related. For example, during an aerobic running event, an individual's performance intensity increases as evidenced by a decrease in minute per mile pace. This increased pace corresponds to increases in both perceptual responses (RPE) and physiological responses, such as HR and VO_2. Knowledge of the functional interdependence of perceptual and physiological responses during exercise can provide valuable information about exercise performance and is the theoretical backbone for applications of RPE research.

The basic tenet underlying the Borg's Range Model makes inter- and intra-individual comparisons of RPE possible. The model describes how the increase in RPE from a very low to a very high level matches the increase in exercise intensity specific to an individual's performance capacity (Borg 1998). In other words, the lowest RPE value matches the lowest exercise intensity and the highest RPE value matches maximal exercise intensity. In addition, 50 % of the RPE range corresponds to approximately 50 % of the individual's exercise intensity range. This holds true whether exercise intensity is expressed in physical units, such as PO, or using a physiological variable such as HR or VO_2. When clinically normal individuals perform exercise at a given intensity, the corresponding level of exertion (RPE) can be compared between clients regardless of aerobic fitness level (Robertson 2004). Likewise, RPE obtained from a single individual can be compared at different time points within an exercise program. If an exercise program results in significant improvements in fitness, the individual's range of possible exercise intensities has increased. However, the RPE range corresponding to these exercise intensities remains the same. Therefore, a given RPE will be attained at higher exercise intensity as training adaption occurs. This can be seen in clinically normal individuals as well as those with various diseases and disorders for which exercise can be beneficial, such as cystic fibrosis.

The Range Model forms the conceptual basis of the standard, pre-exercise instructions to teach an individual how to use an RPE scale and is crucial to establishing category scale anchoring points. In this application, it is recognized that for all clinically normal individuals the level of perceived exertion corresponding to very low intensity and maximal intensity is the same. Such correspondence of perceptual and exercise intensity ranges provides the psychophysical rationale underlying anchoring procedures for a numerical category scale.

To satisfy the requirements of the Borg's Range Model, an individual must be able to link the full range of RPE responses with the full range of physiological responses during exercise (Robertson 2004). Therefore, anchoring procedures should be used to ensure that an individual understands this psychophysiological linkage prior to exercise performance in which RPE will be measured or used as a basis for exercise prescription. It is important to note that the scale anchoring procedures be presented on an individual basis because the physiological range required by the exercise task may vary greatly between individuals.

5.1.2 Memory and Exercise Anchoring Procedures

The most practical method of RPE scale anchoring is the *memory procedure* in which the individual is asked to think about the exertion experienced during previous exercise or physical activity. Using this procedure, the individual is asked to remember when he/she reached levels of exertion equal to the low and high anchor points on the scale. Then, during subsequent bouts of exercise, the individual is asked to rate exertion levels based on memory of exertion at the low and high anchor points. An example of this type of procedure is written into the standard instructions for use of the Adult OMNI-Cycle RPE Scale below.

Following administration of these scaling instructions and anchoring procedures, it is beneficial to ask some simple questions of the individual to determine if he/she understands how to use the scale to rate perceived exertion. Ask the individual to provide an RPE that corresponds to the memory of exertion felt during very light exercise. The individual should respond with a very low number on the scale. If the expected rating is not made, verbally reinforce the individual that perceived exertion is the subjective intensity of effort, strain, discomfort and/or fatigue that is felt during exercise. Ask the individual about various types of exercise or recreational activities he/she performs and what a common RPE value is during those activities. This allows the individual to think about RPE during various exercise intensities that are normally performed during recreation and leisure pursuits. Also, ask the individual to think about and explain the most exhausting exercise he/she has ever performed, and remember the level of exertion experienced during that activity. In this case, if the client rates that activity less than the maximal RPE available on the scale, further explanation of maximal exertion may be necessary.

The second method of RPE scale anchoring is the *exercise procedure*. In this procedure, the individual actually performs exercise, preferably using the same mode as the exercise test or physical activity program that is to be performed. The scale anchor points, once established, ensure the linkage between perceptual and physiological responses during a specific type of exercise. The exercise anchoring procedure begins after reading the standard instructions for the RPE scale and conducting the memory anchoring procedure. First, the client performs 2 min of exercise at a very low intensity. For treadmill exercise, slow walking would be appropriate; for cycle exercise, unloaded (i.e., zero brake resistance) cycling would be appropriate. For resistance exercise, a very light weight that the subject can lift the specified number of repetitions without any fatigue would be appropriate. The number of repetitions used in a resistance exercise anchoring procedure may vary depending on the exercise test or training program to be performed. At the end of the orientation period, instruct the subject to assign the lowest RPE values (0 or 1 on the OMNI Scale) to the level of exertion experienced at that intensity. Next, the client performs load-incremented exercise (i.e., aerobic or resistance) to maximal intensity, which occurs at the point of volitional termination owing to exhaustion. Begin with the intensity that was previously linked to the lowest RPE on the scale and progressively increase intensity until he/she reaches maximal exercise. Immediately following cessation of exercise, instruct the subject to assign a maximal RPE value (10 on the OMNI Scale) to the level of exertion experienced at that intensity.

A load-incremented exercise protocol that employs standard procedures to determine maximal aerobic power, or maximal oxygen uptake (VO_2max), can also be used to establish the high anchor point for aerobic exercise. VO_2max is defined as the maximum amount of oxygen that can be consumed while breathing ambient air during load-incremented aerobic exercise at sea level. Normally, a graded exercise test (GXT) to measure VO_2max involves 2- to 3-min stages with the test ultimately terminating owing to the subjects inability to continue consequent to fatigue. The length of the exercise stage can be shortened to 30 s or 1 min to quickly progress the individual to a very high intensity.

A load-incremented resistance exercise protocol that employs standard procedures to determine maximal muscle strength, or one-repetition maximum (1RM), can also be used to establish the high anchor point for a category perceived exertion metric such as the OMNI-Resistance Exercise Scale. 1RM is defined as the maximal amount of force that can be produced during a single isotonic contraction of a muscle (group) moving through the full range of joint motion.

5.1.3 Undifferentiated Versus Differentiated RPE and the Dominant Signal

The scale anchoring procedures should separately establish low and high perceptual reference points for the *undifferentiated RPE* for the overall body and the *differentiated RPE* for the active limbs and chest/breathing. Rating exertion separately for the chest/breathing (RPE-C), also referred to as respiratory exertion, is appropriate for any type of exercise. In addition, during cycle and treadmill exercise it is appropriate to ask subjects to rate exertion separately for the legs (RPE-L). Other examples of differentiated RPE's include estimating exertion for the arms (RPE-A) during arm ergometry and the back during rowing exercise.

When performing the exercise anchoring procedures, it is appropriate to choose a primary type of RPE to use in establishing the low and high anchor points. For cycle exercise, RPE-L is representative of the major muscle mass being used during exercise and is often the most dominant signal, showing higher values than RPE-O or RPE-C. Therefore, RPE-L can be used as the primary RPE for exercise anchoring and is presented as such in the laboratory procedures that are presented in this manual. For treadmill exercise, RPE-L may be the dominant perceptual signal compared to RPE-O. However, since walking/running exercise is considered as a weight-bearing, total body activity, RPE-O can be used to establish the anchor points. For resistance exercises, it is appropriate to operationally define a specific differentiated RPE that represents the level of exertion for the active muscle mass (RPE-AM). This RPE may be labeled according to the agonist muscle group, or prime movers, for the specific exercise. For example, differentiated RPE for bench press exercise is specific to the chest/pectoral muscles and should be used to set the scale anchor points.

5.1.4 Exercise Anchoring Procedures and the Perceptual Outlier

It is common practice for clinicians and researchers to orient their clients and subjects using memory anchoring procedures only. However, this is not always appropriate, especially for individuals who may not be familiar with a given type of exercise and may not have experienced exercise intensities across their entire physiological response range. It is not possible to ask someone to remember a level of exertion experienced at certain exercise intensity if they have never performed that intensity. For example, asking a child or sedentary adult to assign a maximal RPE value to the memory of the most difficult exercise ever performed would not be appropriate if they had never performed maximal exercise. Therefore, memory anchoring followed by exercise anchoring is most appropriate in these individuals.

It is important to note that, even for extremely active and/or fit individuals, rating perceived exertion is a learned skill (Robertson 2004). Physical activity and fitness levels may not determine one's ability to rate perceived exertion accurately across the full physiological and performance range. Individuals who rate perceived exertion inappropriately and whose responses do not conform to the Borg's Range Model are termed as *perceptual outliers*. Some individuals tend to augment RPE, or provide higher RPE values than expected relative to the measured physiological response (Fig. 5.1, client A). They may even report a maximal RPE when performing submaximal exercise intensity. Likewise, some individuals tend to reduce RPE, or provide lower RPE values than expected relative to the measured physiological response (Fig. 5.1, clients B and C). They may assign a submaximal RPE to maximal exercise intensity. Perceptual reducers seem to be more common than perceptual augmenters, especially among young recreationally active adults. Therefore, the combination of memory and exercise anchoring procedures is recommended for all individuals who are not experienced with RPE procedures in order to identify perceptual outliers who require additional practice, feedback and reinforcement.

There is a more advanced exercise anchoring procedure that has been employed in previous investigations involving an exercise program in which a "target" RPE is used to self-regulate exercise intensity (Higgins et al. 2013). This procedure allows time for additional practice, feedback and reinforcement that is not usually included in the standard exercise anchoring procedure presented in the instructional set. This intensity-specific anchoring procedure may be helpful for any individual having

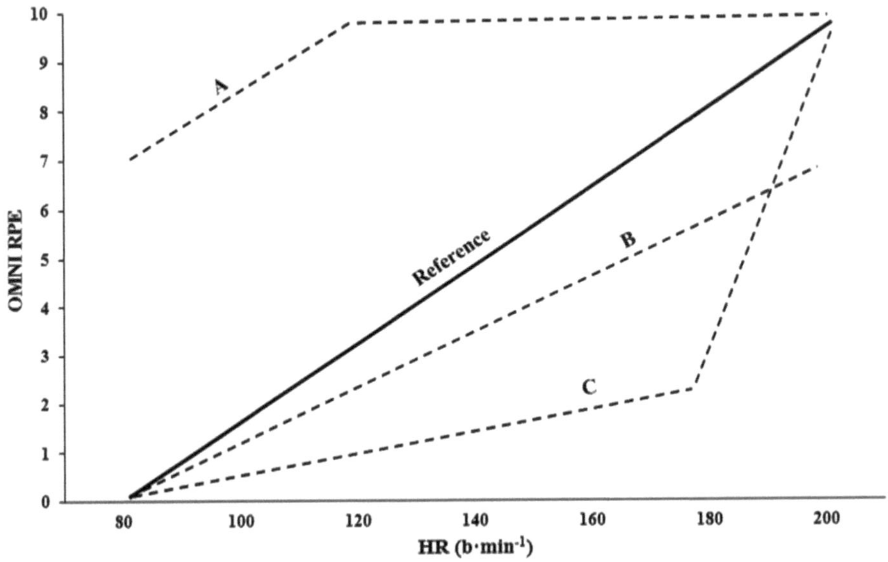

Fig. 5.1 OMNI RPE responses given by three clients (*A*, *B*, *C*) who were having difficulty using the RPE scale in comparison to the expected, i.e., reference, RPE response (Robertson 2004)

difficulty understanding how to use a category scale to rate exertion levels, especially young children. In this procedure, the exercise anchoring is divided into three distinct phases: low, moderate, and high/maximal intensity. Each phase includes a brief, 2- to 4-min bout of load-incremented exercise in which physical intensity is increased and the client provides an RPE every 15 or 30 s. In addition to using the low and moderate intensities for anchoring purposes, these bouts can include a brief, 2- to 4-min perceptual production format in which the client performs exercise that elicits a specific (i.e., target) level of exertion. See Appendix F for a detailed description of this advanced perceived exertion scaling procedure.

5.2 Methods

5.2.1 Treadmill Procedures

5.2.1.1 Equipment

1. Adult OMNI-Walk/Run RPE Scale (Fig. A.2)
2. Treadmill
3. HR monitor (optional)
4. Respiratory-metabolic measurement system (optional)

5.2.1.2 Memory Anchoring Procedure

1. Read the standard instructions for the Adult OMNI-Walk/Run RPE Scale for RPE-O to the subject (Appendix B.1).
2. Following the standard instructions and answering any of the subject's questions, ask the subject this series of questions to check understanding. Take notes about the subject's responses.

   (a) What RPE corresponds to your memory of exertion experienced during light walking activity you performed recently?
   (b) What sport or recreational activity have you performed recently? What RPE corresponds to a preferred level of exertion experienced during that activity?
   (c) What was the most exhausting exercise you remember performing? What RPE would you assign to the level of exertion you experienced during that exercise?

5.2.1.3 Exercise Anchoring Procedure

*Clinical note: During the anchoring procedures, it may be beneficial to have the subject wear the same physiological monitoring equipment that will be worn during the actual exercise test or conditioning program where RPE will be measured. The following instructional set includes procedures for HR and respiratory-metabolic measurement, but these physiological assessment methods are optional for this experiment.

1. Position the HR monitor and respiratory-metabolic mouthpiece (with head support unit and nose clip if applicable) on the subject.
2. Instruct the subject to step onto the treadmill and review test termination procedures: When the subject cannot continue exercise due to exhaustion or discomfort, he/she should grasp the treadmill hand rails, at which time the test administrator will gradually slow the treadmill down for performance of a cool-down. The subject should be reminded not to step off the treadmill belt while it is still in motion.
3. With the treadmill grade set at 0 %, increase the treadmill speed so the subject can walk slowly for 2 min.
4. Establish the *low anchor point.*

   (a) At the end of the 2-min period, with the subject still walking and the Adult OMNI-Walk/Run RPE Scale in full view, instruct the subject that he/she should assign an RPE-O of 0 to the intensity of exertion that is experienced at that moment.
   (b) If using a respiratory-metabolic mouth piece, instruct the subject to point to the numbers on the RPE scale, which should be conveniently positioned within the subject's arm reach. State aloud the numerical ratings for each momentary assessment to which the subject pointed and request a confirmatory nod that the number stated was correct. If incorrect, allow the subject to

point to the appropriate rating on the RPE scale once more. Ask the subject to hold his or her finger on the appropriate scale number for approximately 1 s.

5. Establish the *high anchor point* using an abbreviated version of the Bruce Multistage Treadmill Test Protocol. This can be performed by manually adjusting treadmill speed and grade or using a program on a computer that is interfaced to the treadmill.

   (a) Instruct the subject to face the front of the treadmill, straddle the treadmill belt so the feet are not on the belt and hold onto the hand rails.
   (b) Each exercise test stage will last for 30 s. The stages progress as follows:

      Stage 1—1.7 miles $\cdot$ h^{-1} and 10 % grade
      Stage 2—2.5 miles $\cdot$ h^{-1} and 12 % grade
      Stage 3—3.4 miles $\cdot$ h^{-1} and 14 % grade
      Stage 4—4.2 miles $\cdot$ h^{-1} and 16 % grade
      Stage 5—5.0 miles $\cdot$ h^{-1} and 18 % grade
      Stage 6—5.5 miles $\cdot$ h^{-1} and 20 % grade
      Stage 7—6.0 miles $\cdot$ h^{-1} and 22 % grade
      Stage 8—6.5 miles $\cdot$ h^{-1} and 24 % grade

   (c) When the subject cannot continue exercise any longer due to exhaustion and indicates such by grasping the hand rails, terminate the test. Instruct the subject to assign an RPE-O of 10 to the intensity of exertion experienced at this maximal exercise level.

5.2.2 Cycle Ergometer Procedures

5.2.2.1 Equipment

1. Adult OMNI-Cycle RPE Scale (Fig. 2.4)
2. Treadmill
3. HR monitor (optional)
4. Respiratory-metabolic measurement system (optional)

5.2.2.2 Memory Anchoring Procedure

1. Read the standard instructions for the Adult OMNI-Cycle RPE Scale for RPE-L to the subject (Appendix B.4).
2. Following administration of the standard instructions and answering any of the subject's questions, ask the subject the following to check understanding. Take notes about the subject's responses.

   (a) What RPE corresponds to your memory of exertion experienced during light cycle exercise you performed recently?

(b) What sport or recreational activity have you performed recently? What RPE corresponds to a preferred level of exertion experienced during that activity?
(c) What was the most exhausting exercise you remember performing? What RPE would you assign to the level of exertion you experienced during that exercise?

5.2.2.3 Exercise Anchoring Procedure

*Clinical note: During the anchoring procedures, it may be beneficial to have the subject wear the same physiological monitoring equipment that will be worn during the actual exercise test or conditioning program where RPE will be measured. The following instructional set includes procedures for HR and respiratory-metabolic measurement, but these physiological assessment methods are optional for this experiment.

1. Position the HR monitor and respiratory-metabolic mouthpiece (with head support unit and nose clip if applicable) on the subject.
2. Set the proper seat height on the cycle ergometer according to leg length. When the foot is flat on the right pedal and the pedal is in the extreme down position, there should be a flexion of the right knee should be in 5 degrees of flexion.
3. Instruct the subject to maintain a 50 $rev \cdot min^{-1}$ pedal cadence. Set the metronome to 100 $b \cdot min^{-1}$ so each downward movement of each foot is synchronized with a beat of the metronome. The subject may also use the digital monitor on the cycle control panel to regulate pedal cadence.
4. For electronically braked cycle ergometers (e.g., Lode), with the cycle set at 0 W, instruct the subject to begin unloaded pedaling for 2 min.
5. For friction-braked cycle ergometers (e.g., Monark), with the cycle break resistance set at 0 kg, instruct the subject to begin unloaded pedaling for 2 min.
6. Establish the *low anchor point*.

   (a) At the end of the 2-min period, with the subject still pedaling and the Adult OMNI-Cycle RPE Scale in full view, instruct the subject that he/she should assign an RPE-L of 0 to the intensity of exertion that is experienced at that moment.
   (b) If using a respiratory-metabolic mouth piece, instruct the subject to point to the numbers on the RPE scale, which should be conveniently positioned within the subject's arm reach. State aloud the numerical ratings for each momentary assessment to which the subject pointed and request a confirmatory nod that the number stated was correct. If incorrect, allow the subject to point to the appropriate rating on the RPE scale once more. Ask the subject to hold his or her finger on the appropriate number on the scale for approximately 1 s.

7. Establish the *high anchor point* using an abbreviated version of a load-incremented peak exercise test. This can be performed by manually adjusting cycle resistance or using a program on a computer that is interfaced to the cycle.

8. For electronically braked cycle ergometers (e.g., Lode), increase the resistance 25 W every 30 s.
9. For friction-braked cycle ergometers (e.g., Monark), increase the resistance 0.5 kg every 30 s.
10. When the subject cannot maintain the pedal cadence for 10 consecutive seconds owing to exhaustion in the leg muscles, terminate the exercise test.
11. Instruct the subject to assign an RPE-L of 10 to the level of exertion experienced at the moment of test termination owing to fatigue.

5.2.3 *Resistance Exercise Procedures*

5.2.3.1 Equipment

1. Adult OMNI-Resistance Exercise RPE Scale (Fig. A.5)
2. Resistance exercise equipment of choice

5.2.3.2 Memory Anchoring Procedure

1. Read the standard instructions for the Adult OMNI-Resistance Exercise RPE Scale for RPE-AM to the subject (Appendix B.7).
2. Following administration of the standard instructions and answering any of the subject's questions, ask the subject the following to check understanding of the procedures. Take notes about the subject's responses.
   (a) What RPE corresponds to your memory of exertion experienced during light resistance exercise you performed recently?
   (b) What was the most exhausting resistance exercise you remember performing? What RPE would you assign to the level of exertion you experienced at the point of exhaustion during that exercise?

5.2.3.3 Exercise Anchoring Procedure

1. Prior to resistance exercise, explain and demonstrate proper lifting technique for the isotonic exercise to be performed and discuss how a test administrator will "spot" (i.e., guide) the subject while lifting the weight both concentrically and eccentrically. Then, instruct the subject to take the proper position on the weight bench or resistance exercise machine, if applicable.
2. Establish the *low anchor point*.
   (a) Instruct the subject to perform the lift using an extremely light resistance for 1 repetition. This may involve performing the lift without additional weight beyond the bar or rack. You may even choose to have the subject perform

the repetition with a broom stick or light dumbbells to better simulate the actual lift.

(b) When the repetition is complete, with the Adult OMNI-Resistance RPE Scale in full view, instruct the subject that he/she should assign an RPE (for the active muscle group, e.g., RPE for the chest/pectoral muscles if performing bench press) of 0 to the intensity of exertion that was felt.

3. Establish the *high anchor point* using a 1RM procedure (Baechle and Earle 2008).

   (a) Instruct the subject to warm-up with a light resistance that can be performed in 5–10 repetitions, then provide a 1-min rest.
   (b) Estimate a warm-up load that will allow the subject to complete 3–5 repetitions by adding 10–20 pounds (5–10 % of previous weight lifted) for upper body exercise or 30–40 pounds (10–20 % of previous weight lifted) for lower body exercise, then provide a 2-min rest.
   (c) Estimate a conservative, near maximal load that will allow the subject to complete 2–3 repetitions by adding 10–20 pounds (5–10 % of previous weight lifted) for upper body exercise or 30–40 pounds (10–20 % of previous weight lifted) for lower body exercise, then provide a 2- to 4-min rest.
   (d) Increase the load by 10–20 pounds (5–10 % of previous weight lifted) for upper body exercise or 30–40 pounds (10–20 % of previous weight lifted) for lower body exercise and instruct the subject to attempt a 1RM.
   (e) If the subject successfully completed the lift using proper technique, provide a 2- to 4-min rest and repeat the previous step. If the subject failed to complete the lift using proper technique, provide a 2- to 4-min rest then decrease the resistance by 5–10 pounds (2.5–5 % of previous weight attempted) for upper body exercise or 15–20 pounds (5–10 % of previous weight attempted) for lower body exercise and instruct the subject to attempt a 1RM.
   (f) Continue increasing or decreasing the resistance load until the subject can complete a 1RM with proper exercise technique.
   (g) Following the final set, instruct the subject to assign an RPE of 10 to the feelings of exertion arising from the active muscle group as felt during the 1RM lift.

5.3 Discussion Questions

1. Explain the concept of perceptual scale anchoring using the Borg's Range Model as a theoretical framework.
2. Based on the questions asked about the subject following the memory anchoring procedure, do you believe the procedure would suffice for this individual prior to exercise testing or engage in an exercise program? Why?
3. During the exercise anchoring procedures performed in this laboratory experiment, did your subject conform to the Borg's Range Model? Explain.

4. Based on your knowledge of Borg's Range Model, what should RPE responses be during a maximal graded exercise test that is volitionally terminated by the individual at exhaustion? How could RPE be used as a criterion for attainment of VO_2max?

References

Baechle TR, Earle RW. Essentials of strength training and conditioning. 3rd ed. Champaign, IL: Human Kinetics; 2008.

Borg G. Borg's perceived exertion and pain scales. Champaign, IL: Human Kinetics; 1998.

Higgins LW, Robertson RJ, Kelsey SF, Olson MB, Hoffman LA, Rebovich PJ, Haile L, Orenstein DM. Exercise intensity self-regulation using the OMNI scale in children with cystic fibrosis. Pediatr Pulmonol. 2013;48:497–505.

Robertson RJ. Development of the perceived exertion knowledge base: an interdisciplinary process. Int J Psychol. 2001;12:189–96.

Robertson RJ. Perceived exertion for practitioners: rating effort with the OMNI picture system. Champaign, IL: Human Kinetics; 2004.

Chapter 6
Perceived Exertion Scale Validation

Both concurrent and construct validity are important psychometric properties for perceived exertion scales applied in the clinical and performance settings. A graded exercise test that employs a perceptual estimation protocol is the standard for determination of perceived exertion scale validity for both aerobic and resistance exercise. Both undifferentiated and differentiated RPE can be measured using a scale validity experiment, but special attention should be paid to determination of the dominant RPE signal during exercise. Concurrent and construct validity evidence has been shown for both undifferentiated and differentiated RPE in various sample populations performing aerobic and resistance exercise. A perceived exertion scale that demonstrates concurrent and construct validity can be applied to both exercise testing and prescription in hospital and field settings. Such applications can include the prediction of impending exercise test termination and exercise intensity self-regulation. The primary purpose of this laboratory experiment is to establish concurrent and construct validity for an OMNI RPE Scale. Secondary purposes include the comparison of concurrent validity evidence between the OMNI RPE Scale and the Borg Scale and to determine differentiated RPE signal dominance.

6.1 Background

6.1.1 Validity

Validity can be defined as the degree to which a test or test item measures what it is intended to measure and is the most important characteristic of any specific test (Baechle and Earle 2008). Without validity, test results have no meaning. Measures of basic physical characteristics of an individual (e.g., height and weight) are relatively easy to validate. The validity of metrics to be used during exercise

L. Haile et al., *Perceived Exertion Laboratory Manual*,
DOI 10.1007/978-1-4939-1917-8_6

performance, especially perceived exertion scales, can be more difficult to establish. Therefore, there are two main types of validation experiments used to confirm the validity of a perceived exertion scale: concurrent validity and construct validity.

6.1.2 Concurrent Validity

Concurrent validity is the extent to which test scores are associated with those of other accepted tests when both measures are obtained along a common stimulus range. In the case of a scale that measures RPE, this definition refers to the accuracy of the metric to measure perceived exertion across one's entire physiological range as exercise intensity is systematically increased from low to high levels. A test of concurrent validity involves a statistical calculation of the relation between a criterion variable (the stimulus) and the concurrent variable (the response). This statistical paradigm often uses a Pearson correlation calculation that yields "*r*" values referred to as validity coefficients (Baechle and Earle 2008). To establish concurrent validity of a perceived exertion scale, it is expected that the concurrent variable, RPE, increases concurrently with increases in a physical and/or physiological criterion variable as intensity of exercise increases (Robertson 2004). A statistically significant concurrent validity coefficient indicates a strong relation between the concurrent and criterion variables, often resulting in high *r* values greater than 0.70.

The theoretical framework underlying the assessment of concurrent validity of a perceived exertion scale is derived from the basic tenets of Borg's Effort Continua and Range Models. There are three main effort continua: performance, physiological, and perceptual. An increase in exercise performance, usually denoted as increasing intensity, results in corresponding and interdependent increases in both physiological and perceptual responses. Exercise intensity can be measured as minute per mile pace or PO for aerobic exercise and as absolute weight lifted or %1RM for resistance exercise. Physiological responses are underlying processes that an individual subjectively monitors during exercise to ultimately mediate their RPE response. HR and VO_2 are respiratory-metabolic exertional mediators that are commonly measured during exercise serving as physiological indicators of exercise intensity. Physiological and perceptual responses display a functional interdependence. As such, the model predicts that perceptual responses will increase in correspondence with physiological responses throughout the individual's entire exercise intensity range, from a very low to a maximal level. In addition, the lowest RPE value matches the lowest exercise intensity and the highest RPE value matches maximal exercise intensity. This holds true whether exercise intensity is expressed in physical units, such as PO, or using a physiological variable such as HR or VO_2. In this context, it is the goal of scale anchoring procedures to set the low and high anchor points on an RPE scale, linking them to very low and maximal exercise intensities. Once it is known that an individual conforms to the model following scale anchoring, a concurrent validation experiment can be used to measure the physiological and perceptual effort continua across the full range of possible exercise performance intensities.

6.1.3 Construct Validity

Construct validity can be defined as the ability of a test to represent the underlying construct, or the theory developed to organize and explain aspects of existing knowledge and observations (Baechle and Earle 2008). For perceived exertion scales, construct validity is tested by comparing RPE measured using a newly developed scale for which validity has yet to be established with the RPE derived from a perceived exertion scale having well-established construct validity. In this paradigm, it is expected that both the new (i.e., conditional) scale and the criterion scale have demonstrated a high level of concurrent validity. Traditionally, construct validity of a perceived exertion scale is statistically determined by correlating RPE measured using the conditional metric with RPE measured using the 15-category Borg Scale (i.e., the criterion metric).

6.1.4 Validity Test Protocols

Concurrent and construct validity of a perceived exertion scale can be tested simultaneously using perceptual estimation test protocols. An *estimation protocol* is a GXT during which an individual estimates RPE during each exercise stage. Using commonly employed procedures for determining maximal aerobic or resistance exercise capacity, an estimation protocol allows an individual to rate RPE from a very low exercise intensity to maximal exercise intensity. For example, the Bruce treadmill protocol for the determination of VO_2max employs incremental stages of walking and running exercise. For resistance exercise, variations on 1RM or multiple-RM procedures are used. These procedures must include measurements of physiological exertional mediators (e.g., HR, VO_2) and the recording of physical markers of exercise intensity (e.g., PO, weight lifted, %1RM) necessary for the determination of concurrent validity. As such, a concurrent and construct validation experiment only requires that RPE be measured using the 15-category Borg Scale and the RPE scale for which validity is sought.

6.1.5 Use of Perceived Exertion Scaling Procedures Prior to the Estimation Protocol

It is important to note that the scale anchoring procedures should be presented separately from and prior to the estimation test protocol used for a scale validation experiment. For the concurrent/construct validation experiment to be valid, it must be known that the individual's RPE responses conform to Borg's Range Model. Individuals who have experience using perceived exertion scales and have participated in exercise anchoring procedures in the past are more likely to provide RPE responses that conform to the prediction of Borg's Range Model. However, as discussed previously, some individuals are perceptual outliers, either overestimating or underestimating the RPE response. Such responses usually occur upon initial exposure to RPE assessment and prior to administration of full memory and exercise anchoring procedures. When possible, youth and sedentary adult subjects should undergo both the memory and exercise procedures for perceived exertion scale anchoring. However, even active adults have been known to be perceptual outliers and as such can benefit from additional anchoring and practice in estimating RPE prior to undertaking the actual exercise trial.

6.1.6 Differentiated Versus Undifferentiated Exertional Ratings

Experiments to validate perceived exertion scales can employ both undifferentiated and differentiated RPE. A differentiated RPE is linked to a specific anatomical region of the body. Differentiated RPE specific to the leg muscles (RPE-L) can be measured during cycle ergometry and treadmill exercise. The RPE-L reflects peripheral exertional signals resulting from localized metabolic acidosis, blood glucose level, muscle glycogen content, and muscle blood flow (Robertson 2004). Differentiated RPE specific to the chest and breathing (RPE-C) can be measured during any aerobic activity. The RPE-C reflects respiratory-metabolic exertional mediators such as V_E and total body VO_2. Differentiated RPE rated during resistance exercises are usually specific to the active muscle mass (RPE-AM). Undifferentiated RPE is a measure of the overall body (RPE-O) exertional level. It is formed by integrating the various exertional signals arising from the composite of anatomical regions involved in the exercise task. Many investigations have asked subjects to rate RPE-O only, but important information can be derived by also measuring differentiated exertional ratings.

6.1.7 Concurrent Validity Evidence for Undifferentiated RPE

Numerous investigations have established concurrent validity of various iterations of the OMNI Perceived Exertion Scale using mode-specific estimation protocols. Experiments have included male and female children and adults performing a wide variety of exercise modalities: cycle ergometry, treadmill walking and running, stepping exercise, elliptical ergometry, and resistance exercise. High validity coefficients were reported for male and female children and adults during cycle ergometry and treadmill exercise, with r values ranging from 0.67 to 0.99 for the associations between RPE and HR or VO_2 (Balasekaran et al. 2012; Robertson 2004; Robertson et al. 2000; Utter et al. 2004). High validity coefficients were found in a sample of male and female children and a sample of adult females performing load-incremented stepping exercise. In these stepping experiments, the relation between RPE and VO_2 exhibited r values ranging from 0.87 to 0.96. The relation between RPE and HR exhibited r values ranging from 0.81 to 0.95 (Krause et al. 2012; Robertson et al. 2005b). High validity coefficients were found in adult males and females performing elliptical ergometry. The relation between RPE and VO_2 exhibited r values ranging from 0.93 to 0.95, while the relation between RPE and HR exhibited r values ranging from 0.95 to 0.97 (Mays et al. 2010). High validity coefficients were reported during biceps curl and knee extension exercises, with r values ranging from 0.72 to 0.91 for the association between RPE and total weight lifted in both children and adults (Robertson et al. 2003, 2005a). In addition, an $r = 0.87$ was determined for the association between RPE-AM and blood lactic acid concentration in adults; providing evidence for lactacidemia as a physiological exertional mediator for resistance exercise (Robertson et al. 2003).

Concurrent validation has been tested and confirmed for other perceived exertion scales as well, such as the Children's Effort Rating Table (CERT). CERT, a 10-category scale ranging from 1 to 10, was developed specifically for children to be easily understood with verbal descriptors positioned at each numerical category. CERT, however, does not include pictorial descriptors as does the OMNI Scale. Concurrent validation of CERT has been examined for various youth populations performing stepping and cycle ergometer exercise. Validity coefficients for the relation between RPE and HR ranged from $r = 0.73$ to 0.99 during stepping exercise (Williams et al. 1994) and from $r = 0.70$ to 0.97 for cycle ergometry (Eston et al. 1994; Lamb 1995; Leung et al. 2002). In addition, investigations determined the relations between RPE measured by the CERT and both PO and VO_2 for cycle ergometer exercise. The relation between RPE and power output exhibited r values ranging from 0.70 to 0.98 (Eston et al. 1994; Lamb 1995; Leung et al. 2002). The relation between RPE and VO_2 exhibited r values ranging from 0.85 to 0.91 (Leung et al. 2002).

HR and VO_2 are the most commonly used physiological criterion variables to demonstrate concurrent scale validity for aerobic exercise modalities. They are the most widely used because they increase as a positive function of increases in exercise intensity. However, other physiological criterion variables have been used to study concurrent scale validity during aerobic exercise. Investigations have correlated OMNI Scale RPE with $\%VO_2max$, %HRmax, pulmonary ventilation (V_E), respiratory rate (RR), the respiratory exchange ratio (RER), and the V_E to VO_2 ratio ($V_E \cdot VO_2^{-1}$). All of these physiological variables are expected to increase concurrently with increases in aerobic exercise intensity, demonstrating either linear or polynomial growth functions. High validity coefficients for OMNI Scale responses were found for adolescent girls performing treadmill exercise, as evidenced by the relation between RPE and %HRmax ($r = 0.86$) and the relation between RPE and $\%VO_2max$ ($r = 0.89$) (Pfeiffer et al. 2002). Moderate validity coefficients ranging from $r = 0.33$ to 0.43 were shown between RPE with $\%VO_2max$, V_E, RR, and $V_E \cdot VO_2^{-1}$ for children performing treadmill exercise (Utter et al. 2002). Another study involving children performing treadmill exercise found high validity coefficients ranging from $r = 0.71$ to 0.81 for the relation between RPE with $\%VO_2max$, V_E, RR, and RER using the Spanish version of the Children's OMNI-Walk/Run RPE Scale (Suminski et al. 2008). In addition, high validity coefficients

Table 6.1 Summary of OMNI Scale validation for aerobic and resistance exercise when undifferentiated RPE for the overall body (RPE-O) was the concurrent variable

				Correlation coefficient[a]					
		Scale		HR			VO_2		
Investigation	Mode	Format	Age	F	M	B	F	M	B
Robertson et al. (2000)	Cycle	Cycle	Child	0.94	0.92	0.93	0.93	0.94	0.94
Pfeiffer et al. (2002)	Treadmill	Cycle	Child	0.82	–	–	0.88	–	–
Robertson et al. (2004)	Cycle	Cycle	Adult	0.83	0.90	–	0.88	0.94	–
Utter et al. (2004)	Treadmill	Walk/run	Adult	0.84	0.75	–	0.85	0.86	–
Robertson et al. (2005b)	Step	Step	Child	0.83	0.88	–	0.88	0.93	–
Suminski et al. (2008)[b]	Treadmill	Walk/run	Child	–	–	0.85	–	–	0.85
Mays et al. (2010)	Elliptical	Elliptical	Adult	0.97	0.96	–	0.95	0.95	–
Balasekaran et al. (2012)	Cycle	Cycle	Child	0.99	0.98	0.98	0.99	0.99	0.95
Krause et al. (2012)	Step	Step	Adult	0.95	–	–	0.96	–	–
				WTtot					
				F	**M**				
Robertson et al. (2003)	BC	Resistance	Adult	0.87	0.89				
Robertson et al. (2003)	KE	Resistance	Adult	0.86	0.87				
Robertson et al. (2005a)	BC	Resistance	Child	0.87	0.80				
Robertson et al. (2005a)	KE	Resistance	Child	0.80	0.88				

The criterion variables were: *HR* heart rate, *VO_2* oxygen consumption, *WTtot* total weight lifted. *F* female, *M* male, *B* both males and females, *BC* biceps curl, *KE* knee extension

[a]All correlation coefficients are significant ($p<0.05$)

[b]This study validated a Spanish version of the Children's OMNI-Walk/run Scale

ranging from $r=0.67$ to 0.88 were found for the relation between RPE with $\%VO_2max$, V_E, RR, and RER where data were determined for adults performing treadmill exercise (Utter et al. 2004) (Tables 6.1 and 6.2).

6.1.8 Construct Validity Evidence for Undifferentiated RPE

Construct validity has also been tested and confirmed for various OMNI RPE scales using the 15-category Borg Scale as the criterion metric. Validity coefficients ranged from $r=0.92$ to 0.97 for adults performing cycle ergometry (Robertson et al. 2004),

Table 6.2 Summary of OMNI Scale validation for aerobic and resistance exercise when differentiated RPE for the legs (L), chest/breathing (C) and active muscle (AM) were the concurrent variables

					Correlation coefficient[a]					
			Scale		HR			VO_2		
Investigation	Mode	RPE	Format	Age	F	M	B	F	M	B
Robertson et al. (2000)	Cycle	L	Cycle	Child	0.92	0.93	0.92	0.93	0.93	0.93
Robertson et al. (2000)	Cycle	C	Cycle	Child	0.88	0.90	0.87	0.87	0.90	0.87
Robertson et al. (2004)	Cycle	L	Cycle	Adult	0.81	0.86	–	0.87	0.95	–
Robertson et al. (2004)	Cycle	C	Cycle	Adult	0.82	0.88	–	0.90	0.95	–
Robertson et al. (2005b)	Step	L	Step	Child	0.81	0.89	–	0.87	0.94	–
Robertson et al. (2005b)	Step	C	Step	Child	0.83	0.87	–	0.88	0.92	–
Mays et al. (2010)	Elliptical	L	Elliptical	Adult	0.97	0.96	–	0.93	0.95	–
Mays et al. (2010)	Elliptical	C	Elliptical	Adult	0.96	0.96	–	0.94	0.95	–
Balasekaran et al. (2012)	Cycle	L	Cycle	Child	0.99	0.98	0.99	0.99	0.98	0.98
Balasekaran et al. (2012)	Cycle	C	Cycle	Child	0.99	0.99	0.99	0.99	0.99	0.98
					WTtot		**[Hla]**			
					F	**M**	**B**			
Robertson et al. (2003)	BC	AM	Resistance	Adult	0.89	0.91	0.87			
Robertson et al. (2003)	KE	AM	Resistance	Adult	0.79	0.87	–			
Robertson et al. (2005a)	BC	AM	Resistance	Child	0.88	0.81	–			
Robertson et al. (2005a)	KE	AM	Resistance	Child	0.72	0.75	–			

The criterion variables were: *HR* heart rate, VO_2 oxygen consumption, *WTtot* total weight lifted, *[Hla]* lactic acid concentration. *F* female, *M* male, *B* both males and females, *BC* biceps curl, *KE* knee extension

[a]All correlation coefficients are significant ($p<0.05$)

treadmill exercise (Utter et al. 2002), elliptical ergometry (Mays et al. 2010), and knee extension resistance exercise (Lagally and Robertson 2006). The basis for the development of child versions of perceived exertion metrics, such as the OMNI Scale and CERT, was that children often exhibited semantic limitations in understanding the verbal descriptors employed in adult formatted metrics such as the Borg Scale. The numerical categories of the Borg Scale range from 6 to 20 and many of its verbal descriptors use the word "exertion," a word not typically a part of a younger child's vocabulary. As such, researchers did not believe the Borg Scale to be a valid metric to measure a child's perceived exertion. An investigation by Robertson and colleagues (2005b) conducted a construct validation experiment for the Children's OMNI-Step RPE Scale using the previously validated Children's OMNI-Cycle RPE Scale as the criterion metric. Even though the study involved stepping exercise, the only difference between the two RPE scales was the mode-specific pictorial descriptors. Since the concurrent validity of the Children's OMNI-Cycle Scale was well established by previous investigations, using it as the criterion metric was conceptually similar to previous construct validation experiments in adults that compared OMNI RPE with the well-established Borg Scale RPE. Future investigations seeking to establish construct validity of a new perceived exertion scale for children in future investigations may also consider using as a criterion a well-established children's OMNI scale that has been shown to have a high level of concurrent validity.

6.1.9 *Differentiated RPE in a Validity Experiment*

Previous RPE scale validation studies have employed differentiated RPE during cycle ergometry, elliptical ergometry, and resistance exercise in adults (Lagally and Robertson 2006; Mays et al. 2010; Robertson et al. 2003, 2004); and cycle ergometry, stepping exercise, and resistance exercise in children (Balasekaran et al. 2012; Robertson et al. 2000, 2005a, 2005b). All investigations found evidence of concurrent scale validity where differentiated RPE were employed as well as the undifferentiated, overall-body RPE. In addition, construct scale validity has been confirmed using differentiated RPE measured during cycle ergometry, elliptical ergometry, and resistance exercise in adults, as well as for stepping exercise in children (Lagally and Robertson 2006; Mays et al. 2010; Robertson et al. 2004, 2005b).

It is possible to ask a subject to rate both undifferentiated and differentiated RPE's during each exercise test stage of an aerobic or resistance exercise load-incremented protocol. Three RPE values can easily be rated within a 30-s time frame at the end of each stage of the Bruce treadmill protocol (Robertson 2004). Using this procedure, it can be determined which RPE signal, the overall signal or a differentiated signal, is the dominant perception (i.e., most intense) for a specific mode of exercise. Three primary factors determine the dominant RPE response during exercise: (1) the mode of exercise, (2) the anatomical origin of the differentiated feelings, and (3) the performance environment (i.e., air or water, temperature, humidity) (Robertson 2004). Robertson and colleagues (2001) asked child subjects to rate RPE-O, RPE-L, and RPE-C at the end of each stage of an incremental cycle ergometer exercise test. When the RPE's associated with the ventilatory threshold (VT) were calculated, it was found that RPE-L provided the dominant RPE signal with RPE-C being comparatively less intense (Robertson et al. 2001). Therefore, since RPE-O fell between the two differentiated RPE signals, it generally appeared as a mathematical average of the anatomically regionalized ratings. Such a response confirmed that the undifferentiated RPE is a good overall indicator of total body exertion level and represented an integration of differentiated perceptual signals. Also, it should be noted that differentiated RPE's can be compared with undifferentiated RPE at any intensity of exercise to identify perceptual signal dominance and mode of signal integration.

6.1.10 Application of a Valid Perceived Exertion Scale

Concurrent and construct validity evidence has been shown for scales that measured both undifferentiated and differentiated RPE signals for a wide variety of exercise modalities. This is an important confirmation of the original intent and practical importance of the OMNI perceived exertion scales and the reason for the name OMNI. The name OMNI is an abbreviation of the word omnibus, meaning "of, relating to, or providing for many things at once" (Merriam-Webster Online 2014). Even though the first iteration of the OMNI Scale focused on children's responses during cycle ergometer exercise, it was intended that the original design could be reformatted for use by female and male clients of all ages performing a wide variety of exercise modalities consequent to future scale development. However, RPE scales are not solely restricted for use during incremental exercise, such as that used in this experiment. It was reasoned that as RPE scales showed strong concurrent and construct validity and the perceptual responses conformed to Borg's Range Model, that an individual could self-regulate exercise at a prescribed intensity using a "target" RPE. For example, after an individual successfully performs perceptual scale anchoring and undergoes a separate perceptual estimation test procedure, the exercise professional chooses a specific target RPE that corresponds to an a priori determined physiological intensity, one of the most important of which is the VT. Then, in a separate production procedure, the exercise professional teaches the individual to self-adjust exercise intensity until it feels like the level of exertion equal to the target RPE. The ability to determine an appropriate target RPE and teach an individual how to accurately self-regulate exercise intensity according to the designated perceptual level is one of the most important applications of RPE to exercise prescription and programming for overall health-fitness activities.

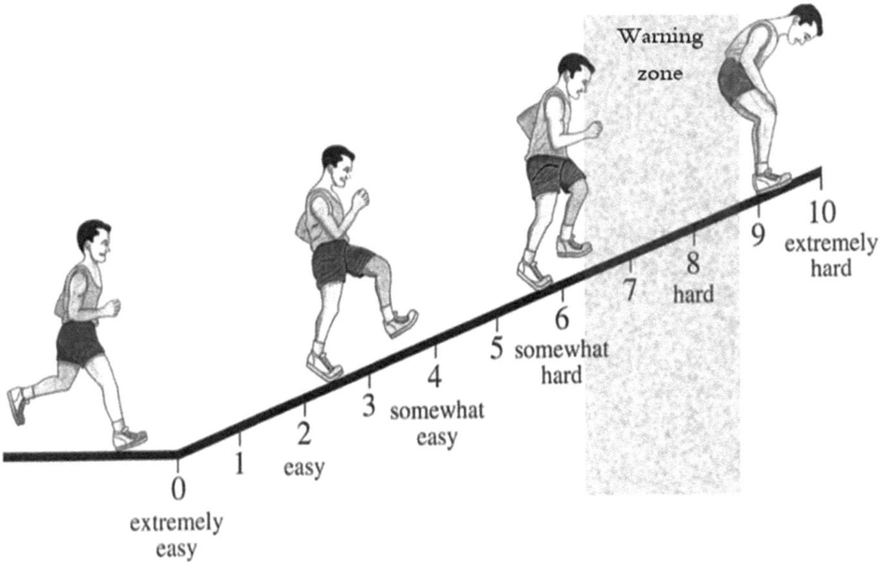

Fig. 6.1 OMNI RPE warning zone that signals impending exercise test termination (Robertson 2004)

6.1.11 *Clinical Application of RPE During Maximal Exercise Testing*

6.1.11.1 Use of RPE to Predict Impending Graded Exercise Test Termination

ACSM recommends the assessment of RPE throughout graded exercise testing to monitor an individual's progress toward maximal exertion in an effort to predict impending fatigue and test termination (ACSM 2013; Morgan and Borg 1976). For this reason, undifferentiated and/or differentiated RPE should be estimated at least at the end of each exercise test stage, preferably at the end of each minute of exercise in more functionally limited individuals whose exercise tolerance could deteriorate rapidly at comparatively higher intensities. Noble and Robertson (1996) identified *RPE warning zones* for graded exercise test termination using both the Borg Scale and OMNI Scales. The warning zones were defined as a range of RPE's, 15–17 on the Borg Scale and 7–8 on the OMNI Scales (Fig. 6.1). Those RPE zones signal impending test termination and as such indicate that it is the time at which procedures to safely end the GXT should be initiated. Goss and colleagues (2010a) identified the mean Borg RPE that indicated a subject would terminate the exercise test during the next 3-min stage of a Bruce treadmill protocol in apparently healthy male and female adults. Using the Borg 6–20 category scale, women terminated the exercise test an average of 142 s after an RPE of 14, while men terminated the test an average of 120 s after an RPE of 15 (Goss et al. 2010a). Using RPE as a determinant of time until test termination can be especially important when a patient is taking a medication that affects the HR response to exercise, such as coronary artery disease patients who take beta-blockers. In such patients, the expected HR response based on a percent of age-predicted maximum is not an accurate predictor of impending test termination because the actual HR response is pharmacologically blunted. However, the RPE response is independent of cardioactive medication effects. As such, RPE will increase appropriately from rest to maximal exertion as it would without medication. Goss and colleagues (2010b) identified the mean Borg RPE that indicated a subject would terminate the exercise test during the next 3-min stage of a Bruce treadmill protocol in men with coronary artery disease. The men, who were all taking beta-blocker medication, terminated the exercise test an average of 153 s after a Borg (6–20) Scale RPE of 14 (Goss et al. 2010b), a perceptual value similar to that reported previously in apparently healthy men and women (Goss et al. 2010a).

6.1.11.2 Use of RPE as a Criterion for the Achievement of VO_2max

The "gold standard" assessment of cardiorespiratory fitness is a GXT used to determine VO_2max by indirect calorimetry. The Bruce treadmill protocol is an ideal example because it involves upright, dynamic, weight-bearing exercise using the total body. It is common for the researcher or health-fitness professional to use defined criteria to determine if the individual has achieved a "true" VO_2max. The "gold standard" criterion for the achievement of VO_2max is a plateau in VO_2 seen at the end of exercise when the individual has terminated the exercise test owing to exhaustion. A VO_2 plateau is defined as less than a 2.1 $ml \cdot kg^{-1} \cdot min^{-1}$ (or 150 $ml \cdot min^{-1}$) increase in VO_2 with an increase in workload (Siconilfi et al. 1982). However, a number of investigations have found that few individuals actually achieve a VO_2 plateau at the end of load-incremented GXTs (Day et al. 2003; Foster et al. 2007; Rossiter et al. 2006). Therefore, other VO_2max criteria that have been used include a respiratory exchange ratio (RER; VCO_2 divided by VO_2) of greater than 1.15, blood lactate concentration greater than 8 $mmol \cdot l^{-1}$, and HR within 10 $b \cdot min^{-1}$ of age-predicted maximum HR (APMHR). Yet many individuals will not achieve these supplementary criteria as well (Powers and Howley 2012).

ACSM (2013) reports that most apparently healthy individuals estimate RPE's for the overall-body from 18 to 19 (using the Borg Scale 6 to 20 format) or 9 to 10 (using a 0 to 10 format such as the OMNI Scales) at exercise test termination. According to the Range Model, if maximal exertion is reached, an individual should report the highest RPE available on the perceived exertion scale. However, since undifferentiated RPE (RPE-O) is an integration of differentiated signals, an OMNI RPE-O of 9 most likely indicates that the dominant differentiated response was a 10, such as RPE for the active musculature. Previous investigations measuring OMNI RPE during maximal or peak aerobic power testing have used the rating of an OMNI RPE ≥ 9 as a primary criterion for attainment of VO_2max/peak in children (Andreacci et al. 2007) and adults (Krause et al. 2012). Therefore, for the purposes of this and the following laboratory experiments, the primary criterion for the successful completion of a GXT should be volitional termination due to exhaustion as indicated by maximal RPE, i.e., OMNI RPE $\geq$ 9 or Borg RPE $\geq$ 19.

6.2 Methods

6.2.1 Treadmill Procedures

6.2.1.1 Equipment

1. Adult OMNI-Walk/Run RPE Scale (Fig. A.2)
2. Borg RPE Scale
3. Treadmill
4. HR monitor
5. Respiratory-metabolic measurement system

6.2.1.2 Pre-exercise Procedures

1. Measure height (cm) and weight (kg) of subject.
2. Read the standard instructions for the Adult OMNI-Walk/Run RPE Scale to the subject for undifferentiated and differentiated RPE (Appendix B.2). Perform the memory anchoring procedure as described in Chap. 5.
3. Read the standard instructions for the Borg Scale during treadmill exercise for RPE-O, which will be used to determine construct validity of the OMNI Scale (Appendix B.3).

6.2.1.3 Graded Exercise Test

1. Position the HR monitor and respiratory-metabolic mouthpiece (with head support unit and nose clip if applicable) on the subject.
2. Instruct the subject to step onto the treadmill and review exercise termination procedures: When the subject cannot continue exercise due to exhaustion or discomfort, he/she should grasp the treadmill hand rails, at which time the test administrator will gradually slow the treadmill down for performance of a cool-down. The subject should be reminded not to step off the treadmill belt while it is still in motion.
3. Bruce Multistage Treadmill Test Protocol: this can be performed by manually adjusting treadmill speed and grade or using a program on a computer that is interfaced to the treadmill.

(a) Begin the warm-up at 1.5 miles $\cdot$ h^{-1} and 0 % grade for 3 min.
(b) Each exercise test stage will last for 3 min. The stages progress as follows:

Stage 1—1.7 miles $\cdot$ h^{-1} and 10 % grade
Stage 2—2.5 miles $\cdot$ h^{-1} and 12 % grade
Stage 3—3.4 miles $\cdot$ h^{-1} and 14 % grade
Stage 4—4.2 miles $\cdot$ h^{-1} and 16 % grade
Stage 5—5.0 miles $\cdot$ h^{-1} and 18 % grade
Stage 6—5.5 miles $\cdot$ h^{-1} and 20 % grade
Stage 7—6.0 miles $\cdot$ h^{-1} and 22 % grade
Stage 8—6.5 miles $\cdot$ h^{-1} and 24 % grade

(c) When the subject cannot continue any longer, terminate the exercise test by initiating the cool-down period at 1.5 miles $\cdot$ h^{-1} and 0 % grade. The cool-down should be 5 min in duration.
(d) Ask the subject to estimate RPE starting at 2:30 of each exercise stage using both the Borg Scale (RPE-O only) and the OMNI Scale (RPE-O, RPE-L, and RPE-C). The RPE's should be rated in a counterbalanced sequence. Because the position of the respiratory-metabolic mouth piece prevents a verbal response, instruct the subject to point to the numbers on the RPE scale, which should be conveniently positioned within the subject's arm reach. State aloud the numerical ratings for each momentary assessment to which the subject pointed and request a confirmatory nod that the number stated was correct. If incorrect, allow the subject to point to the appropriate rating on the RPE scale once more. Ask the subject to hold his or her finger on the appropriate number on the scale for approximately 1 s.
(e) Record HR (b $\cdot$ min^{-1}) at 2:55 of each exercise stage.
(f) Record the final 15-s VO_2 (ml $\cdot$ kg^{-1} $\cdot$ min^{-1}) for each exercise stage.
(g) Record HRmax as the highest HR value recorded during the final exercise stage or immediately post-exercise.
(h) Record VO_2max as the highest 15-s VO_2 value recorded at the end of the test.

6.2.1.4 Data Organization and Analysis

1. In a Microsoft Excel spreadsheet, label columns of data for the following variables: Exercise Stage, VO_2 (ml $\cdot$ kg^{-1} $\cdot$ min^{-1}), Borg RPE-O, OMNI RPE-O, OMNI RPE-L, OMNI RPE-C, HR (b $\cdot$ min^{-1}).
2. Plot of VO_2 and Borg RPE-O for determination of concurrent scale validation.

   (a) Click on the **INSERT** tab and in the **CHARTS** section click on **SCATTER**. Select the first available chart option. A blank or example scatter plot will appear on your screen.
   (b) Click on the **SELECT DATA** tab. Remove any entries found in the **LEGEND ENTRIES** text box then click **ADD**. Under **SERIES NAME**, enter VO_2 and Borg RPE-O. Then click on the icon to the right of the **SERIES X VALUES** text box and highlight the VO_2 values. After the values are highlighted click

the icon on the box that appeared. Then click on the icon to the right of the **SERIES Y VALUES** text box and highlight the Borg RPE-O values. After the values are highlighted click the icon on the box that appeared. Click **OK** on the next two screens.

(c) You should now have a scatter plot with Borg RPE-O on the *y*-axis and VO_2 on the *x*-axis. Create a title for the plot and enter the appropriate axis labels and units of measure.

(d) To determine the validity coefficient, click on one of the data points to highlight the entire data series. Right click on one of the data points and a menu will appear. Click **ADD TRENDLINE** and the **FORMAT TRENDLINE** menu will appear. Select **LINEAR**, **DISPLAY EQUATION ON CHART**, and **DISPLAY R-SQUARED VALUE ON CHART** then click **CLOSE**. The trendline and equation will be displayed on the chart. Take the square root of the R^2 value to determine the Pearson correlation coefficient.

3. Repeat the above steps to plot and determine validity coefficients for the following variable pairs to establish concurrent validity: HR and Borg RPE-O, VO_2 and OMNI RPE-O, HR and OMNI RPE-O, VO_2 and OMNI RPE-L, HR and OMNI RPE-L, VO_2 and OMNI RPE-C, HR and OMNI RPE-C; and for construct validation: Borg RPE-O and OMNI RPE-O.
4. An example of these procedures with a screenshot depicting each step as performed using Microsoft Excel 2013 can be found in Appendix C.

6.2.2 *Cycle Ergometer Procedures*

6.2.2.1 Equipment

1. Adult OMNI-Cycle RPE Scale (Fig. 2.4)
2. Borg RPE Scale
3. Cycle ergometer
4. Metronome
5. HR monitor
6. Respiratory-metabolic measurement system

6.2.2.2 Pre-exercise Procedures

1. Measure height (cm) and weight (kg) of subject.
2. Read the standard instructions for the Adult OMNI-Cycle RPE Scale for undifferentiated and differentiated RPE to the subject (Appendix B.5). Perform the memory anchoring procedure as described in Chap. 5.
3. Read the standard instructions for use of the Borg Scale during cycle exercise emphasizing measurement of RPE-L, which will be used to determine construct validity of the OMNI Scale (Appendix B.6).

6.2.2.3 Graded Exercise Test

1. Position the HR monitor and respiratory-metabolic mouthpiece (with head support unit and nose clip if applicable) on the subject.
2. Set the proper seat height on the cycle ergometer according to leg length. When the foot is flat on the right pedal and the pedal is in the extreme down position, there should be a flexion of the right knee of approximately 5°.
3. Load-incremented protocol for electronically braked and friction-braked cycle ergometers:

   (a) Instruct the subject to maintain a 50 rev · min^{-1} pedal cadence. Set the metronome to 100 b · min^{-1} so each downward movement of each foot is synchronized with a beat of the metronome. The subject may also use the digital monitor on the cycle control panel to regulate pedal cadence.
   (b) For electronically braked cycle ergometers (e.g., Lode), begin stage 1 at 50 W then increase the resistance 25 W every 2 min.
   (c) For friction-braked cycle ergometers (e.g., Monark), begin stage 1 at 1 kg resistance then increase the resistance 0.5 kg every 2 min.
   (d) When the subject cannot maintain the pedal cadence for 10 consecutive seconds, terminate the exercise test. The test may also be volitionally terminated by the subject owing to fatigue.
   (e) Ask the subject to estimate RPE starting at 1:30 of each exercise stage using both the Borg Scale (RPE-O only) and the OMNI Scale (RPE-O, RPE-L, and RPE-C). The RPE's should be rated in a counterbalanced sequence. Because the position of the respiratory-metabolic mouth piece prevents a verbal response, instruct the subject to point to the numbers on the RPE scale, which should be conveniently positioned within the subject's arm reach. State aloud the numerical ratings for each momentary assessment to which the subject pointed and request a confirmatory nod that the number stated was correct. If incorrect, allow the subject to point to the appropriate rating on the RPE scale once more. Ask the subject to hold his or her finger on the appropriate number on the scale for approximately 1 s.
   (f) Record HR (b · min^{-1}) at 1:55 of each exercise stage.
   (g) Record the final 15-s VO_2 (l · min^{-1}) for each exercise stage.
   (h) Record HRpeak as the highest HR value recorded during the final exercise stage or immediately post-exercise.
   (i) Record VO_2peak as the highest 15-s VO_2 value recorded at the end of the test.

6.2.2.4 Data Organization and Analysis

1. In a Microsoft Excel spreadsheet, label columns of data for the following variables: Exercise Stage, VO_2 (l · min^{-1}), Borg RPE-L, OMNI RPE-O, OMNI RPE-L, OMNI RPE-C, HR (b · min^{-1}).

2. Plot of VO_2 and Borg RPE-L for determination of concurrent validity.
   (a) Click on the **INSERT** tab and in the **CHARTS** section click on **SCATTER**. Select the first available chart option. A blank or example scatter plot will appear on your screen.
   (b) Click on the **SELECT DATA** tab. Remove any entries found in the **LEGEND ENTRIES** text box then click **ADD**. Under **SERIES NAME**, enter VO_2 and Borg RPE-L. Then click on the icon to the right of the **SERIES X VALUES** text box and highlight the VO_2 values. After the values are highlighted click the icon on the box that appeared. Then click on the icon to the right of the **SERIES Y VALUES** text box and highlight the Borg RPE-L values. After the values are highlighted click the icon on the box that appeared. Click **OK** on the next two screens.
   (c) You should now have a scatter plot with Borg RPE-L on the *y*-axis and VO_2 on the *x*-axis. Create a title for the plot and enter the appropriate axis labels and units of measure.
   (d) To determine the validity coefficient, click on one of the data points to highlight the entire data series. Right click on one of the data points and a menu will appear. Click **ADD TRENDLINE** and the **FORMAT TRENDLINE** menu will appear. Select **LINEAR**, **DISPLAY EQUATION ON CHART**, and **DISPLAY R-SQUARED VALUE ON CHART** then click **CLOSE**. The trendline and equation will be displayed on the chart. Take the square root of the R^2 value to determine the Pearson correlation coefficient.
3. Repeat the above steps to plot and determine validity coefficients for the following variable pairs for concurrent scale validity: HR and Borg RPE-L, VO_2 and OMNI RPE-O, HR and OMNI RPE-O, VO_2 and OMNI RPE-L, HR and OMNI RPE-L, VO_2 and OMNI RPE-C, HR and OMNI RPE-C; and for construct validation: Borg RPE-L and OMNI RPE-L.
4. An example of these procedures with a screenshot depicting each step as performed using Microsoft Excel 2013 can be found in Appendix C.

6.2.3 *Resistance Exercise Procedures*

6.2.3.1 Equipment

1. Adult OMNI-Resistance Exercise RPE Scale (Fig. A.5)
2. Borg RPE Scale
3. Resistance exercise equipment of choice
4. Metronome

6.2.3.2 Pre-exercise Procedures

1. Read the standard instructions to the subject for use of the Adult OMNI-Resistance Exercise RPE Scale for undifferentiated and differentiated RPE (Appendix B.8). Perform the memory anchoring procedure as described in Chap. 5.

2. Read the standard instructions to the subject for the Borg Scale during resistance exercise emphasizing measurement of RPE-AM, which will be used to determine construct validity of the OMNI Scale (Appendix B.9).

6.2.3.3 Exercise Protocols

1. Administer a 1RM procedure for assessment of muscular strength (Baechle and Earle 2008).
   (a) Instruct the subject to warm-up with a light resistance that can be performed for 5–10 repetitions, then provide a 1-min rest.
   (b) Estimate a warm-up load that will allow the subject to complete 3–5 repetitions by adding 10–20 lb (5–10 % of weight lifted) for upper body exercise or 30–40 lb (10–20 % of weight lifted) for lower body exercise, then provide a 2-min rest.
   (c) Estimate a conservative, near maximal load that will allow the subject to complete 2–3 repetitions by adding 10–20 lb (5–10 % of weight lifted) for upper body exercise or 30–40 lb (10–20 % of weight lifted) for lower body exercise, then provide a 2–4-min rest.
   (d) Make a load increase of 10–20 lb (5–10 % of weight lifted) for upper body exercise or 30–40 lb (10–20 % of weight lifted) for lower body exercise and instruct the subject to attempt a 1RM.
   (e) If the subject successfully completed the lift using proper technique, provide a 2–4-min rest and repeat the previous step. If the subject failed to complete the lift using proper technique, provide a 2–4-min rest then decrease the resistance by 5–10 lb (2.5–5 % of weight lifted) for upper body exercise or 15–20 lb (5–10 % of weight lifted) for lower body exercise and instruct the subject to attempt a 1RM.
   (f) Continue increasing or decreasing the load until the subject can complete a 1RM with proper exercise technique.
   (g) Calculate the weight equal to the following %1RM intensities: 20, 40, 50, 60, 70, 80, and 90 %.
   (h) It may be beneficial to ask the subject to rate Borg RPE-AM, OMNI RPE-O and OMNI RPE-AM in a counterbalanced fashion immediately following each set. This will provide additional practice and feedback prior to undertaking the scale validation protocol.
2. Category scale validation will be assessed using the procedures described by Lagally and Robertson (2006).
   (a) Instruct the subject to warm-up with one set of ten repetitions at 20 % of exercise specific 1RM then provide a 1-min rest.
   (b) Instruct the subject to perform one repetition at 40, 50, 60, 70, 80, and 90 % of 1RM in a random order with a 2-min rest between intensities.
   (c) Repetition speed should be paced by a metronome set at 70 $b \cdot min^{-1}$ so each repetition is performed with a two-count-up, two-count-down pattern.

(d) Instruct the subject to estimate Borg RPE-AM, OMNI RPE-O and OMNI RPE-AM in a counterbalanced sequence immediately following each repetition.

6.2.3.4 Data Organization and Analysis

1. In a Microsoft Excel spreadsheet, label columns of data for the following variables: %1RM, Weight Lifted, Borg RPE-AM, OMNI RPE-O, OMNI RPE-AM.
2. Plot of Weight Lifted and Borg RPE-AM for determination of concurrent validity:
   (a) Click on the **INSERT** tab and in the **CHARTS** section click on **SCATTER**. Select the first available chart option. A blank or example scatter plot will appear on your screen.
   (b) Click on the **SELECT DATA** tab. Remove any entries found in the **LEGEND ENTRIES** text box then click **ADD**. Under **SERIES NAME**, enter Weight Lifted and Borg RPE-AM. Then click on the icon to the right of the **SERIES X VALUES** text box and highlight the Weight Lifted values. After the values are highlighted click the icon on the box that appeared. Then click on the icon to the right of the **SERIES Y VALUES** text box and highlight the Borg RPE-AM values. After the values are highlighted click the icon on the box that appeared. Click **OK** on the next two screens.
   (c) You should now have a scatter plot with Borg RPE-AM on the *y*-axis and Weight Lifted on the *x*-axis. Create a title for the plot and enter the appropriate axis labels and units of measure.
   (d) To determine the validity coefficient, click on one of the data points to highlight the entire data series. Right click on one of the data points and a menu will appear. Click **ADD TRENDLINE** and the **FORMAT TRENDLINE** menu will appear. Select **LINEAR**, **DISPLAY EQUATION ON CHART**, and **DISPLAY R-SQUARED VALUE ON CHART** then click **CLOSE**. The trendline and equation will be displayed on the chart. Take the square root of the R^2 value to determine the Pearson correlation coefficient.
3. Repeat the above steps to plot and determine validity coefficients for the following variable pairs to establish concurrent scale validity: Weight Lifted and OMNI RPE-O, Weight Lifted and OMNI RPE-AM; and for construct validation: Borg RPE-AM and OMNI RPE-AM.
4. An example of these procedures with a screenshot depicting each step as performed using Microsoft Excel 2013 can be found in Appendix C.

6.3 Discussion Questions

1. Define validity, in general, as it applies to the use of a perceived exertion scale during exercise.
2. Explain the differences and similarities between concurrent and construct validation as they apply to a perceived exertion scale.

3. Did your subject's perceived exertion responses conform to Borg's Range Model? Explain your data using the conceptual framework of Borg's Effort Continua Model.
4. Describe why RPE should be used as one of the criteria for the achievement of VO_2max/peak? Should the criterion be based on RPE-O, a differentiated RPE, or both? Explain.
5. Do your results for OMNI Scale concurrent and construct validity agree with previous RPE validation studies? Explain why, citing previous literature.
6. Which RPE scale demonstrated stronger concurrent validity for RPE-O, the OMNI Scale or Borg Scale? Why?
7. Choose a specific exercise test stage (treadmill and cycle) or %1RM from your data sheet.
   (a) Which OMNI RPE was the dominant signal, RPE-O or a specific differentiated RPE?
   (b) If you measured differentiated RPE (L and C) during treadmill or cycle exercise, did RPE-O represent a true integration (i.e., average) of these differentiated signals?
8. Based on your results, how would you use the perceived exertion scale from this experiment to prescribe exercise to the individual you tested?

References

American College of Sports Medicine. ACSM's guidelines for exercise testing and prescription. 9th ed. Philadelphia, PA: Lippincott Williams & Wilkins; 2013.

Andreacci JL, Haile L, Dixon C. Influence of testing sequence on a child's ability to achieve maximal anaerobic and aerobic power. Int J Sports Med. 2007;28(8):673–7.

Baechle TR, Earle RW. Essentials of strength training and conditioning. 3rd ed. Champaign, IL: Human Kinetics; 2008.

Balasekaran G, Loh MK, Govindaswamy VV, Robertson RJ. OMNI scale of perceived exertion: mixed gender and race validation for Singapore children during cycle exercise. Eur J Appl Physiol. 2012;112:3533–46.

Day JR, Rossiter HB, Coats EM, Skasick A, Whipp BJ. The maximally attainable VO_2 during exercise in humans: the peak vs. maximum issue. J Appl Physiol. 2003;95:1901–7.

Eston RG, Lamb KL, Bain A, Williams AM, Williams JG. Validity of a perceived exertion scale for children: a pilot study. Percept Mot Skills. 1994;78:691–7.

Foster C, Kuffel E, Bradley N, Battista RA, Wright G, Porcari JP, Lucia A, deKoning JJ. VO_2max during successive maximal efforts. Eur J Appl Physiol. 2007;102:67–72.

Goss FL, Robertson RJ, Haile L, Krause MP, Nagle EF, Metz KF, Kim K. Identification of a rating of perceived exertion-based warning zone to anticipate graded treadmill test termination. Percept Mot Skills. 2010a;110:213–23.

Goss FL, Robertson RJ, Haile L, Nagle EF, Metz KF, Kim K. Use of ratings of perceived exertion to anticipate treadmill test termination in patients taking beta-blockers. Percept Mot Skills. 2010b;111:310–8.

Krause MP, Goss FL, Robertson RJ, Kim K, Elsangedy HM, Krinski K, da Silva S. Concurrent validity of an OMNI rating of perceived exertion scale for bench stepping exercise. J Strength Cond Res. 2012;26:506–12.

Lagally KM, Robertson RJ. Construct validity of the OMNI resistance exercise scale. J Strength Cond Res. 2006;20:252–6.
Lamb KL. Children's ratings of effort during cycle ergometry: an examination of the validity of two effort rating scales. Pediatr Exerc Sci. 1995;7:407–21.
Leung M, Chung P, Leung RW. An assessment of the validity and reliability of two perceived exertion rating scales among Hong Kong children. Percept Mot Skills. 2002;95:1047–62.
Mays RJ, Goss FL, Schafer MA, Kim KH, Nagle-Stilley EF, Robertson RJ. Validation of adult OMNI perceived exertion scales for elliptical ergometry. Percept Mot Skills. 2010;111:848–62.
Merriam-Webster Online (2014) Dictionary and thesauraus. http://www.merriam-webster.com/netdict.html. Accessed 28 Mar 2014.
Morgan W, Borg GA. Perception of effort in the prescription of physical activity. In: Nelson T, editor. Mental health and emotional aspects of sports. Chicago, IL: American Medical Association; 1976. p. 126–9.
Noble BJ, Robertson RJ. Perceived exertion. Champaign, IL: Human Kinetics; 1996.
Pfeiffer KA, Pivarnik JM, Womack CJ, Reeves MJ, Malina RM. Reliability and validity of the Borg and OMNI rating of perceived exertion scales in adolescent girls. Med Sci Sports Exerc. 2002;34:2057–61.
Powers SK, Howley ET. Exercise physiology: theory and application to fitness and performance. 8th ed. New York, NY: McGraw-Hill; 2012.
Robertson RJ. Perceived exertion for practitioners: rating effort with the OMNI picture system. Champaign, IL: Human Kinetics; 2004.
Robertson RJ, Goss FL, Boer NF, People JA, Foreman AJ, Dabayebeh IM, Millich NB, Balasekaran G, Riechman SE, Gallagher JD, Thompkins T. Children's OMNI scale of perceived exertion: mixed gender and race validation. Med Sci Sports Exerc. 2000;32:452–8.
Robertson RJ, Goss FL, Boer N, Gallagher JD, Thompkins T, Bufalino K, Balasekaran G, Meckes C, Pintar J, Williams A. OMNI scale perceived exertion at ventilatory breakpoint in children: response normalized. Med Sci Sports Exerc. 2001;33:1946–52.
Robertson RJ, Goss FL, Rutkowski J, Lenz B, Dixon C, Timmer J, Frazee K, Dube J, Andreacci J. Concurrent validation of the OMNI perceived exertion scale for resistance exercise. Med Sci Sports Exerc. 2003;35:333–41.
Robertson RJ, Goss FL, Dube J, Rutkowski J, Dupain M, Brennan C, Andreacci J. Validation of the adult OMNI scale of perceived exertion for cycle ergometer exercise. Med Sci Sports Exerc. 2004;36:102–8.
Robertson RJ, Goss FL, Andreacci JL, Dube JJ, Rutkowski JJ, Frazee KM, Aaron DJ, Metz KF, Kowallis RA, Snee BM. Validation of the children's OMNI-resistance exercise scale of perceived exertion. Med Sci Sports Exerc. 2005a;37:819–26.
Robertson RJ, Goss FL, Andreacci JL, Dube JJ, Rutkowski JJ, Snee BM, Kowallis RA, Crawford K, Aaron DJ, Metz KF. Validation of the children's OMNI RPE scale for stepping exercise. Med Sci Sports Exerc. 2005b;37:290–8.
Rossiter HB, Kowalchuk JM, Whipp BJ. A test to establish maximum O_2 uptake despite no plateau in the O_2 uptake response to ramp incremental exercise. J Appl Physiol. 2006;100:764–70.
Siconilfi SF, Culliname EM, Carleton RA, Thompson PD. Assessing VO_2max in epidemiologic studies: modifications of the Astrand-Rhyming test. Med Sci Sports Exerc. 1982;14:335–8.
Suminski RR, Robertson RJ, Goss FL, Olvera N. Validation of the OMNI scale of perceived exertion in a sample of Spanish-speaking youth from the USA. Percept Mot Skills. 2008;107:181–8.
Utter AC, Robertson RJ, Nieman DC, Kang J. Children's OMNI scale of perceived exertion: walking/running evaluation. Med Sci Sports Exerc. 2002;34:139–44.
Utter AC, Robertson RJ, Green JM, Suminski RR, Mcanulty SR, Nieman DC. Validation of the adult OMNI Scale of perceived exertion for walking/running exercise. Med Sci Sports Exerc. 2004;36:1776–80.
Williams JG, Eston RG, Furlong B. CERT: a perceived exertion scale for young children. Percept Mot Skills. 1994;79:1451–8.

Chapter 7
Target RPE at the Ventilatory Threshold

The *ventilatory threshold* (VT) is an important physiological marker of the exercise intensity at which an individual can sustain performance for a prolonged period, and therefore is an important determinant of aerobic exercise performance capacity. Determination of the exercise training intensity equivalent to the VT is an important measure for both elite athletes and sedentary individuals. Training at the intensity equivalent to the individual's VT provides an optimal overload stimulus to achieve both performance and health-fitness benefits. Identification of the VT in the laboratory setting requires experienced personnel and expensive laboratory equipment. Therefore, a surrogate measure of the VT is needed to guide exercise training intensity. This can be achieved by identifying a target HR associated with the VT (HR-VT). However, exercise HR can be affected by both environmental and clinical conditions and requires skill for accurate palpation when a HR monitoring device is unavailable. RPE has been validated for the prescription and regulation of exercise in a variety of settings and subject populations. Owing to its ease of use and cost-efficiency, a target RPE associated with the VT (RPE-VT) presents a practical method for prescribing exercise at VT intensity. The primary purpose of this laboratory experiment is to determine RPE-VT, using both the Borg and OMNI Scales, and HR-VT during load-incremented aerobic exercise.

7.1 Background

7.1.1 Ventilatory Threshold

The VT (also known as the ventilatory breakpoint) can be defined as the point during dynamic exercise of increasing intensity when V_E begins to increase at a rate disproportionately faster than that of VO_2. Some investigations refer to this as the 1stVT inflection point because a further increase in V_E resulting in a 2ndVT inflection point

L. Haile et al., *Perceived Exertion Laboratory Manual*,
DOI 10.1007/978-1-4939-1917-8_7

(i.e., 2ndVT) often follows the 1stVT during load-incremented exercise tests (Alberton et al. 2013). The experiment described later in this chapter employs the 1stVT. This increase in V_E occurs as the body attempts to expel excess carbon dioxide produced as a result of an increased reliance on anaerobic metabolism to meet the energy requirements of higher intensity exercise. The VT occurs at approximately the same exercise intensity or VO_2 as the lactate threshold (LT). The LT can be defined as the point during exercise of increasing intensity when the clearance of lactate from the blood can no longer keep up with the increased rate of lactate production by muscle. During exercise of progressively increasing intensity, blood lactate begins to accumulate above resting values owing to an increased reliance on anaerobic metabolism. In the typical adult, the VT and LT occur between 55 and 70 % of VO_2max (Kenney et al. 2012).

Identifying the VT is very important when developing training programs for elite endurance athletes because it marks the highest exercise intensity that an individual can sustain for a prolonged period. The VT is a better indicator of performance in endurance athletes than VO_2max because one cannot sustain the intensity at which VO_2max occurs for extended periods. In addition, an athlete's VO_2max and ability to increase VO_2max with training have a significant hereditary determinant. Twenty-five to 50 % of the variance in VO_2max can be explained by genetic factors (Kenney et al. 2012). However, once an elite athlete achieves genetically determined VO_2max through aerobic training, the athlete can continue to increase endurance performance through further increases in VT. Elite endurance athletes have been known to increase their VT to 80–90 % of VO_2max.

The VT can also be used for exercise prescription and programming for non-athletes. The exercise training intensity equivalent to the VT provides an optimal overload stimulus to achieve health-fitness benefits, including weight loss and the improvement of cardiorespiratory fitness. When individuals are allowed to self-select aerobic exercise intensity, many choose intensities similar to the VT (Dishman et al. 1994; Ekkekakis and Lind 2006; Lind et al. 2005). However, at exercise intensities above the VT, marked decreases in self-reported pleasure begin to occur that may lead to cessation of exercise and subsequent dropping out from an exercise program (Acevedo et al. 2003; Bixby et al. 2001; Ekkekakis et al. 2004; Hall et al. 2002).

Direct assessment of the VT or LT is often impractical because it requires expert personnel and expensive laboratory equipment. Generally, an individual must undergo a load-incremented exercise test terminating at maximal intensity in order to adequately measure all of the physiological variables necessary to determine the VT or LT. For many with cardiovascular risk factors, such a maximal GXT may require continuous heart monitoring using an electrocardiograph (ECG) and direct physician supervision. To measure the VT, a respiratory-metabolic measurement system is required. This automated system determines the individual's volume of expired air, VO_2, and the volume of CO_2 production (VCO_2). This system requires the individual to wear a respiratory apparatus including a head support, mouthpiece and nose clip. These can be uncomfortable and would not be worn during a normal exercise session. To measure the LT, venous or capillary blood samples are taken at regular intervals throughout the exercise test. These invasive measurement procedures can be painful and psychologically stressful.

7.1.2 Target HR at the VT for Exercise Prescription

The calculation of a target HR associated with the VT (HR-VT) has been proposed as an inexpensive method to estimate this important marker of anaerobic metabolism. It is recognized that the target HR method may be more practical and much less expensive than respiratory-metabolic methodology to identify the VT. However, the HR response to exercise can be affected by psychological stress and the body's need for thermoregulation, especially during exercise in high ambient heat and humidity. When measurements are performed outside of a controlled laboratory environment, there may be considerable variability in HR-VT. In addition, when an exercise prescription is based on a target HR, the individual must be able to measure HR during exercise. This requires skill if measuring HR by palpation, additional cost if using a HR monitoring system, or restriction to a fitness facility where HR can be measured using monitors attached to available ergometers.

7.1.3 Target RPE at the VT for Exercise Prescription

RPE can be used to prescribe and regulate exercise in a variety of athletic, clinical and pedagogical settings (Goss et al. 2003) and can be used in place of or in addition to traditional exercise prescriptions based on HR. Rather than prescribe exercise based on a target HR range, one can prescribe exercise using a target RPE. This application is justified because for most exercise modalities RPE is more closely linked to prescribed levels of VO_2 than HR (Goss et al. 2011; Noble and Robertson 1996).

The RPE used in an exercise prescription is selected based on its correspondence to a specific physiological intensity, such as the VT, a %VT or a %VO_2max. A target RPE at the VT (RPE-VT) can be identified using responses to a perceptual estimation test protocol including measures of VO_2. Once the RPE-VT is calculated, the individual is taught to self-regulate exercise intensity to produce the specified target RPE. It should be noted that when an exercise prescription is based on a target RPE, the client should have a firm knowledge of the RPE scale and its use during exercise. Such familiarization with the RPE scale can be established when the instructional set and anchoring procedures are administered as part of the test orientation.

Goss and colleagues (2003) defined the *group-normalized perceptual response* as a range of RPE's that corresponds to a target physiological outcome during exercise and that is common to a specified group of individuals. The use of group-normalized RPE to prescribe and monitor exercise intensity has application to a variety of activities for a wide range of individuals as it is comparatively easy to determine and is readily understood by the participant (Goss et al. 2003). The group-normalized RPE-VT may be a more practical method to establish an optimal training intensity (i.e., zone) to improve cardiorespiratory fitness than the VT or LT which must be determined by laboratory-based exercise testing (Goss et al. 2011). The measurement of RPE does not require expensive equipment or extensive technical skill as do determination of the VT or LT.

Many individuals choose to exercise at intensities near to or just below their VT. However, studies have shown that a substantial number of individuals would not prefer to exercise at intensities above the VT because these levels induce unpleasant feelings and psychological distress (Lind et al. 2005). As such, prescription of exercise intensities above the VT may lead to a decrease in exercise adherence. Therefore, it is important in the health-fitness setting to have a simple, inexpensive method to identify a physiologically optimal training intensity that is subjectively acceptable to the participant. In this context, the RPE-VT may be useful to identify the upper-limit of a prescribed exercise intensity range, especially for beginning exercisers who are psychologically intolerant of high-intensity aerobic exercise. In addition, the RPE-VT is a perceptual marker that can be used to identify the optimal training intensity for elite endurance performers. Competitive cyclists and runners can use the RPE-VT during training and races alike to produce their optimal performance pace, especially when HR and VO_2 monitoring are not practical or not allowed (Monnier-Benoit et al. 2009).

7.1.4 Evidence for the RPE-VT: Borg and OMNI Scales

Studies have found RPE-VT to range from 11 to 14 using the Borg (6 to 20) Scale in a wide variety of subjects (Alberton et al. 2013; Ekkekakis et al. 2004; Feriche et al. 1998; Hill et al. 1987; Mahon et al. 1998; Purvis and Cureton 1981; Swaine et al. 1995). A recent study by Elsangedy and colleagues (2013) compared RPE-VT between sedentary women who were normal weight, overweight, and obese as classified by body mass index (BMI). RPE-VT was a mean of ~12 on the Borg Scale regardless of BMI classification (Elsangedy et al. 2013). A recent study by Alberton and colleagues (2013) determined Borg Scale RPE at the 1stVT and 2ndVT inflection points during treadmill exercise and three different water aerobic exercises: stationary running, jumping jacks, and forward kicks. Mean RPE-VT for the 1stVT ranged from 12 to 13, while mean RPE-VT for the 2ndVT ranged from 15 to 16 (Alberton et al. 2013).

Studies investigating the RPE-VT using the OMNI Scale have observed values ranging from 5 to 7 (Fig. 7.1). Goss and colleagues (2011) identified a mean RPE-VT of 5.1 in Division I football players performing treadmill exercise. RPE was assessed by the Adult OMNI Walk/Run RPE Scale. Robertson et al. (2001) identified a mean OMNI RPE-VT of 6.1 in children of average and above average fitness levels performing cycle ergometer exercise. In addition, Robertson et al. (2007) identified the RPE-VT in children using direct observation, rather than subject estimation, with OMNI RPE values ranging from 6 to 6.5. These results derived from the OMNI Scale seem to be in agreement with previous research using the Borg Scale. This comparison can be done using Robertson's (2004) table to convert RPE between the Borg and OMNI Scales (Fig. 7.2). Using this table, OMNI RPE values ranging from 5 to 7 correspond to Borg Scale RPE values of 12 to 16.

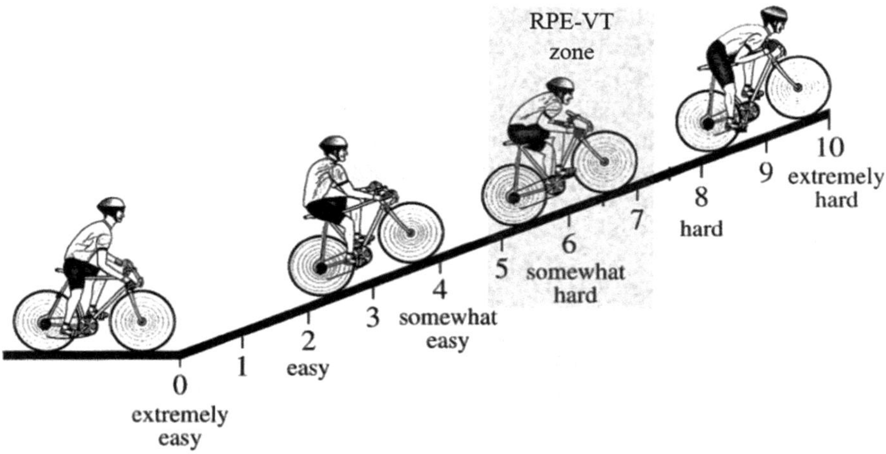

Fig. 7.1 OMNI Scale RPE-VT Zone (Robertson 2004)

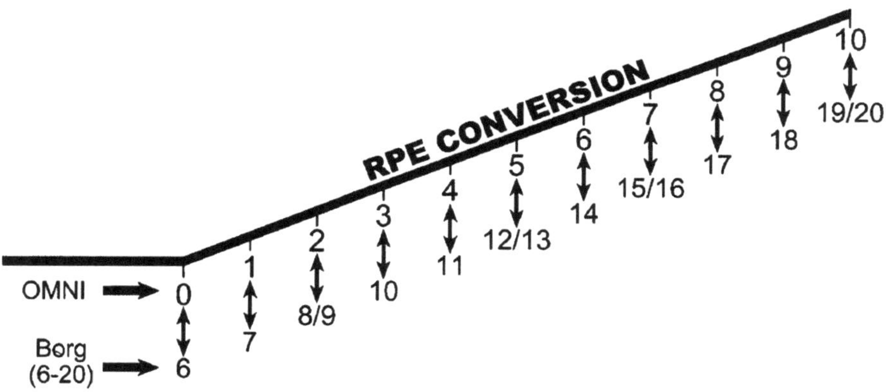

Fig. 7.2 RPE conversions between the OMNI Scale and Borg (6–20) Scale (Robertson 2004)

7.1.5 Case Study

7.1.5.1 Client Information

A 21-year-old male college student comes to your fitness facility. During a pre-participation interview prior to exercise testing, he tells you that he plays recreational basketball once or twice per week. He describes that he gets "winded" easily when playing a full court game. He also tells you that he exercises on a stationary bike at his school's student fitness center once or twice per week for 10–20 min per session. He is moderately overweight and he describes his fitness level as average. His goals are to lose weight and increase his aerobic fitness. He enjoys going to the fitness center because he can go with a friend or watch television while exercising and he prefers the bike over the treadmill. He knows he should go to the fitness center more often. He wants to learn the proper exercise intensity to perform on the bike so he can meet his goals and perform better on the basketball court. Due to the client's age, health status, and current level of PA, a pre-participation GXT is likely not required. Therefore, his exercise prescription could be developed using a group-normalized RPE-VT.

7.1.5.2 Assessments

Perform a graded exercise test on a cycle ergometer or treadmill terminating at maximal exertion to determine CRF (VO_2peak/max), the VT, and RPE-VT. These test responses, when incorporated into an exercise prescription, will identify an effective exercise intensity to achieve his weight loss and aerobic fitness goals.

7.1.5.3 Results and Analysis

Identify HRmax/peak ($b \cdot min^{-1}$):
Identify VO_2max/peak ($ml \cdot kg^{-1} \cdot min^{-1}$ or $l \cdot min^{-1}$):
Identify the VT ($ml \cdot kg^{-1} \cdot min^{-1}$ or $l \cdot min^{-1}$, %VO_2max/peak, PO):
Identify OMNI RPE-VT:
Identify HR-VT ($b \cdot min^{-1}$):

7.2 Methods

7.2.1 Treadmill Procedures

7.2.1.1 Equipment

1. Adult OMNI-Walk/Run RPE Scale (Fig. A.2)
2. Borg RPE Scale
3. Treadmill
4. HR monitor
5. Respiratory-metabolic measurement system

7.2.1.2 Pre-exercise Procedures

1. Measure height (cm) and weight (kg) of subject.
2. Read the standard instructions for use of the Adult OMNI-Walk/Run RPE Scale for RPE-O to the subject (Appendix B.1). If determination of differentiated RPE-VT (RPE-L and RPE-C) is also desired, read the standard instructions for

use of the Adult OMNI-Walk/Run RPE Scale for undifferentiated and differentiated RPE to the subject (Appendix B.2). Perform the memory anchoring procedure as described in Chap. 5.
3. Read the standard instructions for use of the Borg Scale during treadmill exercise for measurement of RPE-O to the subject (Appendix B.3).

7.2.1.3 Graded Exercise Test

1. Position the HR monitor and respiratory-metabolic mouthpiece (with head support unit and nose clip if applicable) on the subject.
2. Instruct the subject to step onto the treadmill and review exercise termination procedures: When the subject cannot continue exercise due to exhaustion or discomfort, he/she should grasp the treadmill hand rails, at which time the test administrator will gradually slow the treadmill down for performance of a cool-down. The subject should be reminded not to step off the treadmill belt while it is still in motion.
3. Bruce Multistage Treadmill Test Protocol: this can be performed by manually adjusting treadmill speed and grade or using a program on a computer that is interfaced to the treadmill.

   (a) Begin the warm-up at 1.5 miles $\cdot$ h^{-1} and 0 % grade for 3 min.
   (b) Each exercise test stage will last 3 min. The stages progress as follows:

   Stage 1—1.7 miles $\cdot$ h^{-1} and 10 % grade
   Stage 2—2.5 miles $\cdot$ h^{-1} and 12 % grade
   Stage 3—3.4 miles $\cdot$ h^{-1} and 14 % grade
   Stage 4—4.2 miles $\cdot$ h^{-1} and 16 % grade
   Stage 5—5.0 miles $\cdot$ h^{-1} and 18 % grade
   Stage 6—5.5 miles $\cdot$ h^{-1} and 20 % grade
   Stage 7—6.0 miles $\cdot$ h^{-1} and 22 % grade
   Stage 8—6.5 miles $\cdot$ h^{-1} and 24 % grade

   (c) When the subject cannot continue any longer owing to fatigue, terminate the exercise test by initiating the cool-down period at 1.5 miles $\cdot$ h^{-1} and 0 % grade. The cool-down is 5 min in duration.
   (d) Ask the subject to estimate RPE starting at 2:30 of each exercise stage using both the Borg Scale (RPE-O only) and the OMNI Scale (RPE-O, RPE-L, and RPE-C). The RPE's should be rated in a counterbalanced sequence. Because a verbal response is inhibited by the position of the respiratory-metabolic mouth piece, instruct the subject to point to the number on the RPE scale, which should be conveniently positioned within the subject's arm reach. State aloud the numerical rating for each momentary assessment to which the subject pointed and request a confirmatory nod that the number stated was correct. If incorrect, allow the subject to point to the appropriate rating on the RPE scale once more. Ask the subject to hold his or her finger on the appropriate number on the scale for approximately 1 s.

(e) Record HR ($b \cdot min^{-1}$) at 2:55 of each exercise stage.
(f) Record the final 15-s VO_2 ($ml \cdot kg^{-1} \cdot min^{-1}$) for each exercise stage.
(g) Record HRmax as the highest HR value recorded during the final exercise stage or immediately post-exercise.
(h) Record VO_2max as the highest 15-s VO_2 value recorded at the end of the test.
(i) Obtain the VO_2 ($ml \cdot kg^{-1} \cdot min^{-1}$) and %$VO_2$max associated with the VT using measurements obtained from the respiratory-metabolic system and calculate the VT using the automated program.

7.2.1.4 Data Organization and Analysis

1. In a Microsoft Excel spreadsheet, label columns of data for the following variables: Exercise Stage, VO_2 ($ml \cdot kg^{-1} \cdot min^{-1}$), Borg RPE-O, OMNI RPE-O, HR ($b \cdot min^{-1}$). Include columns for OMNI RPE-L and OMNI RPE-C if applicable.
2. If the respiratory-metabolic measurement system does not automatically calculate VT or if instruction on manual calculation and visual identification of the VT is desired, refer to Appendix D for detailed instructions for the following:

   (a) Calculation of $V_E \cdot VO_2^{-1}$ and $V_E \cdot VCO_2^{-1}$.
   (b) Plot of $V_E \cdot VO_2^{-1}$ and $V_E \cdot VCO_2^{-1}$ for visual identification of the VT using the ventilatory equivalent method.
   (c) Adjustment of automatic VT calculation using a respiratory-metabolic measurement system if a computer application is available for the system employed.

3. Plot of VO_2 as a function of Borg RPE-O for determination of RPE-VT.

   (a) Click on the **INSERT** tab and in the **CHARTS** section click on **SCATTER**. Select the first available chart option. A blank or example scatter plot will appear on your screen.
   (b) Click on the **SELECT DATA** tab. Remove any entries found in the **LEGEND ENTRIES** text box then click **ADD**. Under **SERIES NAME**, enter VO_2 and Borg RPE-O. Then click on the icon to the right of the **SERIES X VALUES** text box and highlight the VO_2 values. After the values are highlighted click the icon on the box that appeared. Then click on the icon to the right of the **SERIES Y VALUES** text box and highlight the Borg RPE-O values. After the values are highlighted click the icon on the box that appeared. Click **OK** on the next two screens.
   (c) You should now have a scatter plot with Borg RPE-O on the *y*-axis and VO_2 on the *x*-axis. Create a title for the plot and enter the appropriate axis labels and units of measure.
   (d) To determine RPE-VT, click on one of the data points to highlight the entire data series. Right click on one of the data points and a menu will appear. Click **ADD TRENDLINE** and the **FORMAT TRENDLINE** menu will appear. Select **LINEAR** and **DISPLAY EQUATION ON CHART** then click **CLOSE**. The trendline and equation will be displayed on the chart.

(e) Use this linear equation to calculate RPE-VT. Use VO_2 ($ml \cdot kg^{-1} \cdot min^{-1}$) corresponding to the VT as the "x" value in the equation and solve for "y." The calculated "y" value, once rounded to the nearest whole integer, is the Borg RPE-VT.

4. Repeat the above steps for VO_2 and OMNI RPE-O to determine OMNI RPE-VT and VO_2 and HR to determine HR-VT. Repeat the above steps for the following variable pairs to determine differentiated OMNI RPE-VT's: VO_2 and OMNI RPE-L, VO_2 and OMNI RPE-C.
5. An example of these procedures with a screenshot depicting each step as performed using Microsoft Excel 2013 can be found in Appendix D.

7.2.2 Cycle Ergometer Procedures

7.2.2.1 Equipment

1. Adult OMNI-Cycle RPE Scale (Fig. 2.4)
2. Borg RPE Scale
3. Cycle ergometer
4. Metronome
5. HR monitor
6. Respiratory-metabolic measurement system

7.2.2.2 Pre-exercise Procedures

1. Measure height (cm) and weight (kg) of subject.
2. Read the standard instructions for use of the Adult OMNI-Cycle RPE Scale for RPE-L to the subject (Appendix B.4). For determination of undifferentiated RPE-VT (RPE-O) and differentiated RPE-VT for (RPE-L and -C), read the standard instructions for use of the Adult OMNI-Cycle RPE Scale for undifferentiated and differentiated RPE to the subject (Appendix B.5). Perform the memory anchoring procedure as described in Chap. 5.
3. Read the standard instructions for use of the Borg Scale during cycle exercise for RPE-L to the subject (Appendix B.6).

7.2.2.3 Graded Exercise Test

1. Position the HR monitor and respiratory-metabolic mouthpiece (with head support unit and nose clip if applicable) on the subject.
2. Set the proper seat height on the cycle ergometer according to leg length. When the foot is flat on the right pedal and the pedal is in the extreme down position, there should be a flexion of the right knee of approximately 5°.

3. Load-incremented protocol for electronically braked and friction-braked cycle ergometers:
   (a) Instruct the subject to maintain a 50 rev · min^{-1} pedal cadence. Set the metronome to 100 beats · min^{-1} so the downward movement of each foot is synchronized with a beat of the metronome. The subject may also use the digital monitor on the cycle control panel to regulate pedal cadence.
   (b) For electronically braked cycle ergometers (e.g., Lode), begin stage 1 at 50 W then increase the resistance 25 W every 2 min.
   (c) For friction-braked cycle ergometers (e.g., Monark), begin stage 1 at 1 kg resistance then increase the resistance 0.5 kg every 2 min.
   (d) When the subject cannot maintain the pedal cadence for 10 consecutive seconds, terminate the exercise test.
   (e) Ask the subject to estimate RPE starting at 1:30 of each exercise stage using both the Borg Scale (RPE-L only) and the OMNI Scale (RPE-L; RPE-O and –C if desired). The RPE's should be rated in a counterbalanced sequence. Because the position of the respiratory-metabolic mouth piece prohibits a verbal response, instruct the subject to point to the number on the RPE scale, which should be conveniently positioned within the subject's arm reach. State aloud the numerical rating for each momentary assessment to which the subject pointed and request a confirmatory nod that the number stated was correct. If incorrect, allow the subject to point to the appropriate rating on the RPE scale once more. Ask the subject to hold his or her finger on the appropriate number on the scale for approximately 1 s.
   (f) Record HR (b · min^{-1}) at 1:55 of each exercise stage.
   (g) Record the final 15-s VO_2 (l · min^{-1}) for each exercise stage.
   (h) Record HRpeak as the highest HR value recorded during the final exercise stage or immediately post-exercise.
   (i) Record VO_2peak as the highest 15-s VO_2 value recorded at the end of the test.
   (j) Obtain the VO_2 (l · min^{-1}) and %VO_2peak associated with the VT using measurements obtained from the respiratory-metabolic system and calculate the VT using the automated program.

7.2.2.4 Data Organization and Analysis

1. In a Microsoft Excel spreadsheet, label columns of data for the following variables: Exercise Stage, VO_2 (l · min^{-1}), Borg RPE-L, OMNI RPE-L, HR (b · min^{-1}). Include columns for OMNI RPE-O and OMNI RPE-C if applicable.
2. If the respiratory-metabolic measurement system does not automatically calculate VT or if instruction on manual calculation and visual identification of the VT is desired, refer to Appendix D for detailed instructions for the following:
   (a) Calculation of $V_E \cdot VO_2^{-1}$ and $V_E \cdot VCO_2^{-1}$.
   (b) Plot of $V_E \cdot VO_2^{-1}$ and $V_E \cdot VCO_2^{-1}$ for visual identification of the VT using the ventilatory equivalent method.

(c) Adjustment of automatic VT calculation using a respiratory-metabolic system if the computer application is available for the program employed.

3. Plot of VO_2 as a function of Borg RPE-L for determination of RPE-VT.
   (a) Click on the **INSERT** tab and in the **CHARTS** section click on **SCATTER**. Select the first available chart option. A blank or example scatter plot will appear on your screen.
   (b) Click on the **SELECT DATA** tab. Remove any entries found in the **LEGEND ENTRIES** text box then click **ADD**. Under **SERIES NAME**, enter VO_2 and Borg RPE-L. Then click on the icon to the right of the **SERIES X VALUES** text box and highlight the VO_2 values. After the values are highlighted click the icon on the box that appeared. Then click on the icon to the right of the **SERIES Y VALUES** text box and highlight the Borg RPE-L values. After the values are highlighted click the icon on the box that appeared. Click **OK** on the next two screens.
   (c) You should now have a scatter plot with Borg RPE-L on the *y*-axis and VO_2 on the *x*-axis. Create a title for the plot and enter the appropriate axis labels and units of measure.
   (d) To determine Borg RPE-VT, click on one of the data points to highlight the entire data series. Right click on one of the data points and a menu will appear. Click **ADD TRENDLINE** and the **FORMAT TRENDLINE** menu will appear. Select **LINEAR** and **DISPLAY EQUATION ON CHART** then click **CLOSE**. The trendline and equation will be displayed on the chart.
   (e) Use this linear equation to calculate RPE-VT. Use VO_2 ($l \cdot min^{-1}$) corresponding to the VT as the "x" value in the equation and solve for "y." The calculated "y" value, once rounded to the nearest whole integer, is the Borg RPE-VT.
4. Repeat the above steps expressing VO_2 as a function of OMNI RPE-L to determine OMNI RPE-VT. Next, express VO_2 as a function of HR to determine HR-VT. Repeat the above steps for the following variable pairs to determine other OMNI RPE-VT's if desired: VO_2 and OMNI RPE-O; VO_2 and OMNI RPE-C.
5. An example of these procedures with a screenshot depicting each step as performed using Microsoft Excel 2013 can be found in Appendix D.

7.3 Discussion Questions

1. In general, explain the concept of the RPE-VT using the group-normalized RPE model as a theoretical framework.
2. Do the results of this experiment agree with previous literature concerning the RPE-VT? Specifically, are they consistent with previous reports of a group-normalized RPE equivalent to the VT where both the OMNI and Borg (6–20) Scales were separately employed? Please explain why, citing previous literature.

3. How can you use your RPE-VT results to prescribe a self-regulated aerobic exercise intensity for the client described in the case study? To what population can your results be applied?
4. Describe why the VT can be an effective exercise intensity for aerobic conditioning, both physiologically and psychologically.
5. In what type of environments should you avoid using HR-VT to prescribe exercise intensity? What happens to the HR response in these abnormal conditions?
6. Describe the characteristics of another client, different from that identified in the case study in this laboratory module, who would benefit from aerobic exercise at an intensity equivalent to the VT.
   (a) Explain how you would customize the exercise prescription to fit his/her needs and goals?
   (b) What is unique about this other client that you must address in the exercise program?

References

Acevedo EO, Kraemer RR, Haltom RW, Tryniecki JL. Perceptual responses proximal to the onset of blood lactate accumulation. J Sports Med Phys Fitness. 2003;43:267–73.

Alberton CL, Antunes AH, Beilke DD, Pinto SS, Kanitz AC, Tartaruga MP, Kruel LFM. Maximal and ventilatory thresholds of oxygen uptake and rating of perceived exertion responses to water aerobic exercises. J Strength Cond Res. 2013;27:1897–903.

Bixby WR, Spalding TW, Hatfield BD. Temporal dynamics and dimensional specificity of the affective response to exercise of varying intensity: differing pathways to a common outcome. J Sport Exerc Psychol. 2001;23:171–90.

Dishman RK, Farquhar RP, Curetone KG. Responses to preferred intensities of exertion in men differing in activity levels. Med Sci Sports Exerc. 1994;26:783–90.

Ekkekakis P, Hall EE, Petruzzello SJ. Practical markers of the transition from aerobic to anaerobic metabolism during exercise: rationale and a case for affect-based exercise prescription. Prev Med. 2004;35:149–59.

Ekkekakis P, Lind E. Exercise does not feel the same when you are overweight: the impact of self-selected and imposed intensity on affect and exertion. Int J Obes (Lond). 2006;30:652–60.

Elsangedy HM, Krinski K, Costa EC, Haile L, Fonteles AI, Timossi LD, da Silva SG. The rating of perceived exertion is not different at the ventilatory threshold in sedentary women with different body mass indices. J Exerc Sci Fitness. 2013;11:102–6.

Feriche B, Chicharro JL, Vaquero AF, Perez M, Lucia A. The use of a fixed value of RPE during a ramp protocol: comparison with the ventilatory threshold. J Sports Med Phys Fitness. 1998; 38:35–8.

Goss F, Robertson R, DaSilva S, Suminski R, Kang J, Metz K. Ratings of perceived exertion and energy expenditure during light to moderate activity. Percept Mot Skills. 2003;96:739–47.

Goss FL, Robertson RJ, Gallagher Jr M, Piroli A, Nagle EF. Response normalized OMNI rating of perceived exertion at the ventilatory breakpoint in division I football players. Percept Mot Skills. 2011;112:539–48.

Hall EE, Ekkekakis P, Petruzzello SP. The affective beneficence of vigorous exercise revisited. Br J Health Psychol. 2002;7:47–66.

Hill DW, Cureton KJ, Grisham SC, Collins MA. Effect of training on the rating of perceived exertion at the ventilatory threshold. Eur J Appl Physiol. 1987;56:206–11.

Kenney WL, Wilmore JH, Costill DL. Physiology of sport and exercise. 5th ed. Champaign, IL: Human Kinetics; 2012.

Lind E, Joens-Matre RR, Ekkekakis P. What intensity of physical activity do previously sedentary middle-aged women select? evidence of a coherent pattern from physiological, perceptual, and affective markers. Prev Med. 2005;40:407–19.

Mahon AD, Gay JA, Stolen KQ. Differentiated ratings of perceived exertion at ventilatory threshold in children and adults. Eur J Appl Physiol. 1998;18:115–20.

Monnier-Benoit P, Groslambert A, Rouillon J. Determination of the ventilatory threshold with affective valence and perceived exertion in trained cyclists: a preliminary study. J Strength Cond Res. 2009;23:1752–7.

Noble BJ, Robertson RJ. Perceived exertion. Champaign, IL: Human Kinetics; 1996.

Purvis JW, Cureton KJ. Ratings of perceived exertion at the anaerobic threshold. Ergonomics. 1981;24:295–300.

Robertson RJ. Perceived exertion for practitioners: rating effort with the OMNI picture system. Champaign, IL: Human Kinetics; 2004.

Robertson RJ, Goss FL, Boer N, Gallagher JD, Thompkins T, Bufalino K, Balasekaran G, Meckes C, Pintar J, Williams A. OMNI scale perceived exertion at ventilatory breakpoint in children: response normalized. Med Sci Sports Exerc. 2001;33:1946–52.

Robertson RJ, Goss FL, Aaron DJ, Utter AC, Nagle E. OMNI scale rating of perceived exertion at ventilatory breakpoint by direct observation of children's kinematics. Percept Mot Skills. 2007;105:975–84.

Swain IL, Emmet J, Murty D, Dickinson C, Dudfield M. Rating of perceived exertion and heart rate relative to ventilatory threshold in women. Br J Sports Med. 1995;29:57–60.

Chapter 8
Prediction of Maximal Aerobic Power and Dynamic Muscular Strength Using RPE

Cardiorespiratory fitness and muscular strength are important components of health-related physical fitness. The criterion measures for cardiorespiratory fitness and dynamic muscular strength are maximal oxygen uptake (VO_2max) and one-repetition maximum (1RM), respectively. VO_2max testing requires expensive laboratory equipment and expert personnel. Such maximal intensity testing may require physician clearance or supervision for older, sedentary and/or unfit individuals. 1RM testing requires the performance of multiple resistance exercise sets at near-maximal and maximal intensities and may not be safe for many individuals who are unfamiliar with resistance exercise or who have medical contraindications to high intensity exercise. Therefore, administration of maximal fitness test protocols may not be possible, pragmatic or even desirable in health-fitness and clinical settings. As such, submaximal tests have been developed to predict both VO_2max and 1RM. It is proposed that the use of submaximal test protocols is both safe and cost-effective. Statistical models to predict VO_2max have been based on the positive relation between HR and VO_2 during load-incremented exercise. Models to predict 1RM have been based on the inverse relation between weight lifted and repetitions performed to the point of muscular fatigue. However, both VO_2max and 1RM can be predicted from submaximal RPE. The use of RPE as a predictor variable in such tests is valid, technically simple, and easily understood by most individuals. In addition, submaximal fitness tests including the measurement of RPE can be used as an assessment of training-induced fitness changes. The primary purpose of this laboratory experiment is to predict VO_2max/peak and 1RM from submaximal RPE measured during an estimation protocol exercise test.

L. Haile et al., *Perceived Exertion Laboratory Manual*,
DOI 10.1007/978-1-4939-1917-8_8

8.1 Background

8.1.1 Assessment of Cardiorespiratory Fitness

Cardiorespiratory fitness determines an individual's ability to perform dynamic exercise of a moderate to vigorous intensity using large muscle groups for a prolonged period. Cardiorespiratory fitness depends on the functional capacity of the cardiovascular and respiratory systems and the oxidative capacity of skeletal muscle (ACSM 2013). The criterion or "gold-standard" assessment of cardiorespiratory fitness is VO_2max (i.e., maximal aerobic power), defined as the maximal amount of oxygen an individual can use during dynamic exercise while breathing air at sea level. The assessment of VO_2max requires an individual to perform a load-incremented aerobic exercise protocol, or GXT, where pulmonary ventilation (V_E) and expired concentrations of VO_2 and VCO_2 are determined using a respiratory-metabolic measurement system. For most clinically normal individuals, a valid VO_2max can only be achieved during upright, weight-bearing, total body exercise such as uphill walking or running on a treadmill. The term VO_2peak is used rather than VO_2max when the measure is obtained using cycle ergometer exercise and other partial- or non-weight-bearing exercise modalities (e.g., swimming, elliptical ergometer, rowing ergometer, and arm ergometer). However, it is of note that when elite cyclists perform a load-incremented GXT on a cycle ergometer they often demonstrate a higher VO_2max than for uphill treadmill exercise. It is important to assess an individual's VO_2max/peak with a mode-specific GXT that matches their exercise experience and the mode of exercise for which the prescription is intended. For most individuals, a treadmill protocol is appropriate.

VO_2max and VO_2peak tests allow determination of the VT. The VT is an important physiological marker for aerobic training intensity. At test intensities above the VT, an individual can no longer achieve steady state metabolic energy production. HR and/or RPE can be measured throughout a GXT. The HR or RPE values corresponding to specific physiological intensities (e.g., %VO_2max or %VT) can then be used to prescribe exercise intensity. However, the ability to calculate target HR or RPE ranges based on VO_2 responses to an exercise test comes with inherent costs and risks. First, the measurement and interpretation of respiratory-metabolic responses are time-consuming procedures that require expensive laboratory equipment and expert personnel that are not available in many clinics and fitness facilities. Maximal exercise testing often requires physician clearance and test supervision, especially when older, sedentary and/or unfit individuals are evaluated. This imposes additional financial and testing burden on the client and testing facility. Therefore, it is often not possible, practical or even desirable to perform a maximal test with respiratory-metabolic measurement to establish an individual's cardiorespiratory fitness prior to undertaking an aerobic exercise program.

8.1.2 Assessment of Dynamic Muscular Strength

Muscular strength reflects the ability of a muscle or muscle group to exert force. Muscular strength can be assessed statically, involving isometric muscular action with no change in muscle length or joint angle, or dynamically, involving concentric and/or eccentric muscular contractions and changes in muscle length. Since assessments of static muscular strength are only specific to the joint angle used in testing, dynamic muscular strength is considered more ecologically valid. The criterion or "gold-standard" assessment of dynamic muscular strength is the 1RM test. The 1RM is used to establish the maximal amount of force an individual can exert during one repetition of single maximal effort using a defined muscle or muscle group. The lift must be performed in a controlled manner through the full range of motion with proper technique (ACSM 2013). This test requires that an individual perform multiple resistance exercise sets at near-maximal and maximal intensities using a progressive protocol. Unlike VO_2max testing, most clinics and fitness facilities have ample resistance exercise equipment and educated personnel to assess 1RM strength. The primary concern for an individual who is undergoing a 1RM baseline test prior to beginning a resistance exercise program is safety. Many individuals who are beginning a resistance exercise program, including children and adolescents, have little to no experience with resistance exercise. In particular, they lack instruction on proper exercise technique. For these individuals, performing a 1RM test is not practical and may result in musculoskeletal injury. Even a multiple-RM test, where the goal is to perform a set number of repetitions ending at maximal intensity (e.g., 5RM or 10RM), may not be safe for many individuals. If a 1RM or multiple-RM test is employed, it may be best to guide the individual through a brief orientation and practice period that employs the exercise protocol prior to maximal testing.

8.1.3 Submaximal Tests to Predict Maximal Aerobic Power and Muscular Strength

Due to the methodological and safety limitations of measuring maximal aerobic power (i.e., VO_2max) and muscular strength (i.e., 1RM), researchers have designed submaximal exercise tests from which maximal values can be predicted. The methods used to predict VO_2max and 1RM are based on the relations between the criterion variable and predictor variable(s). For aerobic exercise, VO_2 serves as the criterion variable since VO_2max/peak is the unknown that is estimated by prediction models. For resistance exercise, weight lifted serves as the criterion variable since 1RM is the unknown that is estimated by prediction models. The predictor variables are physiological and/or physical markers that rise concurrently with increases in exercise intensity.

Traditional prediction models for aerobic exercise are based on the strong positive correlation between VO_2 and HR that has been consistently demonstrated in the literature. The model of VO_2max using submaximal HR is dependent on the relative accuracy of age-predicted maximal HR (APMHR) equations (e.g., APMHR = 220 − age) and the positive relation between VO_2 and HR as measured during a load-incremented protocol. The prediction can be presented graphically and/or determined using linear regression analysis. For a graphic determination, VO_2 and HR data points from multiple submaximal exercise intensities are plotted on separate axes. HR, the predictor variable, is on the *x*-axis. VO_2, the criterion variable, is on the *y*-axis. A line of best fit that describes the relation between the two variables is drawn by visual determination. The point where the line intercepts APMHR is extended laterally to the VO_2 axis, identifying predicted VO_2max. Using a linear regression model a line of best fit is calculated that describes the relation between HR and VO_2 where these variables are expected to change as a function of increasing exercise intensity. This calculation yields an equation, $VO_2\text{max} = s(\text{APMHR}) + i$. In this equation, s is the slope of the line and i is the *y*-intercept. One solves for VO_2max by entering APMHR into the prediction equation.

Although HR is relatively easy to measure during aerobic exercise and most clinics and fitness facilities have appropriate technology and/or personnel, APMHR is not always the most accurate prediction of actual HRmax. The primary prediction equations used in the health-fitness setting are: APMHR = 220 − age for males; APMHR = 226 − age for females. These equations are based on large sample data with standard deviations of ±11 $b{\cdot}min^{-1}$ (Londeree and Moeschberger 1982). A standard deviation can be defined as the average amount by which the scores in a distribution differ from the mean. Therefore, based on the above APMHR equation for men and women and the reported standard deviation for the derived value, the average 20-year-old will have a HRmax between 189 and 211 $b{\cdot}min^{-1}$, but could even have a HRmax below 189 $b{\cdot}min^{-1}$ or above 211 $b{\cdot}min^{-1}$. As such, APMHR can have a considerable amount of error for an individual resulting in even greater error for predicted VO_2max. A target HR range for a prescribed training program that is based on a VO_2max prediction model using APMHR, could then be either below or above that which provides an optimal overload stimulus for the individual. A target HR range set too low may not provide the individual with an appropriate overload stimulus to achieve health-fitness benefits. A target HR range set too high may not be tolerable for the individual, causing early termination of an exercise session and eventually leading to dropout from the exercise program.

The use of equations that employ APMHR to predict VO_2max/peak as a basis for exercise prescription is not appropriate for many individuals with certain clinical conditions. Individuals taking beta-blocker medication to control hypertension exhibit a blunted HR response, especially during aerobic exercise. In such conditions, HRmax would be much lower than that estimated by the APMHR equation. Individuals with pulmonary limitations to exercise, such as chronic obstructive pulmonary disease (COPD) or cystic fibrosis, often terminate exercise because of dyspnea, or shortness of breath, and cannot reach maximal/peak HR levels similar to clinically normal individuals of the same age. Also, individuals with peripheral artery disease (PAD) experience claudication pain in active limbs, even at submaximal exercise intensities. In these patients, exercise is prescribed based on tolerable levels of perceived pain rather than a target HR range. For any clinical conditions where there may be an increased risk for adverse events during exercise, it is important to obtain physician clearance prior to participation in exercise testing or an exercise program. Often, exercise testing with physician supervision may be indicated.

The only way to ensure accuracy when prescribing aerobic exercise based on the expected positive relation between HR and VO_2 is to actually measure HRmax and VO_2max. This necessitates that the GXT terminates at maximal intensity. However, a maximal GXT may not be possible due to the lack of respiratory-metabolic instrumentation or trained testing personnel. In addition, a maximal GXT may not be appropriate because of time constraints when administering the test protocol and the possible error induced when developing a prediction based on APMHR. VO_2 has shown strong correlations with RPE, a fact well-established in experiments demonstrating the concurrent validity of RPE scales. As such, submaximal RPE expressed as a function of either HR or VO_2 can be used to predict VO_2max/peak. Numerous investigations have shown the validity of RPE-based exercise tests to predict VO_2max/peak (Davies et al. 2008; Eston et al. 2005, 2006, 2008, 2012; Faulkner et al. 2007; Faulkner and Eston 2007; Morris et al. 2009, 2010).

Prior to the initiation of a resistance training program for an untrained individual, the administration of 1RM testing may not be indicated because of safety concerns. The statistical basis for traditional models to predict 1RM is the strong inverse relation between weight lifted per repetition and the number of repetitions performed until fatigue. As weight increases, the number of repetitions that can be performed until fatigue decreases, ultimately resulting in the 1RM value. The linear relation between submaximal weight lifted and repetitions performed can be analyzed to predict the amount of weight lifted for 1RM. Weight lifted has shown strong correlations with RPE, a fact well-established in experiments demonstrating the concurrent validity of RPE scales. As such, the relation between resistance and RPE allows the prediction of 1RM. A number of investigations have shown the validity of RPE-based exercise tests to predict 1RM (Eston and Evans 2009; Gearhart et al. 2008; Robertson et al. 2008).

8.1.4 RPE-Based Models to Predict VO_2max and 1RM

The development of RPE-based models to predict VO_2max and 1RM follow the same design as models based on HR and weight lifted, respectively. Prediction of VO_2max using RPE as the predictor variable is based on the positive relation between VO_2 and RPE that occurs during load-incremented aerobic exercise. Likewise, prediction of 1RM using RPE as the predictor variable is based on the positive relation between weight lifted and RPE that occurs during load-incremented resistance exercise. These relations and their predictive properties can be presented graphically and/or determined using linear regression analysis. For a graphic prediction procedure, VO_2 (or weight lifted) and RPE data points measured at multiple submaximal exercise intensities are plotted on separate axes. A line of best fit that describes the relation between the two variables is drawn by visual determination. When using a HR-based model to predict maximal aerobic power, the point where the line intercepts APMHR is extended laterally to the VO_2 axis, identifying predicted VO_2max. However, when using a category rating scale, the RPE used to predict VO_2max or 1RM is a fixed value, i.e., the maximal RPE (RPEmax) on the category scale. The use of such a fixed upper rating category facilitates the prediction of the corresponding VO_2max/1RM value when employing a graphic procedure.

Statistical prediction models that employ RPE as a predictor variable are based on Borg's Range Model. According to Borg's Range Model, when an individual reaches maximal intensity of the perceptual response range, they should report the highest numerical category, i.e., an RPEmax of 10 using the OMNI Scale, or 20 using the Borg Scale. Therefore, for visual determination of VO_2max using the line of best fit derived from an RPE-based model, the point where the line of best fit intercepts RPEmax is extended laterally to the VO_2 axis, identifying predicted VO_2max. For visual determination of 1RM using submaximal RPE responses, the point where the line of best fit intercepts RPEmax is extended laterally to the y-axis to identify predicted 1RM. Using linear regression, the average positive relation as depicted by the line of best fit is calculated using submaximal VO_2 (or weight lifted) and RPE responses to a load-incremented protocol. For aerobic exercise, this calculation yields an equation, $VO_2\text{max} = s(\text{RPEmax}) + i$, where the predictor variable is RPE and the criterion variable is VO_2. For resistance exercise, 1RM is predicted using the same linear regression model. In these linear regression equations, s is the slope of the line of best fit and i is the y-intercept. Then, one solves for VO_2max or 1RM by entering RPEmax into the prediction equation.

8.1.5 Evaluating the Accuracy of RPE-Based Prediction Models

The accuracy of both HR- and RPE-based statistical models to predict VO_2max can be evaluated in a laboratory setting where standardized testing instrumentation and control and experimental conditions are available. Using a statistical regression procedure where multiple measurements are required, the subject performs three different submaximal exercise intensities. On a treadmill, the intensities are progressively increased by changing speed and/or grade. On a cycle ergometer, the intensities are progressively increased by incremental changes in PO. The intensities on both a treadmill and cycle ergometer are presented in a load-incremented format. However, there are also less frequently employed protocols that present the different intensities in random order. HR and RPE are measured near the end of each 2–3-min exercise stage.

Using a computer program such as Microsoft Excel, plots of VO_2 expressed as a function of HR and as a function of RPE are developed so the relation between variables can be depicted both graphically and using statistical linear regression analysis. VO_2max can be predicted by extrapolating submaximal responses to intercept at APMHR or RPEmax using the graphic procedure. In addition, the computer program employs the linear regression equation to determine the line of best fit. The slope and intercept of this regression line can be used to predict VO_2max as previously described. Alternatively, some computer programs, such as the Statistical Package for the Social Sciences (SPSS), use a linear regression equation to determine the line of best fit without inclusion of the graphic procedure. In order to compare measured VO_2max with that predicted using both the HR and RPE models, the subject should perform an entire GXT terminating at maximal intensity.

For aerobic exercise, both undifferentiated RPE (RPE-O) and the dominant differentiated RPE (i.e., RPE-Legs) have been used in models to predict VO_2max/peak (Faulkner and Eston 2007). In some instances, the dominant RPE for a given exercise mode may explain a greater amount of variance in VO_2max (i.e., is a better predictor) than the undifferentiated RPE-O. For resistance exercise, only differentiated RPE for the active muscle mass (RPE-AM) has been used in statistical models to predict 1RM (Eston and Evans 2009; Robertson et al. 2008).

8.1.6 Cross-Validation of RPE-Based Prediction Models

From a research perspective, statistical prediction models can be developed for the field setting that do not require actual measurement of VO_2. These statistical models must be validated in one sample of subjects then cross-validated in a separate but similar sample of subjects. For the initial validation study, subjects undergo two separate exercise trials: (1) a criterion VO_2max test on a treadmill in a laboratory setting; and (2) a submaximal exercise protocol appropriate for administration in the field setting during which RPE are measured. A statistical regression procedure is used to develop an equation to predict criterion measured VO_2max using submaximal RPE responses measured during the field test. This equation is used to calculate predicted VO_2max for each subject. Criterion measured VO_2max is then compared with predicted VO_2max. If predicted VO_2max is similar to actual measured VO_2max (i.e., exhibiting no statistically significant difference), the field-based prediction model is deemed valid.

The statistical model developed and validated in one subject group is then cross-validated by testing the equation on a separate sample of subjects. For such a cross-validation study, the subjects also undergo two separate exercise trials: (1) a criterion VO_2max test on a treadmill in a laboratory setting; and (2) the same submaximal exercise field protocol used for the initial validation study, during which RPE are measured. This second, cross-validation sample should have similar characteristics (i.e., age, sex, PA level) as the initial validation sample used to develop the RPE-based prediction equation. Generally, the more specific the population for which the prediction equation is designed, the less error there will be in the predicted VO_2max using criterion variables measured under field conditions. Predicted VO_2max, calculated using the previously developed equation, is compared with actual measured

VO_2max using the cross-validation subject sample. If predicted VO_2max is similar to actual measured VO_2max, then the field-based prediction equation has been successfully cross-validated using an independently selected subject sample. The equation could then be used to predict VO_2max for similar individuals in a health-fitness facility where respiratory-metabolic instrumentation is unavailable.

8.1.7 A Perceptually Regulated Exercise Test to Predict VO_2peak

Eston and colleagues (2005) developed an alternative type of submaximal exercise test to predict cycle ergometer VO_2peak in healthy adults. This exercise is in contrast to commonly used protocols that employ specific increments in power output to produce a systematic increase in exercise intensity. This exercise test is perceptually regulated using step-wise increments in RPE that are produced sequentially to progressively increase exercise intensity. Using the Borg (6–20) Scale, at each 2-min test stage subjects were asked to self-regulate exercise intensity by producing the following RPE's: 9, 11, 13, 15, and 17. VO_2 was recorded near the end of each stage. Linear regression was used to develop equations for the relation between RPE and the corresponding submaximal VO_2 response. To predict VO_2peak, RPEmax (i.e., Borg Scale RPE of 20) was entered into the regression equations. The perceptually regulated exercise test and subsequent prediction equation d procedures were repeated two additional times to test the effects of protocol familiarity on VO_2peak prediction. The results of the investigation found that equations developed using RPE's 9 through 15 and RPE's 9 through 17 both predicted VO_2peak with reasonable accuracy. However, the equation developed using RPE's 9 through 17 was more precise, predicting most subjects' VO_2peak within 5–7 $ml{\cdot}kg^{-1}\ min^{-1}$. In addition, the accuracy of such prediction was improved with practice, evidenced by a closer agreement of predicted and measured VO_2peak between trials two and three than between trials one and two (Eston et al. 2005). Subsequent investigations confirmed the ability of perceptually regulated exercise tests to predict VO_2max/peak in both active and sedentary adults performing cycle ergometer exercise (Eston et al. 2006, 2008; Faulkner et al. 2007; Morris et al. 2009) and treadmill exercise (Eston et al. 2012; Morris et al. 2010), as well as in able-bodied individuals and paraplegics performing arm ergometery (Al-Rahamneh and Eston 2011).

8.1.8 Submaximal Tests to Assess Training-Induced Fitness Changes

An advantage of exercise tests used to predict VO_2max/peak and 1RM is that they can also use submaximal end-points to assess training induced fitness changes. If the same procedures are performed prior to the initiation of an exercise training program and after a designated amount of time participating in the exercise program, pre- and post-training submaximal values can be compared. This can be accomplished using either load-incremented estimation protocols or perceptually regulated protocols. For an estimation protocol, HR and RPE measured during a given workload (i.e., exercise intensity) can be compared before and after training. For a given exercise intensity, decreases in HR reveal a lowered physiological strain to perform the fixed intensity and an improvement in cardiorespiratory fitness. Pre- to post-training decreases in RPE when measured at a given PO (intensity) indicates that the perception of effort associated with a given level of exercise is comparatively lower following a training program. A lower perception of exertion fatigue at a given workload indicates improved tolerance of that exercise intensity and as such an improvement in cardiorespiratory fitness. In addition, this perceptual training adaptation allows the individual to perform the given exercise intensity for a greater amount of time.

This perceptual training adaptation holds true for resistance exercise training programs as well. That is, for a given weight lifted, the perception of exertion decreases from pre- to post-training. A decrease in RPE for a given resistance load indicates improved muscular strength and the ability to perform a lift of that weight for a higher number of repetitions after training. Likewise, for a perceptually regulated aerobic exercise protocol, if a given RPE is produced by self-regulating exercise at a higher workload following training, this decreased perception of exertion indicates an improved exercise tolerance.

8.1.9 Case Study

8.1.9.1 Client Information

A 55-year-old female who works in a local office building comes to your fitness facility. During a pre-participation interview prior to exercise testing, she tells you that she walks her dog each evening after dinner. She describes how she recently completed an office fitness challenge that involved counting steps each week and learned that she does not exercise nearly as much as others in the office nor does she meet recommended guidelines for regular PA participation. She is slightly overweight and believes her fitness level to be below average. Her goals are to lose weight, increase aerobic fitness and improve muscular strength. She is very busy at the office and cannot increase her time available to exercise. She knows she should try to exercise at a higher intensity but is not sure what level is appropriate for her. Also, she wants to learn how she can keep her muscle tone using the resistance equipment that is available through her office wellness center.

8.1.9.2 Assessments, Results and Analysis

Perform submaximal exercise protocols to predict VO_2max/peak and/or muscular strength (1RM) using RPE from which appropriate exercise intensities can be determined to achieve weight loss, aerobic fitness, and muscular strength goals.

1. Calculate predicted VO_2max/peak:
2. Determine the appropriate exercise intensity for her initial aerobic exercise prescription.
3. Calculate predicted 1RM:
4. Determine an appropriate load for the resistance exercises to be performed for her initial resistance exercise prescription.

8.2 Methods

8.2.1 Treadmill Procedures: Prediction of VO_2max

8.2.1.1 Equipment

1. Adult OMNI-Walk/Run RPE Scale (Fig. A.2)
2. Treadmill
3. HR monitor
4. Respiratory-metabolic measurement system

8.2.1.2 Pre-exercise Procedures

1. Measure height (cm) and weight (kg) of subject.
2. Read the standard instructions for the Adult OMNI-Walk/Run RPE Scale for RPE-O to the subject (Appendix B.1). For prediction of VO_2max using differentiated RPE (RPE-L and RPE-C) as well, read the standard instructions for the Adult OMNI-Walk/Run RPE Scale for undifferentiated and differentiated RPE (Appendix B.2). Perform the memory anchoring procedure as described in Chap. 5.

8.2.1.3 Graded Exercise Test to Measure VO_2max

1. Position the HR monitor and respiratory-metabolic mouthpiece (with head support unit and nose clip if applicable) on the subject.
2. Instruct the subject to step onto the treadmill and review exercise termination procedures: When the subject cannot continue exercise due to exhaustion or discomfort, he/she should grasp the treadmill hand rails, at which time the test administrator will gradually slow the treadmill down for performance of a cool-down. The subject should be reminded not to step off the treadmill belt while it is still in motion.
3. Bruce Multistage Treadmill Test Protocol: this can be performed by manually adjusting treadmill speed and grade or using a program on a computer that is interfaced to the treadmill.

(a) Begin the warm-up at 1.5 miles·h^{-1} and 0% grade for 3 min.
(b) Each exercise test stage will last for 3 min. The stages progress as follows:

Stage 1—1.7 miles·h^{-1} and 10 % grade
Stage 2—2.5 miles·h^{-1} and 12 % grade
Stage 3—3.4 miles·h^{-1} and 14 % grade
Stage 4—4.2 miles·h^{-1} and 16 % grade
Stage 5—5.0 miles·h^{-1} and 18 % grade
Stage 6—5.5 miles·h^{-1} and 20 % grade
Stage 7—6.0 miles·h^{-1} and 22 % grade
Stage 8—6.5 miles·h^{-1} and 24 % grade

(c) When the subject cannot continue any longer, terminate the exercise test by initiating the cool-down period at 1.5 miles h^{-1} and 0 % grade. The cool-down should be 5 min in duration.
(d) Instruct the subject to estimate RPE starting at 2:30 of each exercise stage using the OMNI Scale (RPE-O; RPE-L and RPE-C if desired). The RPE's should be rated in a counterbalanced sequence. Because the position of respiratory-metabolic mouth piece inhibits a verbal response, instruct the subject to point to the numbers on the RPE scale, which should be conveniently positioned within the subject's arm reach. State aloud the numerical ratings for each momentary assessment to which the subject pointed and request a confirmatory nod that the number stated was correct. If incorrect, allow the subject to point to the appropriate rating on the RPE scale once more. Ask the subject to hold his or her finger on the appropriate number on the scale for approximately 1 s.
(e) Record HR (b·min^{-1}) at 2:55 of each exercise stage.
(f) Record the final 15-s VO_2 (ml·kg^{-1}·min^{-1}) for each exercise stage.
(g) Record HRmax as the highest HR value recorded during the final exercise stage or immediately post-exercise.
(h) Record VO_2max as the highest 15-s VO_2 value recorded at the end of the test.

8.2.1.4 Submaximal Protocol to Predict VO_2max

1. Position the HR monitor and respiratory-metabolic mouthpiece (with head support unit and nose clip if applicable) on the subject.
2. Instruct the subject to step onto the treadmill and review termination procedures. The subject should be reminded not to step off the treadmill belt while it is still in motion.
3. Three submaximal exercise intensities will be performed. Select the intensity sequence from Table 8.1 that is consistent with the subject's training status.

(a) The subject will perform each exercise intensity (A, B, and C) for 4 min with a 5-min seated recovery between each exercise.
(b) Instruct the subject to estimate RPE-O (RPE-L and RPE-C are optional) at 1:30 and 3:30 of each exercise intensity using the OMNI Scale. The RPE's

Table 8.1 Exercise intensities for submaximal treadmill protocol

Exercise intensity	Mode	Trained (speed/grade)	Sedentary or untrained (speed/grade)
A	Walk	2.5 miles·h^{-1}/0.0 %	1.5 miles·h^{-1}/0.0 %
B	Walk	3.5 miles·h^{-1}/5.0 %	2.5 miles·h^{-1}/5.0 %
C	Run	5.0 miles·h^{-1}/2.5 %	4.0 miles·h^{-1}/2.5 %

should be rated in a counterbalanced sequence. Because the position of the respiratory-metabolic mouth piece inhibits a verbal response, instruct the subject to point to the numbers on the RPE scale, which should be conveniently positioned within the subject's arm reach. State aloud the numerical ratings for each momentary assessment to which the subject pointed and request a confirmatory nod that the number stated was correct. If incorrect, allow the subject to point to the appropriate rating on the RPE scale once more. Ask the subject to hold his or her finger on the appropriate number on the scale for approximately 1 s.

(c) Record HR (b·min^{-1}) every 2 min of each exercise intensity.

(d) Record the final 15-s VO_2 (ml·kg^{-1}·min^{-1}) for each 2-min segment of exercise.

8.2.1.5 Data Organization and Analysis

1. In a Microsoft Excel spreadsheet, label columns of data for the following variables: Exercise Intensity (A, B, C), VO_2 (ml·kg^{-1}·min^{-1}), OMNI RPE-O, HR (b·min^{-1}). Include columns for OMNI RPE-L and OMNI RPE-C if applicable.
2. Plot of VO_2 and OMNI RPE-O for prediction of VO_2max.

   (a) Click on the **INSERT** tab and in the **CHARTS** section click on **SCATTER**. Select the first available chart option. A blank or example scatter plot will appear on your screen.

   (b) Click on the **SELECT DATA** tab. Remove any entries found in the **LEGEND ENTRIES** text box then click **ADD**. Under **SERIES NAME**, enter VO_2 and OMNI RPE-O. Then click on the icon to the right of the **SERIES X VALUES** text box and highlight the OMNI RPE-O values. After the values are highlighted click the icon on the box that appeared. Then click on the icon to the right of the **SERIES Y VALUES** text box and highlight the VO_2 values. After the values are highlighted click the icon on the box that appeared. Click **OK** on the next two screens.

   (c) You should now have a scatter plot with OMNI RPE-O on the *x*-axis and VO_2 on the *y*-axis. Create a title for the plot and enter the appropriate axis labels and units of measure.
   *Methodological note: During aerobic exercise, RPE increases linearly with increases in both physical (PO, speed, grade) and physiological (HR, VO_2) analogs of exercise intensity. Therefore, this laboratory experiment uses a linear equation to predict VO_2max.

(d) To determine the equation from which VO_2max will be predicted, click on one of the data points to highlight the entire data series. Right click on one of the data points and a menu will appear. Click **ADD TRENDLINE** and the **FORMAT TRENDLINE** menu will appear. Select **LINEAR** and **DISPLAY EQUATION ON CHART** then click **CLOSE**. The trendline and equation will be displayed on the chart.

(e) Use this linear equation to calculate predicted VO_2max. Use the OMNI RPE-O of 10 as the "x" value in the equation and solve for "y." The calculated "y" value is the predicted VO_2max ($ml \cdot kg^{-1} \cdot min^{-1}$).

3. Repeat the above steps for VO_2 and HR to determine VO_2max predicted from submaximal HR, using APMHR as the "x" value in the prediction equation. You may also determine predicted VO_2max using differentiated OMNI RPE's (RPE-L and RPE-C), also using the OMNI RPE-L and -C of 10.
4. An example of these procedures with a screenshot depicting each step as performed using Microsoft Excel 2013 can be found in Appendix E.

8.2.2 *Cycle Ergometer Procedures: Prediction of VO_2peak*

8.2.2.1 Equipment

1. Adult OMNI-Cycle RPE Scale (Fig. 2.4)
2. Cycle ergometer
3. Metronome
4. HR monitor
5. Respiratory-metabolic measurement system

8.2.2.2 Pre-exercise Procedures

1. Measure height (cm) and weight (kg) of subject.
2. Read the standard instructions for the Adult OMNI-Cycle RPE Scale for RPE-L to the subject (Appendix B.4). For prediction of VO_2peak using undifferentiated RPE (RPE-O) and differentiated RPE for chest/breathing (RPE-C) as well, read the standard instructions for the Adult OMNI-Cycle RPE Scale for undifferentiated and differentiated RPE to the subject (Appendix B.5). Perform the memory anchoring procedure as described in Chap. 5.

8.2.2.3 Graded Exercise Test to Measure VO_2peak

1. Position the HR monitor and respiratory-metabolic mouthpiece (with head support unit and nose clip if applicable) on the subject.
2. Set the proper seat height on the cycle ergometer according to leg length. When the foot is flat on the right pedal and the pedal is in the extreme down position, there should be a flexion of the right knee of approximately 5°.

3. Load-incremented protocol for electronically braked and friction-braked cycle ergometers:

   (a) Instruct the subject to maintain a 50 rev min^{-1} pedal cadence. Set the metronome to 100 b·min^{-1} so each downward movement of each foot is synchronized with a beat of the metronome. The subject may also use the digital monitor on the cycle control panel to regulate pedal cadence.
   (b) For electronically braked cycle ergometers (e.g., Lode), begin stage 1 at 50 W then increase the resistance 25 W every 2 min.
   (c) For friction-braked cycle ergometers (e.g., Monark), begin stage 1 at 1 kg resistance then increase the resistance 0.5 kg every 2 min.
   (d) When the subject cannot maintain the pedal cadence for 10 consecutive seconds, terminate the exercise test.
   (e) Instruct the subject to estimate RPE starting at 1:30 of each exercise stage using the OMNI Scale (RPE-L; RPE-O, and RPE-C if desired). The RPE's should be rated in a counterbalanced sequence. Because the position of the respiratory-metabolic mouth piece inhibits a verbal response, instruct the subject to point to the numbers on the RPE scale, which should be conveniently positioned within the subject's arm reach. For each momentary assessment, state aloud the numerical ratings to which the subject pointed and request a confirmatory nod that the number stated was correct. If incorrect, allow the subject to point to the appropriate rating on the RPE scale once more. Ask the subject to hold his or her finger on the appropriate number on the scale for approximately 1 s.
   (f) Record HR (b·min^{-1}) at 1:55 of each exercise stage.
   (g) Record the final 15-s VO_2 (l·min^{-1}) for each exercise stage.
   (h) Record HRpeak as the highest HR value recorded during the final exercise stage or immediately post-exercise.
   (i) Record VO_2peak as the highest 15-s VO_2 value recorded at the end of the test.

8.2.2.4 Submaximal Protocol to Predict VO_2peak

1. Position the HR monitor and respiratory-metabolic mouthpiece (with head support unit and nose clip if applicable) on the subject.
2. Instruct the subject to maintain a 50 rev min^{-1} pedal cadence. Set the metronome to 100 b·min^{-1} so each downward movement of each foot is synchronized with a beat of the metronome. The subject may also use the digital monitor on the cycle control panel to regulate pedal cadence.
3. Three submaximal exercise intensities will be performed. Select the intensity sequence from Tables 8.2 or 8.3 that is consistent with the subject's training status.

   (a) The subject will perform each exercise intensity (A, B, and C) for 4 min with a 5-min seated recovery between each exercise. Terminate exercise if the subject cannot maintain the 50 rev min^{-1} pedal cadence for 10 consecutive seconds owing to fatigue.

Table 8.2 Exercise intensities for submaximal, electronically braked cycle ergometer protocol

Exercise intensity	Trained males (W)	Trained females (W)	Sedentary/untrained males (W)	Sedentary/untrained females (W)
A	50	25	50	25
B	100	75	75	50
C	150	125	100	75

Table 8.3 Exercise intensities for submaximal, friction-loaded cycle ergometer protocol

Exercise intensity	Trained males (kg m min^{-1})	Trained females (kg m min^{-1})	Sedentary/untrained males (kg m min^{-1})	Sedentary/untrained females (kg m min^{-1})
A	300	150	300	150
B	600	450	450	300
C	900	750	600	450

(b) Instruct the subject to estimate RPE-L at 1:30 and 3:30 of each exercise intensity using the OMNI Scale (RPE-O and RPE-C are optional). The RPE's should be rated in a counterbalanced sequence. Because the position of the respiratory-metabolic mouth piece inhibits a verbal response, instruct the subject to point to the numbers on the RPE scale, which should be conveniently positioned within the subject's arm reach. State aloud the numerical ratings for each momentary assessment to which the subject pointed and request a confirmatory nod that the number stated was correct. If incorrect, allow the subject to point to the appropriate rating on the RPE scale once more. Ask the subject to hold his or her finger on the appropriate number on the scale for approximately 1 s.

(c) Record HR (b·min^{-1}) every two minutes of each exercise intensity.

(d) Record the final 15-s VO_2 (l·min^{-1}) for each two minute segment of exercise.

8.2.2.5 Data Organization and Analysis

1. In a Microsoft Excel spreadsheet, label columns of data for the following variables: Exercise Intensity (A, B, C), VO_2 (1 min^{-1}), OMNI RPE-L, HR (b·min^{-1}). Include columns for OMNI RPE-O and OMNI RPE-C if applicable.
2. Plot of VO_2 and OMNI RPE-L for prediction of VO_2peak.

   (a) Click on the **INSERT** tab and in the **CHARTS** section click on **SCATTER**. Select the first available chart option. A blank or example scatter plot will appear on your screen.

   (b) Click on the **SELECT DATA** tab. Remove any entries found in the **LEGEND ENTRIES** text box then click **ADD**. Under **SERIES NAME**, enter VO_2 and OMNI RPE-L. Then click on the icon to the right of the **SERIES X VALUES** text box and highlight the OMNI RPE-L values.

After the values are highlighted click the icon on the box that appeared. Then click on the icon to the right of the **SERIES Y VALUES** text box and highlight the VO_2 values. After the values are highlighted click the icon on the box that appeared. Click **OK** on the next two screens.

(c) You should now have a scatter plot with OMNI RPE-L on the *x*-axis and VO_2 on the *y*-axis. Create a title for the plot and enter the appropriate axis labels and units of measure.

*Methodological note: During aerobic exercise, RPE increases linearly with increases in both physical (PO, speed, grade) and physiological (HR, VO_2) analogs of exercise intensity. Therefore, this laboratory experiment uses a linear equation to predict VO_2max.

(d) To determine the equation from which VO_2peak will be predicted, click on one of the data points to highlight the entire data series. Right click on one of the data points and a menu will appear. Click **ADD TRENDLINE** and the **FORMAT TRENDLINE** menu will appear. Select **LINEAR** and **DISPLAY EQUATION ON CHART** then click **CLOSE**. The trendline and equation will be displayed on the chart.

(e) Use this linear equation to calculate predicted VO_2peak. Use the OMNI RPE-L of 10 as the "x" value in the equation and solve for "y." The calculated "y" value is the predicted VO_2peak (l min^{-1}).

3. Repeat the above steps using submaximal VO_2 and HR to predict VO_2peak. Then calculated regression model uses APMHR as the "x" value in the prediction equation. You may also determine predicted VO_2peak from OMNI RPE-O and RPE-C. This procedure uses the OMNI RPE of 10 as the "x" value.
4. An example of these procedures with a screenshot depicting each step as performed using Microsoft Excel 2013 can be found in Appendix E.

8.2.3 Resistance Exercise Procedures: Prediction of 1RM

8.2.3.1 Equipment

1. Adult OMNI-Resistance Exercise RPE Scale (Fig. A.5)
2. Resistance exercise equipment for performance of flat bench press
3. Metronome

8.2.3.2 Pre-exercise Procedures

Read the standard instructions for the Adult OMNI-Resistance Exercise RPE Scale for RPE-AM to the subject (Appendix B). Perform the memory anchoring procedure as described in Chap. 5.

8.2.3.3 1RM Maximum Strength Test

Administer a 1RM procedure for assessment of muscular strength according to the procedures of Baechle and Earle (2008).

1. Instruct the subject to warm-up with a light resistance that can be performed for 5–10 repetitions, then provide a 1-min rest.
2. Estimate a warm-up load that will allow the subject to complete 3–5 repetitions by adding 10–20 lb (5–10% of previous weight lifted) for upper body exercise or 30–40 lb (10–20% of previous weight lifted) for lower body exercise. At the conclusion of the warm-up, provide a 2-min rest.
3. Estimate a conservative, near maximal load that will allow the subject to complete 2–3 repetitions by adding 10–20 lb (5–10% of previous weight lifted) for upper body exercise or 30–40 lb (10–20% of previous weight lifted) for lower body exercise. Following completion of the lift, provide a 2-minute rest.
4. Increase the resistance load by 10–20 lb (5–10% of previous weight lifted) for upper body exercise or 30–40 lb (10–20% of previous weight lifted) for lower body exercise and instruct the subject to attempt a 1RM.
5. If the subject successfully completes the lift using proper technique, provide a 2–4-min rest and repeat the previous step. If the subject failed to complete the lift owing to improper technique or having reached maximum muscle tension production, provide a 2–4-min rest then decrease the resistance by 5–10 lb (2.5–5% of previous weight lifted) for upper body exercise or 15–20 lb (5–10% of previous weight lifted) for lower body exercise and instruct the subject to attempt a 1RM using the adjusted weight.
6. Continue increasing or decreasing the load until the subject can complete a 1RM with proper exercise technique.
7. Instruct the subject to estimate OMNI RPE-AM at the end of each resistance exercise set.

8.2.3.4 Submaximal Protocol to Predict 1RM

1. Three submaximal exercise intensities will be performed. Select the intensity sequence from Table 8.4 that is consistent with the subject's resistance training status.
2. The subject will perform each exercise intensity (A, B, and C) for one set of five repetitions with a 5-min seated recovery between sets.

Table 8.4 Exercise intensities for submaximal resistance exercise protocol for bench press

Exercise intensity	Trained male	Trained female	Recreational male	Recreational female	Untrained male	Untrained female
A	120	60	80	40	40	20
B	150	80	110	60	60	40
C	180	100	140	80	80	60

3. Remind the subject to rate the intensity of feelings of exertion for the active muscle(s) during the concentric phase of the final repetition of each set.
4. Instruct the subject to estimate OMNI RPE-AM at the end of each resistance exercise set.

8.2.3.5 Data Organization and Analysis

1. In a Microsoft Excel spreadsheet, label columns of data for the following variables: Exercise Intensity (A, B, C), Weight Lifted, and OMNI RPE-AM.
2. Plot of Weight Lifted and OMNI RPE-AM for prediction of 1RM.

   (a) Click on the **INSERT** tab and in the **CHARTS** section click on **SCATTER**. Select the first available chart option. A blank or example scatter plot will appear on your screen.
   (b) Click on the **SELECT DATA** tab. Remove any entries found in the **LEGEND ENTRIES** text box then click **ADD**. Under **SERIES NAME**, enter Weight Lifted and OMNI RPE-AM. Then click on the icon to the right of the **SERIES X VALUES** text box and highlight the OMNI RPE-AM values. After the values are highlighted click the icon on the box that appeared. Then click on the icon to the right of the **SERIES Y VALUES** text box and highlight the Weight Lifted values. After the values are highlighted click the icon on the box that appeared. Click **OK** on the next two screens.
   (c) You should now have a scatter plot with OMNI RPE-AM on the x-axis and Weight Lifted on the y-axis. Create a title for the plot and enter the appropriate axis labels and units of measure.
   *Methodological note: During resistance exercise, RPE increases linearly with increases in both physical (absolute weight lifted, %1RM) and physiological (blood lactate concentration) analogs of exercise intensity. Therefore, this laboratory experiment uses a linear equation to predict 1RM.
   (d) To determine the equation from which 1RM will be predicted, click on one of the data points to highlight the entire data series. Right click on one of the data points and a menu will appear. Click **ADD TRENDLINE** and the **FORMAT TRENDLINE** menu will appear. Select **LINEAR** and **DISPLAY EQUATION ON CHART** then click **CLOSE**. The trendline and equation will be displayed on the chart.
   (e) Use this linear regression equation to calculate predicted 1RM. Use the OMNI RPE-AM of 10 as the "x" value in the equation and solve for "y." The calculated "y" value is the predicted 1RM (lbs).

3. An example of these procedures with a screenshot depicting each step as performed using Microsoft Excel 2013 can be found in Appendix E.

8.3 Discussion Questions

1. How does VO_2max/peak or 1RM that is predicted using responses from a submaximal exercise protocol compare to the actual measured VO_2max/peak or 1RM? List some possible reasons why predicted values may differ from actual measured values.
2. How can fitness be tracked over time using a VO_2max/peak or 1RM prediction protocol? What must be done to ensure the methods are sensitive to changes in aerobic or muscular fitness?
3. How could RPE be used to prescribe aerobic and resistance exercise intensity based on the results from a submaximal protocol to predict VO_2max/peak or 1RM?

References

Al-Rahamneh H, Eston RG. Prediction of maximal oxygen uptake from the ratings of perceived exertion during a graded and ramp exercise test in able-bodied and persons with paraplegia. Arch Phys Med Rehabil. 2011;92:277–83.

American College of Sports Medicine. ACSM's guidelines for exercise testing and prescription. 9th ed. Philadelphia, PA: Lippincott Williams & Wilkins; 2013.

Baechle TR, Earle RW. Essentials of strength training and conditioning. 3rd ed. Champaign, IL: Human Kinetics; 2008.

Davies RC, Rowlands AV, Eston RG. The prediction of maximal oxygen uptake from sub-maximal ratings of perceived exertion elicited during the multistage fitness test. Br J Sports Med. 2008;42:1006–10.

Eston RG, Lamb KL, Parfitt G, King N. The validity of predicting maximal oxygen uptake from a perceptually-regulated graded exercise test. Eur J Appl Physiol. 2005;94:221–7.

Eston RG, Faulkner J, Mason E, Parfitt G. The validity of predicting maximal oxygen uptake from perceptually-regulated exercise tests of different durations. Eur J Appl Physiol. 2006;97: 535–41.

Eston R, Lambrick D, Sheppard K, Parfitt G. Prediction of maximal oxygen uptake in sedentary males from a perceptually regulated, sub-maximal graded exercise test. J Sports Sci. 2008; 26:131–9.

Eston R, Evans HJL. The validity of submaximal ratings of perceived exertion to predict one repetition maximum. J Sports Sci Med. 2009;8:567–73.

Eston R, Evans H, Faulkner J, Lambrick D, Al-Rahamneh H, Parfitt G. A perceptually-regulated, graded exercise test predicts peak oxygen uptake during treadmill exercise in active and sedentary participants. Eur J Appl Physiol. 2012;112:3459–68.

Faulkner J, Eston R. Overall and peripheral ratings of perceived exertion during a graded exercise test to volitional exhaustion in individuals of high and low fitness. Eur J Appl Physiol. 2007;101:613–20.

Faulkner J, Parfitt G, Eston R. Prediction of maximal oxygen uptake from the ratings of perceived exertion and heart rate during a perceptually-regulated sub-maximal exercise test in active and sedentary participants. Eur J Appl Physiol. 2007;101:397–407.

Gearhart Jr RF, Lagally KM, Riechman SE, Andrews RD, Robertson RJ. RPE at relative intensities after 12 weeks of resistance-exercise training by older adults. Percept Motor Skills. 2008;106: 893–903.

Londeree B, Moeschberger ML. Effect of age and other factors on maximal heart rate. Res Q Exerc Sport. 1982;53:297–304.
Morris M, Lamb KL, Cotterell D, Buckley J. Predicting maximal oxygen uptake via a perceptually regulated exercise test (PRET). J Exerc Sci Fit. 2009;7:122–8.
Morris M, Lamb KL, Hayton J, Cotterell D, Buckley J. The validity and reliability of predicting maximal oxygen uptake from a treadmill-based sub-maximal perceptually regulated exercise test. Eur J Appl Physiol. 2010;109:983–8.
Robertson RJ, Goss FL, Aaron DJ, Gairola A, Kowallis RA, Liu Y, Randall CR, Tessmer KA, Schnorr TL, Schroeder AE, White B. One repetition maximum prediction models for children using the OMNI RPE scale. J Strength Cond Res. 2008;22:196–201.

Part III
Exercise Prescription and Program Evaluation

Chapter 9
The Estimation–Production Paradigm for Exercise Intensity Self-Regulation

The *estimation–production paradigm* is a set of exercise test procedures designed to assess the validity of using RPE to prescribe and self-regulate exercise intensity. The paradigm is intended to evaluate an individual's ability to accurately self-regulate exercise intensity according to a specified *target RPE* or *target RPE range*. The target RPE is produced and maintained by self-regulating exercise intensity. The target RPE is prescribed using the data derived from a pre-participation GXT also referred to as the *estimation trial*. In a subsequent exercise training bout called the *production trial*, the individual is instructed to produce the target RPE by self-adjusting exercise intensity in order to attain the target level of exertion. In an assessment of prescription congruence, physiological responses such as VO_2 and HR corresponding to the target RPE are compared between estimation and production trials. This cross-trial comparison is used to determine the validity of exercise intensity self-regulation using a target RPE. Evidence for prescription congruence has been shown for adults and children performing various exercise modalities using both the Borg and OMNI Scales. The primary purpose of this laboratory experiment is to use an estimation–production paradigm to determine an individual's ability to self-regulate exercise intensity using a target RPE range.

9.1 Background

9.1.1 The Estimation Protocol

The importance of the estimation protocol alone has been discussed in previous chapters. When proper anchoring procedures are performed and it is confirmed that the individual's perceptual responses conform to Borg's Range Model, the estimation protocol allows the measurement of RPE from very low to maximal exercise intensity. The estimation protocol can be used to test the concurrent

L. Haile et al., *Perceived Exertion Laboratory Manual*,
DOI 10.1007/978-1-4939-1917-8_9

validity of an RPE Scale. In this type of validity experiment, RPE is correlated with such physiological variables as VO_2 and HR that increase concurrently with increasing aerobic exercise intensity.

The measurement of RPE and corresponding physiological variables during an estimation protocol allows the prescription of an exercise intensity that provides an overload training stimulus for cardiorespiratory conditioning. This prescribed exercise intensity can be based on a specific physiological marker such as the VT. The RPE-VT is then used as the target RPE for exercise prescription. Previous studies have shown RPE-VT to range from 5 to 7 on the OMNI Scale and 11 to 16 on the Borg Scale. Recently, Parfitt and colleagues (2012) guided sedentary adults through an 8-week perceptually regulated exercise program. Using an estimation–production paradigm, subjects were taught to self-regulate exercise intensity to produce an exertional level equivalent to 13 on the Borg Scale. The perceptually regulated training program resulted in significant improvements in mean arterial pressure, total cholesterol, and body mass index over the 8-week period (Parfitt et al. 2012).

An estimation protocol that employs corresponding physiological monitoring also allows the determination of an appropriate exercise intensity range to promote cardiorespiratory fitness. This range can be based on specific VO_2 or HR values corresponding to a percent of maximum level or a percent of the VT. The appropriate exercise intensity range to elicit improvement in cardiorespiratory fitness depends on the training level of the individual (Garber et al. 2011). Regular aerobic exercise at intensities between 70 and 85 % of VO_2max is an accepted range to provide an overload stimulus to enhance cardiorespiratory fitness (Robertson 2004), even in trained individuals (Midgley et al. 2006). Therefore, the RPE's corresponding to 70 and 85 % of VO_2max can serve as an appropriate target perceptual range for exercise prescription. However, intensities as low as 30 % of VO_2 reserve may be sufficient to improve cardiorespiratory fitness in low fit individuals (Swain and Franklin 2002) and very high exercise intensities may be necessary for trained runners to improve cardiorespiratory fitness (Midgley et al. 2006). Individual aerobic fitness level must be taken into account prior to exercise intensity prescription, even when the overload stimulus is expressed in relative terms.

9.1.2 The Production Protocol

The production exercise protocol is administered after the estimation exercise protocol. When the individual begins exercise, he/she is asked to self-regulate exercise intensity to produce the target RPE or target RPE range that has been prescribed based on the responses to the estimation protocol. The individual is instructed to adjust exercise intensity throughout the production protocol in order to continually produce the prescribed target RPE (range). VO_2 and/or HR are monitored just as during the estimation protocol so that physiological values can be compared between trials to document validity of the self-regulation procedures. If the individual accurately self-regulated exercise intensity by producing the target RPE(s), the physiological responses from the production trial should be similar to those corresponding to the same target RPE(s) derived from the estimation trial. This is termed *prescription congruence* (Robertson et al. 2002). If the physiological values are different between the estimation and production protocols when measured at the prescribed target RPE, then it is said that the individual is exhibiting *exercise intensity self-regulation error.* The validity of exercise intensity self-regulation using a target RPE can be tested with an assessment of prescription congruence. Evidence for prescription congruence has been shown in studies of cycle ergometry, arm ergometry and treadmill exercise for children and adults (Dunbar et al. 1992, 1994; Dunbar and Kalinski 2004; Kang et al. 1998, 2003, 2009; Parfitt et al. 2007; Robertson et al. 2002), including children with cystic fibrosis (Higgins et al. 2013) and adults with cardiovascular disease (Weiser et al. 2007).

Robertson and colleagues (2002) confirmed prescription congruence in 8–12 year-old children during cycle ergometer exercise. During two separate 6-min production trials, the children were instructed to produce the target RPE's 2 and 6 from the Children's OMNI Cycle Scale in either ascending or descending order. Prescription congruence was exhibited for both target RPE's using HR and VO_2. Neither the order of target RPE production nor gender had an effect on the accuracy of exercise intensity self-regulation (Robertson et al. 2002).

Weiser and colleagues (2007) presented evidence for prescription congruence for cycle ergometer exercise in cardiovascular disease patients participating in cardiac rehabilitation. During the first 6-min production trial, patients produced a target Borg Scale RPE of 13. During the second 6-min production trial, patients were instructed to begin exercise by producing an RPE of 11 then adjust intensity to produce a target RPE of 13 for minutes 3 through 6. The researchers termed this procedure an RPE "step-up" procedure, positing that it would reduce the likelihood of overshoot, or producing an intensity higher than the target. Overshoot production could be potentially hazardous in this population, putting them at risk of an untoward cardiovascular event during exercise. Prescription congruence was confirmed since HR corresponding to an RPE of 13 measured during the estimation trial was similar to HR at the end of both production trials. In addition, the RPE step-up procedure resulted in significantly less patients producing a HR that was higher than the target intensity (Weiser et al. 2007).

9.1.3 Intramodal Versus Intermodal Prescription Congruence

The aforementioned studies examined prescription congruence using a single exercise mode. Thus, the experimental paradigm involved *intramodal* prescription congruence, where estimation and production protocols employed the same exercise modality. For example, target RPE's obtained from a cycle ergometer estimation protocol were used to self-regulate exercise intensity during cycle ergometer production protocols. Normally, target RPE's are based on a predetermined $\%VO_2max$/peak. However, VO_2max/peak varies with differing exercise modalities. As such, it is necessary to examine the validity of intermodal estimation–production paradigms to prescribe and self-regulate exercise intensity using a target RPE. It is possible that an exercise prescription could involve the production of a target RPE using multiple modes of exercise. In this instance, multiple estimation protocols should be administered employing exercise modes that match those used in the production protocols. However, this poses a problem from a practical standpoint since multiple maximal GXTs would require additional burden for both client and exercise professional. Nevertheless there is an advantage in employing multiple exercise modes in a single conditioning session. In particular, such an approach to exercise programming could improve PAE and increase participation. Therefore, some investigations have examined an assessment of prescription congruence between modes, i.e., *intermodal* or *cross-modal* prescription congruence. This paradigm involves estimation and production protocols of differing modes of exercise. For example, target RPE's obtained from a cycle ergometer estimation protocol can be used to self-regulate exercise intensity during a treadmill exercise production protocol or vice versa. The significance of intermodal prescription congruence is the ability to require the performance of only one pre-participation GXT prior to a multimodal exercise prescription.

A study by Kang et al. (2003) tested intramodal and intermodal prescription congruence for target OMNI Scale RPE's corresponding to 50 and 70 % VO_2max/peak in young physically active men and women. Estimation protocols and 20-min production trials were performed for both cycle ergometer and treadmill exercise. Subjects were assigned to one of four groups: estimation and production protocols on a treadmill, estimation and production protocols on a cycle, estimation protocol on a treadmill and production protocol on a cycle, estimation protocol on a cycle and production protocol on a treadmill. Intramodal prescription congruence was confirmed for treadmill and cycle ergometer exercise at 50 and 70 % VO_2max/peak using HR and VO_2 as criterion variables. At only one time-point during the treadmill exercise production protocol was HR higher than the treadmill estimation protocol for a given target RPE. However, intermodal prescription congruence was not confirmed in this paradigm. For the treadmill estimation-cycle ergometer production group, VO_2 and HR were significantly lower during the production protocol for both intensities. For the cycle ergometer estimation-treadmill production group, VO_2 and HR were significantly higher during the production protocol for both intensities (Kang et al. 2003).

The results of the Kang et al. (2003) investigation reveal a problem that can arise when using an intermodal estimation–production paradigm to prescribe exercise intensity according to a target RPE. Physiological responses (VO_2, HR) compared between treadmill and cycle ergometer exercise at the same level of exertion will be higher during treadmill exercise due to a higher metabolic rate and, subsequently, a higher HR and VO_2 (Robertson et al. 1990). Kang and colleagues (2003) conducted a post hoc comparison that normalized physiological variables measured during production protocols to mode-specific estimation trials performed by subjects in other groups. In that analysis, HR and VO_2 values were similar to 50 and 70 % VO_2max/peak as expected (Kang et al. 2003). From a practical standpoint, an estimation protocol employing a single exercise mode (i.e., treadmill or cycle) may be used to identify a target RPE to self-regulate exercise intensity during production protocols of various exercise modalities. It should be expected that physiological responses may change with the metabolic demands of different types of exercise. Regardless, a number of investigations have found evidence for intermodal prescription congruence as evidenced by similar physiological responses between estimation and production modes with assessment undertaken at the same target RPE.

Dunbar and colleagues (1992) presented evidence for intramodal and intermodal prescription congruence in 17–35-year-old men ranging from sedentary to very active. Subjects performed two estimation protocols, one on a cycle and one on a treadmill. These trials were used to calculate target RPE's from the Borg Scale corresponding to 50 and 70 % VO_2max/peak. Subjects then performed four production protocols; two on a cycle and two on a treadmill, each involving self-regulation of exercise intensity at target RPE's corresponding to 50 and 70 % VO_2max/peak. Each exercise bout was 8 min in duration with 5 min of rest between intensities. Both intramodal and intermodal prescription congruence was confirmed for the production trials using target RPE's derived from the cycle ergometer estimation trial. VO_2 and HR values corresponding to 50 and 70 % VO_2peak measured during the cycle ergometer estimation trial were similar to VO_2 and HR measured during the production trials performed on both the cycle and treadmill. Intermodal prescription congruence was confirmed for the cycle ergometer production trial using target RPE's derived from the treadmill estimation trial. However, intramodal prescription congruence for treadmill exercise was confirmed only using the target RPE corresponding to 50 % VO_2max. During the production protocol using the target RPE corresponding to 70 % VO_2max, subjects selected a lower treadmill intensity resulting in lower VO_2 and HR values (Dunbar et al. 1992).

In a similar design, Dunbar and colleagues (1994) tested intramodal and intermodal prescription congruence in active college-aged men. Following a cycle ergometer estimation protocol, four 25-min production protocols were performed at a target RPE (Borg Scale) corresponding to 60 % VO_2peak. Two production protocols were performed on a cycle ergometer to test intramodal prescription congruence and the reproducibility of cycle ergometer exercise intensity self-regulation. Two production protocols were performed on a treadmill to test intermodal prescription congruence and the reproducibility of treadmill exercise intensity self-regulation. Interestingly, intermodal prescription congruence was confirmed but intramodal prescription congruence was not. VO_2 and HR values were similar at 60 % VO_2peak during the cycle ergometer estimation trial and throughout both treadmill production protocols. VO_2 values were similar for only one of the cycle ergometer production protocols, but significantly lower for the other. HR values were significantly lower than the estimation protocol during both cycle ergometer production protocols (Dunbar et al. 1994). These results indicate a somewhat better ability of the subjects tested to self-regulate exercise intensity during treadmill exercise than cycle ergometer exercise.

Higgins and colleagues (2013) presented evidence for prescription congruence in 10–17 year-old children with cystic fibrosis during cycle ergometer and treadmill exercise. First, the subjects performed an estimation protocol on a cycle ergometer. Then, two separate 10-min production protocols were employed. The first protocol was performed on a cycle ergometer to assess intramodal prescription congruence, while the second was performed on a treadmill to assess intermodal prescription congruence. The children were instructed to self-regulate exercise intensity at target RPE's of 4 and 7 using the Children's OMNI Cycle RPE Scale. Both protocols were performed using an interval exercise format during which the children alternated between the target RPE's of 4 and 7 performed for 2-min intervals. Intervals 1, 3 and 5 were performed at an RPE of 4, while intervals 2 and 4 were performed at an RPE of 7. VO_2 and HR from both the cycle ergometer and treadmill production protocols were compared to the values measured during the cycle ergometer GXT. Prescription congruence was not confirmed at an RPE of 4 during the cycle production protocol. VO_2 and HR were significantly higher during the production protocol than the estimation protocol. Prescription congruence was confirmed at an RPE of 4 during the treadmill production protocol and at an RPE of 7 for both the cycle and treadmill production protocols (Higgins et al. 2013).

9.1.4 Teleoanticipation to Improve Prescription Congruence

The studies by Dunbar and colleagues (1992, 1994) and, more recently, Higgins et al. (2013) largely support prescription congruence, but some inconsistencies have been shown for both intramodal and intermodal perceptual prescription procedures. Prescription congruence data were also inconsistent in a study of cycle ergometry and outdoor track walking/running in overweight children (Ward and Bar-Or 1990). Therefore, additional instruction, practice, and feedback may be necessary for some individuals to accurately self-regulate exercise intensity using a target RPE, whether the estimation trial is mode-specific or not.

The physiological values corresponding to target RPE(s) should be measured during the estimation protocol prior to performance of the production protocol. These same physiological values can be monitored during the production protocol to provide feedback to the individual regarding the accuracy of exercise intensity self-regulation according to a target RPE. If the individual is accurately self-regulating exercise intensity, positive reinforcement can be given. Evidence that the individual is exhibiting exercise intensity self-regulation error by either overshooting or undershooting the target level is provided by the VO_2 and HR responses. When such self-regulation error is present, feedback can be provided that exercise intensity should be either decreased or increased to attain the desired level. This would be considered a form of *teleoanticipation*, whereby feedback is given during practice exercise trials prior to participation in an actual exercise program (Ulmer 1996). Providing an individual with multiple production protocols, or pre-participation practice trials, and simultaneously giving appropriate feedback regarding correction of self-regulation error may improve prescription congruence, i.e., avoid exercise intensity self-regulation error.

Dunbar and Kalinski (2004) conducted a 20-week cycle ergometer exercise training study in postmenopausal women. Target RPE's were identified for intensities corresponding to 40, 50 and 60 % of VO_2peak during an initial pre-participation GXT. This testing protocol allowed the training intensity to be increased throughout the first 5 weeks of the exercise program. Specifically, exercise intensity was self-regulated at the target RPE corresponding to 40 % VO_2peak during the first 2 weeks, 50 % VO_2peak during weeks 3 and 4, and 60 % VO_2peak from weeks 5 through 20. Prescription congruence was tested by comparing the average HR achieved during exercise training bouts to the HR corresponding to target RPE's achieved during the pre-participation GXT. At week 2, prescription congruence was confirmed for the target RPE corresponding to 40 % VO_2peak. At week 4, prescription congruence was not confirmed for the target RPE corresponding to 50 % VO_2peak since HR values were significantly lower than the target values. At weeks 6 and 10, HR values were again significantly lower than the target values corresponding to 60 % VO_2peak. At week 20, prescription congruence was confirmed for the target RPE corresponding to 60 % VO_2peak. These results indicate that the women in the study were able to accurately self-regulate exercise intensity at an RPE corresponding to 40 % VO_2peak, but that several weeks of practice might be necessary to accurately self-regulate exercise intensity above 40 % VO_2peak where a target RPE is employed. The researchers provided the subjects with no feedback regarding exercise intensity self-regulation, i.e., the women were not instructed to either increase or decrease exercise intensity when they were respectively under- or over-producing intensity at RPE's corresponding to 50 and 60 % VO_2peak (Dunbar and Kalinski 2004). In this case, teleoanticipation administered during the pre-participation period may have improved the women's accuracy in exercise intensity self-regulation at intensities above 40 % VO_2peak.

9.1.5 *Differentiated RPE for Exercise Intensity Self-Regulation*

The undifferentiated perceptual rating for the overall body may not always be the best choice to prescribe a target RPE to self-regulate exercise intensity. It has been shown that an RPE differentiated to the legs (RPE-L) often provides the dominant perceptual signal during treadmill and cycle ergometer exercise. Therefore, prescribing and self-regulating exercise intensity using a target RPE-L may be preferable under some conditions. That is, choosing the comparatively more intense differentiated RPE signal for exercise intensity self-regulation can help the individual stay within functionally and perceptually tolerable limits (Robertson 2004).

Using a differentiated RPE for exercise prescription can be particularly useful when developing an exercise program for individuals with certain clinical disorders. Self-regulating exercise intensity based on RPE differentiated to the chest and breathing (RPE-C) is appropriate for those who experience exertional dyspnea, including individuals with pulmonary limitations such as (exercise-induced) asthma, chronic obstructive pulmonary disorders, or cystic fibrosis. In addition, the use of RPE-L or RPE differentiated to the arms is appropriate in the rehabilitation setting for limb-specific exercises following (neuro)muscular or articular injury (Robertson 2004).

9.1.6 Exercise Intensity Self-Regulation Using RPE Versus HR

A unique advantage of an RPE-based exercise prescription compared to traditional exercise prescriptions that are based on absolute exercise intensity and/or HR can be seen in the perceptual production protocol. A traditional HR-based exercise prescription involves determining a target HR or HR range. Similar to a target RPE, target HR's are prescribed because they correspond to specific physiological intensities shown to provide an overload stimulus and elicit physiological benefit when performed as part of a regular exercise program. In this procedure the training intensity is set at the VT or a certain percent of maximum. If the individual is adherent to the HR-based exercise prescription, aerobic fitness level improves over time. As aerobic fitness improves, the individual becomes more metabolically efficient at any given submaximal aerobic exercise intensity. Subsequently, less cardiorespiratory work is required to perform the same exercise intensity. Some exercise prescriptions employ an absolute intensity, such as a specific power output setting on an ergometer. In the presence of a training induced increase in maximal aerobic power, the prescribed absolute intensity will no longer serve as an overload stimulus and physiological benefits will plateau. As aerobic fitness increases, changes occur in the individuals overall HR range. Resting HR decreases as parasympathetic (vagal) tone increases. The HR required to produce a specific aerobic metabolic demand may be different. In addition, HR is sensitive to environmental extremes, such as high heat and humidity, which will not cause concomitant changes in VO_2. Therefore, periodic exercise testing will be needed to reevaluate the individual's maximal aerobic power and the HR values corresponding to the target physiological intensity to ensure the continued effectiveness of the exercise program.

With an RPE-based exercise prescription, periodic reevaluation of maximal aerobic power over the time course of the conditioning program is not necessary. If the individual's perceptual responses to the pre-participation GXT conform to predictions of Borg's Range Model, it is assumed that the RPE response range will proportionately redistribute over the expected physiological range. In this case, the target RPE will continue to correspond with the desired physiological intensity required to produce an overload stimulus in the presence of either an increase or decrease in aerobic fitness. For example, if the prescribed RPE corresponds to the VT, the exertional level remains constant even though the absolute work and associated physiological response at the VT will increase consequent to training adaptation. In this context it is important that the RPE scale anchoring procedures are appropriately administered, providing ample practice, feedback, and reinforcement. When these pre-participation conditions are met, exercise intensity self-regulation is accurate and the desired effects of the exercise program are achieved.

9.1.7 Case Study

9.1.7.1 Client Information

A 40-year-old male who is already a member of your fitness facility inquires about personal training. He tells you that he recently lost about 30 lb of body weight by participating in a nutrition program and using the elliptical trainer at the gym. He is happy with the weight loss and is now classified as normal weight. However, he worries about being able to maintain the weight loss because he gets bored exercising on the elliptical trainer. He has recently started to run on the treadmill to improve his aerobic fitness, but he would like to be able to run outside on a nice day. Using exercise tips provided as part of the nutrition program, he learned to exercise at an intensity based on HR measured on the elliptical trainer. However, he would like to learn a method of regulating exercise intensity that would be appropriate no matter where he exercises, be it at the fitness facility, around his neighborhood, or on the new "rails to trails" course located outside of town.

9.1.7.2 Assessments, Results and Analysis

Administer the estimation–production paradigm to the client to determine the appropriate intensity for exercise prescription and test the subject's ability to self-regulate exercise intensity using a target RPE range.

Using the estimation protocol:

1. Determine VO_2max.
2. Determine the appropriate exercise intensity range for the subject's initial aerobic exercise prescription.
3. Calculate the corresponding physiological values that define the exercise intensity range.
4. Identify the target RPE's corresponding to the lower and upper limits of the prescribed exercise intensity range.

Using the production trial, test the subject's ability to self-regulate exercise intensity within the prescribed target RPE range by comparing physiological values collected during the production protocol to those corresponding to the target RPE's measured during the estimation protocol.

9.2 Methods

9.2.1 *Treadmill Procedures*

9.2.1.1 Equipment

1. Adult OMNI-Walk/Run RPE Scale (Fig. A.2)
2. Treadmill
3. HR monitor
4. Respiratory-metabolic measurement system

9.2.1.2 Pre-exercise Procedures

1. Measure height (cm) and weight (kg) of subject.
2. Read the standard instructions for the Adult OMNI-Walk/Run RPE Scaleto the subject to measure RPE-O (Appendix B.1). If measurement of differentiated RPE (RPE-L and/or RPE-C) is desired, read the standard instructions for the Adult OMNI-Walk/Run RPE Scale for undifferentiated and differentiated RPE (Appendix B.2). Perform the memory anchoring procedure as described in Chap. 5

9.2.1.3 Graded Exercise Test: Estimation Protocol

1. Position the HR monitor and respiratory-metabolic mouthpiece (with head support unit and nose clip if applicable) on the subject.
2. Instruct the subject to step onto the treadmill and review exercise termination procedures: When the subject cannot continue exercise due to exhaustion or discomfort, he/she should grasp the treadmill hand rails, at which time the test administrator will gradually slow the treadmill down for performance of a cool-down. The subject should be reminded not to step off the treadmill belt while it is still in motion.
3. Bruce Multistage Treadmill Test Protocol: this can be administered by manually adjusting treadmill speed and grade or using a program on a computer that is interfaced to the treadmill.

   (a) Begin the warm-up at 1.5 miles $\cdot$ h^{-1} and 0 % grade for 3 min.
   (b) Each exercise test stage will last for 3 min. The stages progress as follows:

      Stage 1—1.7 miles $\cdot$ h^{-1} and 10 % grade
      Stage 2—2.5 miles $\cdot$ h^{-1} and 12 % grade
      Stage 3—3.4 miles $\cdot$ h^{-1} and 14 % grade
      Stage 4—4.2 miles $\cdot$ h^{-1} and 16 % grade
      Stage 5—5.0 miles $\cdot$ h^{-1} and 18 % grade
      Stage 6—5.5 miles $\cdot$ h^{-1} and 20 % grade
      Stage 7—6.0 miles $\cdot$ h^{-1} and 22 % grade
      Stage 8—6.5 miles $\cdot$ h^{-1} and 24 % grade

(c) When the subject cannot continue any longer, terminate the exercise test by initiating the cool-down period at 1.5 miles · h^{-1} and 0 % grade. The cool-down should be 5 min in duration.
(d) Ask the subject to estimate RPE starting at 2:30 of each exercise stage. Because the position of the respiratory-metabolic mouth piece inhibits a verbal response, instruct the subject to point to the numbers on the RPE scale, which should be conveniently positioned within the subject's arm reach. State aloud the numerical ratings for each momentary assessment to which the subject pointed and request a confirmatory nod that the number stated was correct. If incorrect, allow the subject to point to the appropriate rating on the RPE scale once more. Ask the subject to hold his or her finger on the appropriate number on the scale for approximately 1 s.
(e) Record HR (b·min^{-1}) at 2:55 of each exercise stage.
(f) Record the final 15-s VO_2 (ml · kg^{-1} · min^{-1}) for each exercise stage.
(g) Record HRmax as the highest HR value recorded during the final exercise stage or immediately post-exercise.
(h) Record VO_2max as the highest 15-s VO_2 value recorded at the end of the test.
(i) Calculate the VO_2 (ml · kg^{-1} · min^{-1}) associated with 70 and 85 % of VO_2max. These values will be used to obtain the RPE's corresponding to the upper and lower limits of the overload training zone to be performed during the production trial. An explanation of how to calculate this target RPE range will be presented in the data organization and analysis section below.

9.2.1.4 Estimation Protocol: Data Organization and Analysis

1. In a Microsoft Excel spreadsheet, label columns of data for the following variables: Exercise Stage, VO_2 (ml · kg^{-1} · min^{-1}), OMNI RPE-O, HR (b·min^{-1}).
2. Plot of VO_2 and OMNI RPE-O for determination of the RPE corresponding to 70 % VO_2max, i.e., the lower limit of the target RPE range.

   (a) Click on the **INSERT** tab and in the **CHARTS** section click on **SCATTER**. Select the first available chart option. A blank or example scatter plot will appear on your screen.
   (b) Click on the **SELECT DATA** tab. Remove any entries found in the **LEGEND ENTRIES** text box then click **ADD**. Under **SERIES NAME**, enter VO_2 and OMNI RPE-O. Then click on the icon to the right of the **SERIES X VALUES** text box and highlight the VO_2 values. After the values are highlighted click the icon on the box that appeared. Then click on the icon to the right of the **SERIES Y VALUES** text box and highlight the OMNI RPE-O values. After the values are highlighted click the icon on the box that appeared. Click **OK** on the next two screens.
   (c) You should now have a scatter plot with OMNI RPE-O on the *y*-axis and VO_2 on the *x*-axis. Create a title for the plot and enter the appropriate axis labels and units of measure.

(d) To determine the RPE corresponding to 70 % VO_2max, click on one of the data points to highlight the entire data series. Right click on one of the data points and a menu will appear. Click **ADD TRENDLINE** and the **FORMAT TRENDLINE** menu will appear. Select **LINEAR** and **DISPLAY EQUATION ON CHART** then click **CLOSE**. The trendline and equation will be displayed on the chart.

(e) Use this linear equation to calculate the RPE corresponding to 70 % VO_2max. Use VO_2 ($ml \cdot kg^{-1} \cdot min^{-1}$) corresponding to 70 % VO_2max as the "x" value in the equation and solve for "y." The calculated "y" value, once rounded to the nearest whole integer, is the RPE corresponding to the lower limit of the target RPE range.

3. Use the forgoing linear equation procedures to determine the RPE corresponding to 85 % VO_2max, i.e., the upper limit of the target RPE range.
4. An example of the procedures to determine an RPE corresponding to a specific %VO_2max with a screenshot depicting each step as performed using Microsoft Excel 2013 can be found in Appendix D since these procedures are the same as those used for determination of RPE-VT.

9.2.1.5 Production Protocol

1. Position the HR monitor and respiratory-metabolic mouthpiece (with head support unit and nose clip if applicable) on the subject.
2. Instruct the subject to step onto the treadmill and review exercise termination procedures: When the subject cannot continue exercise due to exhaustion or discomfort, he/she should grasp the treadmill hand rails, at which time the test administrator will slow down or stop the treadmill. The subject should be reminded not to step off the treadmill belt while it is still in motion.
3. Start the treadmill at 3 $miles \cdot h^{-1}$ and 0 % grade for a 2-min warm-up.
4. Following the warm-up, instruct the subject to exercise for 10 min within the predetermined target RPE range by adjusting the speed of the treadmill until the desired intensity is met.
5. Record HR ($b \cdot min^{-1}$) every 2 min.
6. Record the final 15-s VO_2 ($ml \cdot kg \cdot min^{-1}$) for each 1-min segment of exercise.
7. Following the 10-min exercise, the treadmill should be set to 3 $miles \cdot h^{-1}$ and 0 % grade for a 2-min cool-down period.

9.2.1.6 Production Protocol: Data Organization and Analysis

1. In a Microsoft Excel spreadsheet, label columns of data for the following variables: Exercise Time (minutes), VO_2 ($ml \cdot kg^{-1} \cdot min^{-1}$), HR ($b \cdot min^{-1}$).
2. Count the number of minutes in which the subject's VO_2 was between 70 and 85 % of VO_2max. Divide that number by 10 then multiple by 100 to determine the percent of time the subject exercised in the overload training zone.

3. Count the number of minutes in which the subject's VO_2 was lower than 70 % VO_2max. Divide that number by 10 then multiple by 100 to determine the percent of time the subject exercised below the overload training zone.
4. Count the number of minutes in which the subject's VO_2 was higher than 85 % VO_2max. Divide that number by 10 then multiple by 100 to determine the percent of time the subject exercised above the overload training zone.

9.2.2 *Cycle Ergometer Procedures*

9.2.2.1 Equipment

1. Adult OMNI-Cycle RPE Scale (Fig. 2.4)
2. Cycle ergometer
3. Metronome
4. HR monitor
5. Respiratory-metabolic measurement system

9.2.2.2 Pre-exercise Procedures

1. Measure height (cm) and weight (kg) of subject.
2. Read the standard instructions for the Adult OMNI-Cycle RPE Scale to the subject to measure RPE-L (Appendix B.4). If measurement of undifferentiated RPE (RPE-O) or differentiated RPE for the chest/breathing (RPE-C) is desired, read the standard instructions for the Adult OMNI-Walk/Run RPE Scale for undifferentiated and differentiated RPE (Appendix B.5). Perform the memory anchoring procedure as described in Chap. 5.

9.2.2.3 Graded Exercise Test: Estimation Protocol

1. Position the HR monitor and respiratory-metabolic mouthpiece (with head support unit and nose clip if applicable) on the subject.
2. Set the proper seat height on the cycle ergometer according to leg length. When the foot is flat on the right pedal and the pedal is in the extreme down position, the right knee should be in approximately 5° of flexion.
3. Load-incremented protocol for electronically braked and friction-braked cycle ergometers:

   (a) Instruct the subject to maintain a 50 $rev \cdot min^{-1}$ pedal cadence. Set the metronome to 100 $beats \cdot min^{-1}$ so each downward movement of each foot is synchronized with a beat of the metronome. The subject may also use the digital monitor on the cycle control panel to regulate pedal cadence.
   (b) For electronically braked cycle ergometers (e.g., Lode), begin stage 1 at 50 W then increase the resistance 25 W every 2 min.

(c) For friction-braked cycle ergometers (e.g., Monark), begin stage 1 at 1 kg resistance then increase the resistance 0.5 kg every 2 min.
(d) When the subject cannot maintain the pedal cadence for 10 consecutive seconds, terminate the exercise test.
(e) Ask the subject to estimate RPE starting at 1:30 of each exercise stage. Because the position of the respiratory-metabolic mouth piece inhibits a verbal response, instruct the subject to point to the numbers on the RPE scale, which should be conveniently positioned within the subject's arm reach. State aloud the numerical ratings for each momentary assessment to which the subject pointed and request a confirmatory nod that the number stated was correct. If incorrect, allow the subject to point to the appropriate rating on the RPE scale once more. Ask the subject to hold his or her finger on the appropriate number on the scale for approximately 1 s.
(f) Record HR ($b{\cdot}min^{-1}$) at 1:55 of each exercise stage.
(g) Record the final 15-s VO_2 ($l \cdot min^{-1}$) for each exercise stage.
(h) Record HRpeak as the highest HR value recorded during the final exercise stage or immediately post-exercise.
(i) Record VO_2peak as the highest 15-s VO_2 value recorded at the end of the test.
(j) Calculate the VO_2 ($l \cdot min^{-1}$) associated with 70 and 85 % of VO_2peak, which will be used to obtain the RPE's corresponding to the upper and lower limits of the overload training zone performed during the production trial. An explanation for the calculation of this target RPE range will be presented in the data organization and analysis section below.

9.2.2.4 Estimation Protocol: Data Organization and Analysis

1. In a Microsoft Excel spreadsheet, label columns of data for the following variables: Exercise Stage, VO_2 ($l \cdot min^{-1}$), OMNI RPE-L, HR ($b{\cdot}min^{-1}$).
2. Plot of VO_2 and OMNI RPE-L for determination of the RPE corresponding to 70 % VO_2peak, i.e., the lower limit of the target RPE range.

   (a) Click on the **INSERT** tab and in the **CHARTS** section click on **SCATTER**. Select the first available chart option. A blank or example scatter plot will appear on your screen.
   (b) Click on the **SELECT DATA** tab. Remove any entries found in the **LEGEND ENTRIES** text box then click **ADD**. Under **SERIES NAME**, enter VO_2 and OMNI RPE-L. Then click on the icon to the right of the **SERIES X VALUES** text box and highlight the VO_2 values. After the values are highlighted click the icon on the box that appeared. Then click on the icon to the right of the **SERIES Y VALUES** text box and highlight the OMNI RPE-L values. After the values are highlighted click the icon on the box that appeared. Click **OK** on the next two screens.
   (c) You should now have a scatter plot with OMNI RPE-L on the *y*-axis and VO_2 on the *x*-axis. Create a title for the plot and enter the appropriate axis labels and units of measure.

(d) To determine the RPE corresponding to 70 % VO_2peak, click on one of the data points to highlight the entire data series. Right click on one of the data points and a menu will appear. Click **ADD TRENDLINE** and the **FORMAT TRENDLINE** menu will appear. Select **LINEAR** and **DISPLAY EQUATION ON CHART** then click **CLOSE**. The trendline and equation will be displayed on the chart.

(e) Use this linear equation to calculate the RPE corresponding to 70 % VO_2peak. Use VO_2 ($l \cdot min^{-1}$) corresponding to 70 % VO_2peak as the "x" value in the equation and solve for "y." The calculated "y" value, once rounded to the nearest whole integer, is the RPE corresponding to the lower limit of the target RPE range.

3. Use the linear equation described above to determine the RPE corresponding to 85 % VO_2peak, i.e., the upper limit of the target RPE range.
4. An example of the procedures to determine an RPE corresponding to a specific %VO_2peak with a screenshot depicting each step as performed using Microsoft Excel 2013 can be found in Appendix D since these procedures are the same as those used for determination of RPE-VT.

9.2.2.5 Production Protocol

1. Position the HR monitor and respiratory-metabolic mouthpiece (with head support unit and nose clip if applicable) on the subject.
2. Set the proper seat height on the cycle ergometer according to leg length. When the foot is flat on the right pedal and the pedal is in the extreme down position, the right knee should be in approximately 5° of flexion.
3. Instruct the subject to maintain a 50 $rev \cdot min^{-1}$ pedal cadence. Set the metronome to 100 $beats \cdot min^{-1}$ so each downward movement of each foot is synchronized with a beat of the metronome. The subject may also use the digital monitor on the cycle control panel to regulate pedal cadence.
4. Instruct the subject to perform a 2-min warm-up.

   (a) For electronically braked cycle ergometers (e.g., Lode) set the PO at 25 W.
   (b) For friction-braked cycle ergometers (e.g., Monark), set the brake resistance at 0.5 kg.

5. Following the warm-up, instruct the subject to exercise for 10 min within the predetermined target RPE range by adjusting the power output or break resistance until the desired intensity is met.
6. Record HR ($b \cdot min^{-1}$) every 2 min.
7. Record the final 15-s VO_2 ($l \cdot min^{-1}$) for each 1-min segment of exercise.
8. Following the 10-min exercise bout, instruct the subject to perform a 2-min cool-down period.

   (a) For electronically braked cycle ergometers (e.g., Lode) set the PO at 25 W.
   (b) For friction-braked cycle ergometers (e.g., Monark), set the brake resistance at 0.5 kg.

9.2.2.6 Production Protocol: Data Organization and Analysis

1. In a Microsoft Excel spreadsheet, label columns of data for the following variables: Exercise Time (minutes), VO_2 ($l \cdot min^{-1}$).
2. Count the number of minutes in which the subject's VO_2 was between 70 and 85 % of VO_2peak. Divide that number by 10 then multiple by 100 to determine the percent of time the subject exercised in the overload training zone.
3. Count the number of minutes in which the subject's VO_2 was lower than 70 % VO_2peak. Divide that number by 10 then multiple by 100 to determine the percent of time the subject exercised below the overload training zone.
4. Count the number of minutes in which the subject's VO_2 was higher than 85 % VO_2peak. Divide that number by 10 then multiple by 100 to determine the percent of time the subject exercised above the overload training zone.

9.3 Laboratory Discussion Questions

1. What information can be collected during an estimation protocol that can be used to determine an appropriate intensity for a self-regulated exercise prescription? List various ways that exercise intensity or an exercise intensity range can be calculated using the information collected during a perceptual estimation protocol.
2. Describe how the production protocol in an estimation–production paradigm provides data to test the validity of exercise intensity self-regulation using a target RPE.
3. How did your subject's VO_2 and HR responses compare between the estimation and production protocols when measurements were obtained within the target intensity range of 70–85 % VO_2max/peak? Do the results of this perceptual estimation–production paradigm provide evidence for prescription congruence?
4. Describe methods that can be employed to assist the subject with exercise intensity self-regulation during a perceptual production protocol.
5. Describe the comparative advantages of using a target RPE rather than a target HR to self-regulate exercise intensity.

9.4 Laboratory Addendum: Exercise Prescription for Resistance Exercise

Although the procedures used to determine 1RM can be considered a multi-stage GXT for resistance exercise and were presented as such regarding perceived exertion scale validation in Chap. 6, there is no production trial for resistance exercise that is analogous to a production trial used for aerobic exercise. The use of RPE to prescribe a resistance exercise program recognizes that, for a given intensity

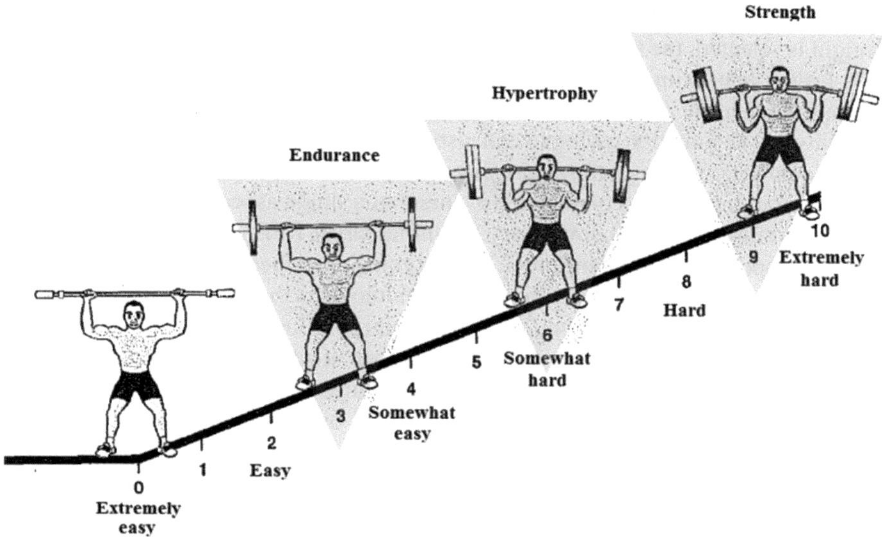

Fig. 9.1 Sliding RPE Zone for Resistance Exercise Training (Robertson 2004)

(i.e., resistance or weight), perceived exertion increases as the number of repetitions increase. A set of resistance exercise begins by identifying a predetermined submaximal weight that produces an a priori determined initial RPE. As the resistance exercise set continues (i.e., the number of repetitions increases), RPE increases until the point of exhaustion causing the individual to terminate exercise. For example, an individual could begin an exercise set using a resistance of 60 % 1RM. Initially, the individual might rate this intensity a 6 on the Adult OMNI-Resistance Exercise RPE Scale. As the individual performs the repetitions of that set, RPE increases until maximal exertion is achieved, i.e., a 10 on the OMNI Scale.

The systematic change in perceived exertion induced by a constant submaximal resistance exercise intensity allows for a unique prescriptive paradigm that is specific to the desired neuromuscular training outcomes: endurance, hypertrophy, strength. Robertson (2004) developed the Sliding RPE Zone System (Fig. 9.1) for individualized resistance exercise training using this paradigm. The system is ideally suited to provide a progressive training stimulus that promotes muscular endurance, hypertrophy, and/or strength, depending on the initial target RPE designated for each resistance exercise set. When muscular endurance is the goal of the resistance exercise prescription, an OMNI RPE of 3 is an ideal initial target, most likely resulting in a set of greater than 12 repetitions. When muscular hypertrophy is the goal, an OMNI RPE of 6 is an ideal initial target, most likely resulting in a set of 8–12 repetitions. When muscular strength is the goal, an OMNI RPE of 9 is an ideal initial target, most likely resulting in a set of 3–4 repetitions. For each resistance exercise set, the individual is instructed to use trial and error to select a weight that produces the initial target RPE during a single lift. This process may require several attempts to identify the initial RPE for a given exercise type. Once the weight is selected, the individual begins the resistance exercise set with each repetition lasting approximately 4 s: 2 s in the concentric phase (i.e., lifting or pushing), 2 s in the eccentric phase (i.e., lowering). Repetitions are performed until the point of volitional termination owing to exhaustion (i.e., an OMNI RPE of 10 is reached). Additional sets can be performed using the same initial target RPE for resistance exercise training to focus on the improvement of a single aspect of muscular fitness. However, subsequent sets could vary the initial target RPE for the improvement of multiple aspects of muscular fitness. The forgoing described method to prescribe an RPE-based resistance exercise program accommodates the individual's unique levels of strength, fatigue, and comfort. This occurs by allowing the individual to control the amount of weight lifted and the number of repetitions performed, which may improve adherence to such a resistance training program.

Since the Sliding RPE Zone System is based on initial target RPE's for each resistance exercise set, the assessment and reevaluation of 1RM is not necessary. The individual automatically increases the weight that corresponds to each initial target RPE as muscular strength and endurance increase. Therefore, the Sliding RPE Zone method continually provides the optimal overload training stimulus to enhance muscular fitness. In addition, although it may be desirable to use the measurement of 1RM to track individual improvement, progression can also be tracked by documenting the weight selected and repetitions performed throughout the training program.

References

Dunbar CC, Robertson RJ, Baun R, Blandin MF, Metz K, Burdett R, Goss FL. The validity of regulating exercise intensity by ratings of perceived exertion. Med Sci Sports Exerc. 1992; 24:94–9.

Dunbar CC, Goris C, Michielli DW, Kalinski MI. Accuracy and reproducibility of an exercise prescription based on ratings of perceived exertion for treadmill and cycle ergometer exercise. Percept Mot Skills. 1994;78:1335–44.

Dunbar CC, Kalinski MI. Using RPE to regulate exercise intensity during a 20-week training program for postmenopausal women: a pilot study. Percept Mot Skills. 2004;99:688–90.

Garber CE, Blissmer B, Deschenes MR, Franklin BA, Lamonte MJ, Lee I, Nieman DC, Swain DP. Quantity and quality of exercise for developing and maintaining cardiorespiratory, musculoskeletal, and neuromotor fitness in apparently healthy adults: guidance for prescribing exercise. Med Sci Sports Exerc. 2011;43:1334–59.

Higgins LW, Robertson RJ, Kelsey SF, Olson MB, Hoffman LA, Rebovich PJ, Haile L, Orenstein DM. Exercise intensity self-regulation using the OMNI scale in children with cystic fibrosis. Pediatr Pulmonol. 2013;48:497–505.

Kang J, Chaloupka EC, Mastrangelo MA, Donnelly MS, Martz WP, Robertson RJ. Regulating exercise intensity using ratings of perceived exertion during arm and leg ergometry. Eur J Appl Physiol. 1998;78:241–6.

Kang J, Hoffman JR, Walker H, Chaloupka EC, Utter AC. Regulating intensity using perceived exertion during extended exercise periods. Eur J Appl Physiol. 2003;89:475–82.

Kang J, Chaloupka EC, Biren GB, Mastrangelo MA, Hoffman JR. Regulating intensity using perceived exertion: effect of exercise duration. Eur J Appl Physiol. 2009;105:445–51.

Midgley AW, McNaughton LR, Wilkinson M. Is there an optimal training intensity for enhancing the maximal oxygen uptake of distance runners?: empirical research findings, current opinions, physiological rationale and practical recommendations. Sports Med. 2006;36:117–32.

Parfitt G, Shepherd P, Eston RG. Reliability of effort perception using the children's CALER and BABE perceived exertion scales. J Exerc Sci Fitness. 2007;5:49–55.

Parfitt G, Evans H, Eston R. Perceptually-regulated training at RPE13 is pleasant and improves physical health. Med Sci Sports Exerc. 2012;44:1613–8.

Robertson RJ. Perceived exertion for practitioners: rating effort with the OMNI picture system. Champaign, IL: Human Kinetics; 2004.

Robertson RJ, Goss FL, Auble TE, Cassinelli DA, Spina RJ, Glickman EL, Galbreath RW, Silberman RM, Metz KF. Cross-modal exercise prescription at absolute and relative oxygen uptake using perceived exertion. Med Sci Sports Exerc. 1990;22:653–9.

Robertson RJ, Goss FL, Bell JA, Dixon CB, Gallagher KI, Lagally KM, Timmer JM, Abt KL, Gallagher JD, Thompkins T. Self-regulated cycling using the children's OMNI scale of perceived exertion. Med Sci Sports Exerc. 2002;34:1168–75.

Swain DP, Franklin BA. VO_2 reserve and the minimal intensity for improving cardiorespiratory fitness. Med Sci Sports Exerc. 2002;34:152–7.

Ulmer HV. Concept of an extracellular regulation of muscular metabolic rate during heavy exercise in humans by psychophysiological feedback. Experientia. 1996;52:416–20.

Ward DS, Bar-Or O. Use of the Borg scale in exercise prescription for overweight youth. Can J Sport Sci. 1990;15:120–5.

Weiser PC, Wojciechowicz V, Funck A, Robertson RJ. Perceived effort step-up procedures for self-regulating stationary cycle exercise intensity by patients with cardiovascular disease. Percept Mot Skills. 2007;104:236–53.

Chapter 10
Exercise Intensity Self-Regulation for Interval Exercise

The estimation–production paradigm can be used to test an individual's ability to self-regulate exercise intensity at multiple target RPE's, a skill that is necessary to perform RPE-based aerobic *interval* exercise. Prior to implementing an interval exercise prescription, an evaluation of intensity discrimination is warranted. This evaluation determines the individual's ability to perceptually differentiate between separate levels of exertion such as are prescribed in an interval training program. Intensity discrimination can be tested using a separate production trial for each target RPE or using an aerobic interval exercise protocol during which exercise intensity self-regulation alternates between two different target RPE's in the same production trial. Intensity discrimination has been confirmed using both production formats such that physiological responses differed between self-regulated target RPE's. Aerobic interval training may have advantages over a traditional exercise prescription that employs a continuous moderate intensity protocol. Such advantages may be a comparatively greater improvement of VO_2max and its rate-limiting physiological factors (i.e., stroke volume and muscle cell oxidation) in healthy and clinical populations. In addition, interval exercise training may minimize the time necessary to achieve health-fitness goals, a major barrier to regular PA participation. The primary purpose of this laboratory experiment is to determine an individual's ability to self-regulate interval exercise intensity using multiple target RPE's according to an estimation–production paradigm.

L. Haile et al., *Perceived Exertion Laboratory Manual*,
DOI 10.1007/978-1-4939-1917-8_10

10.1 Background

10.1.1 The Estimation–Production Paradigm for the Assessment of Intensity Discrimination

The estimation–production paradigm is used to assess the validity of self-regulating exercise intensity using a prescribed target RPE. First, the target RPE and associated physiological responses are derived from the estimation test protocol. The production protocol is then presented, allowing an assessment of an individual's ability to accurately self-regulate exercise intensity at a specific target RPE or within a specific target RPE range. Evidence that the individual can accurately self-regulate exercise intensity is obtained by comparing the physiological responses (e.g., VO_2, HR) determined during the production trial to those corresponding to the target RPE(s) determined during the estimation trial. If the physiological responses are the same between the estimation and production trials, it can be concluded that the exercise intensity prescribed using a target RPE is valid. Such a response established prescription congruence. In contrast, if the physiological responses differ between estimation and production test protocols, then the individual is exhibiting exercise intensity self-regulation error.

The validity of exercise intensity self-regulation can be tested one step further by asking the subject to produce multiple target RPE's during single or multiple production trials. This is a method to evaluate the validity of *intensity discrimination* during interval exercise. In this instance, the test protocol determines the ability to perceptually differentiate between separate target RPE's such that physiological responses differ between different self-regulated intensities (Robertson et al. 2002). Intensity discrimination during interval exercise can be tested using two different formats. The first format uses a separate production trial for each target RPE, similar to the laboratory procedures in the previous chapter. The second format requires the subject to self-regulate exercise intensities corresponding to multiple target RPE's in the same production trial using an interval exercise protocol.

Using separate production trials for different target RPE's, intensity discrimination has been confirmed in children during cycle ergometer exercise (Robertson et al. 2002) and adults during cycle ergometer (Weiser et al. 2007), treadmill and field running exercise (Ceci and Hassmen 1991). Robertson and colleagues (2002) confirmed exercise intensity discrimination between OMNI RPE's 2 and 6 in children performing cycle ergometry. Weiser et al. (2007) confirmed intensity discrimination between Borg RPE's 11 and 13 for cycle ergometer exercise in adults participating in a cardiac rehabilitation program. Ceci and Hassmen (1991) confirmed intensity discrimination between Borg RPE's 11, 13, and 15 for both treadmill and running on an outdoor track in active adult males. Intensity discrimination data were inconsistent in a study of overweight youth performing cycle ergometry and outdoor track walking/running (Ward and Bar-Or 1990). Intensity discrimination was confirmed during cycle ergometry but not for walking/running exercise, indicating that additional instruction and practice may be necessary for overweight children and adolescents when self-regulating exercise intensity using multiple target RPE's (Ward and Bar-Or 1990).

10.1.2 *Assessment of Intensity Discrimination Using Interval Exercise*

When performing an interval exercise protocol, the individual is instructed to begin the exercise trial by self-regulating intensity to produce a specific target RPE. After self-regulating intensity for a specified time interval, the individual is instructed to adjust the exercise intensity to produce a different target RPE for another time interval. The length and number of intervals can be adjusted to accommodate the desired amount of time to complete the entire exercise bout. A simple study of intensity discrimination could involve only one interval at each target RPE. However, multiple intervals at each target RPE are more appropriate when examining the ability of an individual to self-regulate exercise intensity. This is especially important when preparing an individual prior to performing of an interval exercise program. Additional practice, feedback, and reinforcement may be necessary when teaching an individual to perform RPE-based interval exercise compared to continuous exercise.

Higgins and colleagues (2013) used the interval production format to examine intensity discrimination and prescription congruence for both cycle ergometer and treadmill exercise. The intent of the paradigm was to assess the validity of interval exercise intensity self-regulation in children with cystic fibrosis. The research design included an estimation trial on a cycle ergometer and two interval production trials, one on a cycle ergometer and another on a treadmill. Each interval production trial simulated an actual aerobic exercise trial that could be performed during an interval exercise program. Subjects alternated between 3-min intervals of self-regulated exercise intensity at OMNI RPE's 4 and 7 without rest between intervals. Intensity discrimination was confirmed between OMNI RPE's 4 and 7 for cycle ergometer and treadmill exercise when both HR and VO_2 were compared between the two target RPE's (Higgins et al. 2013). In addition, both intramodal and intermodal prescription congruence was confirmed for both RPE's, indicating the validity of exercise intensity self-regulation during interval exercise.

10.1.3 *Advantages of Aerobic Interval Exercise for Health-Fitness and Performance Outcomes*

For healthy adult subjects (Daussin et al. 2007; Esfandiari et al. 2014; Helgerud et al. 2007; Matsuo et al. 2013), patients with coronary artery disease and heart failure (Fu et al. 2011; Guirard et al. 2012; Haykowski et al. 2013; Rognmo et al. 2004; Wisloff et al. 2007), as well as type 2 diabetics (Mitranum et al. 2013), aerobic interval exercise has resulted in more positive health-fitness and performance-related outcomes when compared with a continuous moderate intensity program. Improvement in the rate-limiting physiological factors that influence VO_2max as well as VO_2max itself has been shown to be greater following interval training when compared to moderate intensity continuous exercise. In a study by Daussin and colleagues (2007), previously sedentary subjects performed 8 weeks each of aerobic interval exercise and continuous moderate-intensity exercise with programs presented in random order and separated by 12 weeks of detraining. These cycle ergometer exercise programs were matched for energy expenditure and exercise duration. Both programs resulted in improved arterial-venous oxygen difference, but only the interval exercise program resulted in improved VO_2max and maximal cardiac output (Daussin et al. 2007). In a study by Esfandiari and colleagues (2014), previously untrained but healthy men increased plasma volume, end-diastolic volume, stroke volume and cardiac output after only 6 sessions of cycle ergometer exercise involving either high-intensity intervals or continuous moderate-intensity training. However, only the interval exercise group significantly improved VO_2max (Esfandiari et al. 2014). In a study of previously sedentary men performing an 8-week cycle ergometer exercise program, Matsuo and colleagues (2013) found that interval training resulted in a greater improvement in VO_2max compared to continuous aerobic training. In addition, interval training resulted in significant improvements in left ventricular mass, stroke volume and resting HR as compared to continuous exercise training (Matsuo et al. 2013). Helgerud and colleagues (2007) conducted a study in which moderately trained men performed one of four, 8-week treadmill exercise programs. The protocol employed two different continuous exercise intensities (70 or 85 % of VO_2max) and two different interval exercise protocols (15-s intervals with 15-s recovery periods or 4-min intervals with 4-min recovery periods). All training programs significantly improved running economy and velocity at the lactate threshold, but only the interval exercise programs significantly improved maximal stroke volume and VO_2max. Interestingly, both the interval durations of 15 s and 4 min were equally effective (Helgerud et al. 2007).

A recent meta-analysis investigated the effect of aerobic interval training compared with moderate-intensity continuous exercise on VO_2peak and left ventricular ejection fraction in heart failure patients (Haykowski et al. 2013). Seven randomized trials were identified. Collectively, the studies indicated that interval exercise was significantly more effective at improving VO_2peak compared with continuous exercise. However, observed improvements in ejection fraction were not significantly different between training protocols (Haykowski et al. 2013). Rognmo and colleagues (2004) compared the effects of aerobic interval exercise with moderate intensity continuous exercise in coronary artery disease patients. After 10 weeks of treadmill exercise, patients who performed interval exercise experienced a significantly greater increase in VO_2peak (Rognmo et al. 2004). In a study of diabetic patients, Mitranum et al. (2013) found that 12 weeks of either aerobic interval training or continuous aerobic training conferred various improvements in health, including reductions in body fat, resting HR, and fasting blood glucose, as well as increases in leg muscle strength. However, only the group of patients who performed the interval exercise program significantly decreased glycosylated hemoglobin values, an indication of improved blood glucose control over a prolonged period. In addition, although both groups improved VO_2max, the interval exercise group had a greater increase (Mitranum et al. 2013).

10.1.4 Duration of the Interval Exercise Bout

High-intensity interval training may be an effective method to achieve health-fitness benefits while performing for a comparatively shorter time period during each exercise session. Time commitment is a common barrier to exercise participation (Trost et al. 2002). Minimizing the time required for a given exercise session could be a major factor to promote program adherence in some individuals (Kessler et al. 2012; Reichert et al. 2007). In a study by Matsuo et al. (2013), sedentary men who performed 8 weeks of aerobic interval training achieved significantly greater beneficial changes in VO_2max, left ventricular mass, stroke volume, and resting HR compared with those who performed a continuous aerobic exercise training program. The duration of each interval exercise session was 13 min, compared to the 40-min continuous exercise session. The shorter time period required for an interval program could facilitate an individual's ability to complete exercise sessions within the time they have available for daily PA. However, average caloric expenditure for the interval exercise sessions was 180 kilocalories (kcals) compared to the 360 kcals required during continuous exercise (Matsuo et al. 2013). Therefore, an increased exercise duration or a greater focus on dietary changes may be necessary to achieve weight loss goals when employing an interval program.

10.1.5 Appropriate Intensity of Exercise Intervals

The intensity of the exercise intervals may influence the type and size of physiological training adaptations. In a study by Higgins and colleagues (2013), children with cystic fibrosis performed interval exercise by alternating intensity between target OMNI RPE's 4 and 7. For most individuals, an OMNI RPE 4 represents low to moderate exercise intensity, usually falling below the VT. An OMNI RPE 7 represents high exercise intensity falling above the VT. Studies have examined high-intensity intervals at prescribed HR or VO_2 values of 80–100 % of maximal intensity. This comparatively high intensity range is not only safe and effective for healthy individuals but may also be safe for most patients with coronary artery disease or heart failure who are participating in cardiac rehabilitation programs (Helgerud et al. 2007; Moholdt et al. 2014; Rognmo et al. 2004; Wisloff et al. 2007). Moholdt and colleagues (2013) employed an interval exercise cardiac rehabilitation program where patients performed within an intensity range of 85–95 % of HRmax. They found that those coronary heart disease patients who performed the exercise intensity intervals at the higher end of the prescribed range had greater improvements in VO_2peak. A target RPE of 7 for high-intensity intervals may not be high enough to elicit HR responses within the range of 85–95 % of HRmax. In addition, a target RPE of 4 may not be low enough to allow proper active recovery between high-intensity intervals. Therefore, to ensure optimal target RPE's that require the production of high-intensity exercise intervals and low-intensity active recovery periods, it may be best to determine which target RPE's should be used on an individual basis.

10.1.6 Advanced Perceived Exertion Scaling Procedures for Interval Exercise

Prior to undertaking an interval exercise program, it is important to ensure that the participants are adequately oriented to the scaling procedures necessary to self-regulate exercise intensity using multiple target RPE's in a single exercise bout. Therefore, a more advanced perceived exertion scale anchoring procedure was developed and has been used in previous investigations involving such an interval type exercise program (Higgins et al. 2013). This procedure allows time for additional practice, feedback, and reinforcement that is not included in the standard scale anchoring procedures already presented in Chap. 5. This advanced anchoring procedure may be helpful for any individual having difficulty understanding how to use a category scale to rate exertion levels, especially young children. In this procedure, the scale anchoring is divided into three distinct phases: low, moderate, and high/maximal exercise intensity. Each phase includes a brief, 2–4-min bout of load-incremented exercise in which physical intensity is increased and the client estimates his or her RPE every 15 or 30 s. In addition, the low and moderate intensity phases include a brief, 2–4-min production bout in which the individual is instructed to perform exercise that elicits a specific level of exertion as indicated by a target RPE. See Appendix F for a detailed description of this advanced perceived exertion scaling procedure.

Even after a comprehensive exercise anchoring procedure has been administered, additional feedback may be helpful during production trials. The physiological values corresponding to the multiple target RPE's should be measured during the estimation trial. These values should be monitored during the production trial(s) to provide appropriate feedback and reinforcement regarding the accuracy of intensity self-regulation according to a target RPE. Exercise intensity self-regulation error may be more common when the individual is asked to titrate exercise intensity between different target RPE's a number of times in a single exercise bout. Pre-participation practice exercise trials, or teleoanticipation, may improve prescription congruence and intensity discrimination (Ulmer 1996).

10.1.7 Case Study

10.1.7.1 Client Information

A 63-year-old male with coronary artery disease (CAD) entered your cardiac rehabilitation program after undergoing a surgical procedure that included angioplasty and the placement of a stent in one of his coronary arteries. Prior to the procedure, he would walk his dog through his hilly housing development on a regular basis. Sometimes he would even jog for an extra workout, which is what caused him to experience angina pectoris (i.e., chest pain) symptomatic of CAD. He has been in the rehabilitation program for 4 weeks and is tolerating moderate intensity exercise very well. He has been cleared by his physician to engage in more vigorous intensity exercise under supervision. He looks forward to being able to progress to harder workouts and feeling comfortable enough to exercise at home again.

10.1.7.2 Assessments, Results and Analysis

Administer the estimation–production paradigm using an RPE-based interval exercise protocol. Determine the appropriate intensities for an aerobic interval exercise prescription and determine the subject's ability to self-regulate exercise intensity using multiple target RPE's within the same exercise bout.

Using the estimation trial:

1. Determine VO_2max/peak.
2. Determine the target RPE's corresponding to the high-intensity intervals and the low-intensity active recovery periods.
3. Calculate the corresponding physiological values that correspond to these target RPE's.

Using the production trial, test the ability of the subject to self-regulate exercise intensity at multiple target RPE's by comparing physiological values measured during the production protocol to those corresponding to the target RPE's determined from the estimation protocol.

10.2 Methods

10.2.1 Treadmill Procedures

10.2.1.1 Equipment

1. Adult OMNI-Walk/Run RPE Scale (Fig. A.2)
2. Treadmill
3. HR monitor
4. Respiratory-metabolic measurement system

10.2.1.2 Pre-Exercise Procedures

1. Measure height (cm) and weight (kg) of subject.
2. Perform the memory anchoring procedure as described in Chap. 5, reading the standard instructions for the Adult OMNI-Walk/Run RPE Scale for RPE-O to the subject (Appendix B.1). If measurement of differentiated RPE (RPE-L and/ or RPE-C) is desired, read the standard instructions for the Adult OMNI-Walk/ Run RPE Scale for undifferentiated and differentiated RPE (Appendix B.2).

10.2.1.3 Graded Exercise Test

1. Position the HR monitor and respiratory-metabolic mouthpiece (with head support unit and nose clip if applicable) on the subject.
2. Instruct the subject to step onto the treadmill and review exercise termination procedures: When the subject cannot continue exercise due to exhaustion or discomfort, he/she should grasp the treadmill hand rails, at which time the test administrator will gradually slow the treadmill down for performance of a cool-down. The subject should be reminded not to step off the treadmill belt while it is still in motion.
3. Bruce Multistage Treadmill Test Protocol: this can be performed by manually adjusting treadmill speed and grade or using a program on a computer that is interfaced to the treadmill.

   (a) Begin the warm-up at 1.5 miles $\cdot$ h^{-1} and 0 % grade for 3 min.
   (b) Each exercise test stage will last for 3 min. The stages progress as follows:

   Stage 1—1.7 miles $\cdot$ h^{-1} and 10 % grade
   Stage 2—2.5 miles $\cdot$ h^{-1} and 12 % grade
   Stage 3—3.4 miles $\cdot$ h^{-1} and 14 % grade
   Stage 4—4.2 miles $\cdot$ h^{-1} and 16 % grade
   Stage 5—5.0 miles $\cdot$ h^{-1} and 18 % grade
   Stage 6—5.5 miles $\cdot$ h^{-1} and 20 % grade
   Stage 7—6.0 miles $\cdot$ h^{-1} and 22 % grade
   Stage 8—6.5 miles $\cdot$ h^{-1} and 24 % grade

   (c) When the subject cannot continue any longer, terminate the exercise test by initiating the cool-down period at 1.5 miles $\cdot$ h^{-1} and 0 % grade. The cool-down should be 5 min in duration.
   (d) Ask the subject to estimate RPE starting at 2:30 of each exercise stage. Because of the respiratory-metabolic mouth piece, instruct the subject to point to the numbers on the RPE scale, which should be conveniently positioned within the subject's arm reach. State aloud the numerical ratings for each momentary assessment to which the subject pointed and request a confirmatory nod that the number stated was correct. If incorrect, allow the subject to point to the appropriate rating on the RPE scale once more. Ask the subject to hold his or her finger on the appropriate number on the scale for approximately 1 s.
   (e) Record HR (b $\cdot$ min^{-1}) at 2:55 of each exercise stage.
   (f) Record the final 15-s VO_2 (ml $\cdot$ kg^{-1} $\cdot$ min^{-1}) for each exercise stage.
   (g) Record HRmax as the highest HR value recorded during the final exercise stage or immediately post-exercise.
   (h) Record VO_2max as the highest 15-s VO_2 value recorded at the end of the test.
   (i) Calculate the VO_2 (ml $\cdot$ kg^{-1} $\cdot$ min^{-1}) associated with 40 and 80 % of VO_2max. These values will be used to obtain the RPE's corresponding to the lower and higher exercise intensity intervals that will be performed during the production trial. An explanation of how to calculate the target RPE's will be presented in the data organization and analysis section below.

10.2.1.4 Estimation Protocol: Data Organization and Analysis

1. In a Microsoft Excel spreadsheet, label columns of data for the following variables: Exercise Stage, VO_2 ($ml \cdot kg^{-1} \cdot min^{-1}$), OMNI RPE-O, HR ($b \cdot min^{-1}$).
2. Plot of VO_2 and OMNI RPE-O for determination of the RPE corresponding to 40 % VO_2max, the target exercise intensity for the lower intensity interval.
   (a) Click on the **INSERT** tab and in the **CHARTS** section click on **SCATTER**. Select the first available chart option. A blank or example scatter plot will appear on your screen.
   (b) Click on the **SELECT DATA** tab. Remove any entries found in the **LEGEND ENTRIES** text box then click **ADD**. Under **SERIES NAME**, enter VO_2 and OMNI RPE-O. Then click on the icon to the right of the **SERIES X VALUES** text box and highlight the VO_2 values. After the values are highlighted click the icon on the box that appeared. Then click on the icon to the right of the **SERIES Y VALUES** text box and highlight the OMNI RPE-O values. After the values are highlighted click the icon on the box that appeared. Click **OK** on the next two screens.
   (c) You should now have a scatter plot with OMNI RPE-O on the *y*-axis and VO_2 on the *x*-axis. Create a title for the plot and enter the appropriate axis labels and units of measure.
   (d) To determine the RPE corresponding to 40 % VO_2max, click on one of the data points to highlight the entire data series. Right click on one of the data points and a menu will appear. Click **ADD TRENDLINE** and the **FORMAT TRENDLINE** menu will appear. Select **LINEAR** and **DISPLAY EQUATION ON CHART** then click **CLOSE**. The trendline and equation will be displayed on the chart.
   (e) Use this linear equation to calculate the RPE corresponding to 40 % VO_2max. Use VO_2 ($ml \cdot kg^{-1} \cdot min^{-1}$) corresponding to 40 % VO_2max as the "x" value in the equation and solve for "y." The calculated "y" value, once rounded to the nearest whole integer, is the RPE corresponding to the lower limit of the target RPE range.
3. Use the linear equation as explained above to determine the RPE corresponding to 80 % VO_2max, the target exercise intensity for the higher intensity interval.
4. An example of the procedures to determine an RPE corresponding to a specific %VO_2max with a screenshot depicting each step as performed using Microsoft Excel 2013 can be found in Appendix D since these procedures are the same as those used for determination of RPE-VT.

10.2.1.5 Production Protocol

1. Position the HR monitor and respiratory-metabolic mouthpiece (with head support unit and nose clip if applicable) on the subject.
2. Instruct the subject to step onto the treadmill and review exercise termination procedures: When the subject cannot continue exercise due to exhaustion or

discomfort, he/she should grasp the treadmill hand rails, at which time the test administrator will slow down or stop the treadmill. The subject should be reminded not to step off the treadmill belt while it is still in motion.
3. Start the treadmill at 3 miles $\cdot$ h^{-1} and 0 % grade for a 2-min warm-up.
4. Following the warm-up, instruct the subject to exercise for 2 min at the predetermined target RPE associated with 40 % VO_2max by adjusting the speed of the treadmill until the desired perceptual intensity is met. This is the first lower intensity interval. Repeat these instructions at the 4 and 8-min time points.
5. At the 2-min time point, instruct the subject to exercise for 2 min at the predetermined target RPE associated with 80 % VO_2max by adjusting the speed of the treadmill until the desired perceptual intensity is met. This is the first higher intensity interval. Repeat these instructions at the 6-min time point.
6. Record HR (b $\cdot$ min^{-1}) every 2 min.
7. Record the final 15-s VO_2 (ml $\cdot$ kg $\cdot$ min^{-1}) for each 2-min segment of exercise.
8. Following the 10-min exercise, the treadmill should be set to 3 miles $\cdot$ h^{-1} and 0 % grade for a 2-min cool-down period.

10.2.2 Cycle Ergometer Procedures

10.2.2.1 Equipment

1. Adult OMNI-Cycle RPE Scale (Fig. 2.4)
2. Cycle ergometer
3. Metronome
4. HR monitor
5. Respiratory-metabolic measurement system

10.2.2.2 Pre-exercise Procedures

1. Measure height (cm) and weight (kg) of subject.
2. Perform the memory anchoring procedure as described in Chap. 5, reading the standard instructions for the Adult OMNI-Cycle RPE Scale for RPE-L to the subject (Appendix B.4). If measurement of undifferentiated RPE (RPE-O) or differentiated RPE for the chest/breathing (RPE-C) is desired, read the standard instructions for the Adult OMNI-Walk/Run RPE Scale for undifferentiated and differentiated RPE (Appendix B.5).

10.2.2.3 Graded Exercise Test

1. Position the HR monitor and respiratory-metabolic mouthpiece (with head support unit and nose clip if applicable) on the subject.

2. Set the proper seat height on the cycle ergometer according to leg length. When the foot is flat on the right pedal and the pedal is in the extreme down position, the right knee should be in approximately 5 degrees of flexion.
3. Load-incremented protocol for electronically braked and friction-braked cycle ergometers:

   (a) Instruct the subject to maintain a 50 rev · min^{-1} pedal cadence. Set the metronome to 100 beats · min^{-1} so each downward movement of each foot is synchronized with a beat of the metronome. The subject may also use the digital monitor on the cycle control panel to regulate pedal cadence.
   (b) For electronically braked cycle ergometers (e.g., Lode), begin stage 1 at 50 W then increase the resistance 25 W every 2 min.
   (c) For friction-braked cycle ergometers (e.g., Monark), begin stage 1 at 1 kg resistance then increase the resistance 0.5 kg every 2 min.
   (d) When the subject cannot maintain the pedal cadence for 10 consecutive seconds, terminate the exercise test.
   (e) Ask the subject to estimate RPE starting at 1:30 of each exercise stage. Because the position of the respiratory-metabolic mouth piece inhibit a verbal response, instruct the subject to point to the numbers on the RPE scale, which should be conveniently positioned within the subject's arm reach. State aloud the numerical ratings for each momentary assessment to which the subject pointed and request a confirmatory nod that the number stated was correct. If incorrect, allow the subject to point to the appropriate rating on the RPE scale once more. Ask the subject to hold his or her finger on the appropriate number on the scale for approximately 1 s.
   (f) Record HR (b · min^{-1}) at 1:55 of each exercise stage.
   (g) Record the final 15-s VO_2 (l · min^{-1}) for each exercise stage.
   (h) Record HRpeak as the highest HR value recorded during the final exercise stage or immediately post-exercise.
   (i) Record VO_2peak as the highest 15-s VO_2 value recorded at the end of the test.
   (j) Calculate the VO_2 (l · min^{-1}) associated with 40 and 80 % of VO_2peak. These values will be used to obtain the RPE's corresponding to the lower and higher exercise intensity intervals that will be performed during the production trial. An explanation of how to calculate the target RPE's will be presented in the data organization and analysis section below.

10.2.2.4 Estimation Protocol: Data Organization and Analysis

1. In a Microsoft Excel spreadsheet, label columns of data for the following variables: Exercise Stage, VO_2 (l · min^{-1}), OMNI RPE-L, HR (b · min^{-1}).
2. Plot of VO_2 and OMNI RPE-L for determination of the RPE corresponding to 40 % VO_2peak, the lower limit of the target RPE range.

   (a) Click on the **INSERT** tab and in the **CHARTS** section click on **SCATTER**. Select the first available chart option. A blank or example scatter plot will appear on your screen.

(b) Click on the **SELECT DATA** tab. Remove any entries found in the **LEGEND ENTRIES** text box then click **ADD**. Under **SERIES NAME**, enter VO_2 and OMNI RPE-L. Then click on the icon to the right of the **SERIES X VALUES** text box and highlight the VO_2 values. After the values are highlighted click the icon on the box that appeared. Then click on the icon to the right of the **SERIES Y VALUES** text box and highlight the OMNI RPE-L values. After the values are highlighted click the icon on the box that appeared. Click **OK** on the next two screens.

(c) You should now have a scatter plot with OMNI RPE-L on the *y*-axis and VO_2 on the *x*-axis. Create a title for the plot and enter the appropriate axis labels and units of measure.

(d) To determine the RPE corresponding to 40 % VO_2peak, click on one of the data points to highlight the entire data series. Right click on one of the data points and a menu will appear. Click **ADD TRENDLINE** and the **FORMAT TRENDLINE** menu will appear. Select **LINEAR** and **DISPLAY EQUATION ON CHART** then click **CLOSE**. The trendline and equation will be displayed on the chart.

(e) Use this linear equation to calculate the RPE corresponding to 40 % VO_2peak. Use VO_2 ($l \cdot min^{-1}$) corresponding to 40 % VO_2peak as the "x" value in the equation and solve for "y." The calculated "y" value, once rounded to the nearest whole integer, is the RPE corresponding to the lower limit of the target RPE range.

3. Use the linear equation explained above to determine the RPE corresponding to 80 % VO_2peak, the target exercise intensity for the higher intensity interval.
4. An example of the procedures to determine an RPE corresponding to a specific %VO_2peak with a screenshot depicting each step as performed using Microsoft Excel 2013 can be found in Appendix D. It should be noted that these procedures are the same as those used for determination of RPE-VT.

10.2.2.5 Production Protocol

1. Position the HR monitor and respiratory-metabolic mouthpiece (with head support unit and nose clip if applicable) on the subject.
2. Set the proper seat height on the cycle ergometer according to leg length. When the foot is flat on the right pedal and the pedal is in the extreme down position, the right knee should be in approximately 5 degrees of flexion.
3. Instruct the subject to maintain a 50 $rev \cdot min^{-1}$ pedal cadence. Set the metronome to 100 $beats \cdot min^{-1}$ so each downward movement of each foot is synchronized with a beat of the metronome. The subject may also use the digital monitor on the cycle control panel to regulate pedal cadence.
4. Instruct the subject to perform a 2-min warm-up.

(a) For electronically braked cycle ergometers (e.g., Lode), set the power output at 25 W.

   (b) For friction-braked cycle ergometers (e.g., Monark), set the resistance at 0.5 kg.

5. Following the warm-up, instruct the subject to exercise for 2 min at the predetermined target RPE associated with 40 % VO_2max by adjusting the speed of the treadmill until the desired perceptual intensity is met. This target RPE identifies the first lower intensity interval. Repeat these instructions at the 4- and 8-min time points.
6. At the 2-min time point, instruct the subject to exercise for 2 min at the predetermined target RPE associated with 80 % VO_2max by adjusting the speed of the treadmill until the desired perceptual intensity is met. This target RPE identifies the first higher intensity interval. Repeat these instructions at the 6-min time point.
7. Record HR every 2 min.
8. Record the final 15-s VO_2 in $l \cdot min^{-1}$ for each 1-min segment of exercise.
9. Following the 10-min exercise period, instruct the subject to perform a 2-min cool-down period using the following intensities:

   (a) For electronically braked cycle ergometers (e.g., Lode), set the power output at 25 W.
   (b) For friction-braked cycle ergometers (e.g., Monark), set the break resistance at 0.5 kg.

10.3 Laboratory Discussion Questions

1. Describe the concept of prescription congruence and how it can be tested using an estimation–production procedure in an aerobic interval exercise format.
2. How did your subject's VO_2 and HR responses associated with the lower target RPE interval measured during the production protocol compare with the VO_2 and HR corresponding to the lower target RPE measured during the estimation protocol? How did your subject's VO_2 and HR responses associated with the higher target RPE interval measured during the production protocol compare with the VO_2 and HR corresponding to the higher target RPE measured during the estimation protocol? Do the results of this interval exercise prescription protocol that employs an estimation–production paradigm provide evidence for prescription congruence?
3. Describe the concept of intensity discrimination and how it can be tested using an estimation–production procedure in an aerobic interval exercise format?
4. How did your subject's VO_2 and HR responses associated with the lower target RPE interval measured during the production protocol compare with those associated with the higher target RPE interval during the same protocol? Do the results of this RPE-regulated interval exercise protocol provide evidence for perceptual intensity discrimination?

5. What are the reasons why an RPE-based exercise prescription be more cost-efficient compared with a traditional HR-based exercise prescription?
6. How has high-intensity interval exercise training been found to be superior to continuous moderate-intensity aerobic training regarding physiological outcomes in healthy participants?
7. How has high-intensity interval exercise training been found to be superior to continuous moderate-intensity aerobic training regarding health-related outcomes in individuals with certain chronic diseases?

References

Ceci R, Hassmen P. Self-monitored exercise at three different RPE intensities in treadmill vs field running. Med Sci Sports Exerc. 1991;23:732–8.

Daussin FN, Ponsot E, Dufour SP, Lonsdorfer-Wolf E, Doutrelaue S, Geny B, Piquard F, Richard R. Improvement of VO2max by cardiac output and oxygen extraction adaptation during intermittent versus continuous endurance training. Eur J Appl Physiol. 2007;101:377–83.

Esfandiari S, Sasson Z, Goodman JM. Short-term high-intensity interval and continuous moderate-intensity training improve maximal aerobic power and diastolic filling during exercise. Eur J Appl Physiol. 2014;114:331–43.

Fu TC, Wang CH, Lin PS, Hsu CC, Cherng WJ, Huang SC, Liu MH, Chiang CL, Wang JS. Aerobic interval training improves oxygen uptake efficiency by enhancing cerebral and muscular hemodynamics in patients with heart failure. Int J Cardiol. 2011;167:41–50.

Guirard T, Nigam A, Gremeaux V, Meyer P, Juneau M, Bosquet L. High-intensity interval training in cardiac rehabilitation. Sports Med. 2012;42:587–605.

Haykowski MJ, Timmons MP, Kruger C, McNeely M, Taylor DA, Clark AM. Meta-analysis of aerobic interval training on exercise capacity and systolic function in patients with heart failure and reduced ejection fraction. Am J Cardiol. 2013;111:1466–9.

Helgerud J, Hoydal K, Wang E, Karlsen T, Berg P, Bjerkaas M, Simonsen T, Helgesen C, Hjorth N, Bach R, Hoff J. Aerobic high-intensity intervals improve VO_2max more than moderate training. Med Sci Sports Exerc. 2007;39:665–71.

Higgins LW, Robertson RJ, Kelsey SF, Olson MB, Hoffman LA, Rebovich PJ, Haile L, Orenstein DM. Exercise intensity self-regulation using the OMNI scale in children with cystic fibrosis. Pediatr Pulmonol. 2013;48:497–505.

Kessler HS, Sisson SB, Short KR. The potential for high-intensity interval training to reduce cardiometabolic disease risk. Sports Med. 2012;42:489–509.

Matsuo T, Saotome K, Seino S, Shimojo N, Matsushita A, Iemitsu M, Ohshima H, Tanaka K, Mukai C. Effects of a low-volume aerobic-type interval exercise on VO2max and cardiac mass. Med Sci Sports Exerc. 2013;46:42–50.

Mitranum W, Deerochanawong C, Tanaka H, Suksom D. Continuous vs interval training on glycemic control and macro- and microvascular reactivity in type 2 diabetics. Scand J Med Sci Sports. 2013;24:e69–76.

Moholdt T, Madssen E, Rognmo O, Aamot IL. The higher the better? Interval training intensity in coronary heart disease. J Sci Med Sport. 2014;17:506–10.

Reichert FF, Barros AJ, Domingues MR, Hallal PC. The role of perceived personal barriers to engagement in leisure-time physical activity. Am J Public Health. 2007;97:515–9.

Robertson RJ, Goss FL, Bell JA, Dixon CB, Gallagher KI, Lagally KM, Timmer JM, Abt KL, Gallagher JD, Thompkins T. Self-regulated cycling using the children's OMNI scale of perceived exertion. Med Sci Sports Exerc. 2002;34:1168–75.

Rognmo O, Hetland E, Helgerud J, Hoff J, Slordahl SA. High intensity aerobic interval exercise is superior to moderate intensity exercise for increasing aerobic capacity in patients with coronary artery disease. Eur J Cardiovasc Prev Rehabil. 2004;11:216–22.

Trost SG, Owen N, Bauman AE, Sallis JF, Brown W. Correlates of adults' participation in physical activity: review and update. Med Sci Sports Exerc. 2002;34:1996–2001.

Ulmer HV. Concept of an extracellular regulation of muscular metabolic rate during heavy exercise in humans by psychophysiological feedback. Experientia. 1996;52:416–20.

Ward DS, Bar-Or O. Use of the Borg scale in exercise prescription for overweight youth. Can J Sport Sci. 1990;15:120–5.

Weiser PC, Wojciechowicz V, Funck A, Robertson RJ. Perceived effort step-up procedures for self-regulating stationary cycle exercise intensity by patients with cardiovascular disease. Percept Mot Skills. 2007;104:236–53.

Wisloff U, Stoylen A, Loennechen JP, Bruvold M, Rognmo O, Haram PM, Tjonna AE, Helgerud J, Slordahl SA, Lee SJ, Videm V, Bye A, Smith GL, Najjar SM, Ellingsen O, Skjaerpe T. Superior cardiovascular effect of aerobic interval training versus moderate continuous training in heart failure patients: a randomized study. Circulation. 2007;115:3086–94.

Chapter 11
Exercise Intensity Self-Regulation Using the Perceived Exertion JND

The achievement of optimal performance during exercise is not a new area of interest among athletes, trainers, and coaches. Many tend to employ external performance enhancing devices or substances, such as improved clothing and footwear, nutritional beverages and high caloric (i.e., energy) shakes. However, those that solely rely on such external mechanisms to improve performance may still be limiting the potential of their athletes to achieve the absolute upper limits of a specific exercise endeavor. Regardless of the type of clothing, footwear, or ergogenic beverage, the athlete's psychological and/or physiological capacities to attain and maintain maximal loads provides the greatest contribution to reach the upper limits of exercise performance. Information gained from these interrelated psychophysiological mechanisms, known collectively as teleoanticipation, incorporate memory from previously completed exercise along with both centrally generated efferent signals and peripheral afferent signals during exercise in order to minimize fatigue and obtain optimal performance. The assessment of an individual's just noticeable difference (JND) in perceived exertion may assist in providing appropriate feedback during exercise intensity self-regulation to improve pacing strategy. The perceived exertion JND is a measure of an individual's perceptual acuity, i.e., the smallest change in exercise intensity (expressed as VO_2 or PO) that elicits a noticeable change in perceived exertion. The perceived exertion JND can be applied to the improvement and refinement of teleoanticipation to attain optimal exercise performance. The primary purpose of this laboratory experiment is to examine the use of the perceived exertion JND in a traditional RPE-based exercise prescription. Using the JND as identified in an estimation–production paradigm can in turn provide feedback during the production protocol regarding the accuracy of intensity self-regulation according to a target RPE.

L. Haile et al., *Perceived Exertion Laboratory Manual*,
DOI 10.1007/978-1-4939-1917-8_11

11.1 Background

11.1.1 Pacing Strategy and Teleoanticipation

The study of exercise performance pacing, and associated pacing strategies, arose from research into the mechanisms of fatigue during exercise. It was concluded that an optimal pacing strategy should reduce muscular fatigue by setting performance intensities that optimize both physiological and psychological responses to muscular exertion. It is on this previously established reference point that teleoanticipation plays a critical role in dictating exercise performance in the upcoming and/or ongoing task. Teleoanticipation uses the individual's cognitive ability to recall the stored information regarding metabolic and biomechanical responses to a previously performed exercise task (Ulmer 1996). This recalled information provides a cognitive reference for periodic adjustment in exercise pace throughout the performance, especially at the outset. Once exercise is underway, afferent signals provide feedback on the "state of the body" to obtain optimal performance for that given task, signaling the need to maintain or adjust performance pace. The closer to optimal performance an athlete is at the beginning of exercise, the better the overall performance for that task throughout the duration of the event.

The teleoanticipation process is analogous to looking at someone to provide a general estimate of their age. The evaluator's collective memory of individuals, their age, and appearance accrued over the lifespan are used to provide a cognitive estimate of the age of the target individual. Feedback and subsequent storage of this information is obtained through a series of queries aimed at identifying the exact age of the target individual. This process of using stored information provides a more realistic estimate of age for future estimates of individuals possessing the same general characteristics.

Much of the research and support for teleoanticipation as a mechanism for determining a pacing strategy utilized a perceptual estimation–production paradigm. Additionally, performance pacing strategies are partially dependent on the duration of the task and follow one of three possibilities: positive, negative, or even pacing (Abbiss and Laursen 2008). A positive pacing strategy is commonly observed in events lasting less than 2 min in which an athlete initially starts out at an intensity faster than metabolically optimal, but then the speed or PO is systematically decreased with time. A negative pacing strategy employs an increase in performance pace or producing a higher PO as the task progresses. An even pacing strategy occurs when the athlete performs the task at a preset constant speed or PO throughout the duration of the task.

11.1.2 Mechanisms of Different Pacing Strategies

The three commonly employed pacing strategies may be dependent upon the availability, type and magnitude of metabolic and biomechanical feedback. Tasks that are short-duration may not last long enough to provide sufficient time for the afferent signals to be processed and thereby relying more on the stored memories of that task. These events indicate positive pacing strategies, i.e., a sprinter "bursts" out of the blocks at high speed and the speed tends to slow down towards the end of the sprint. Additionally, the decline in substrate availability of ATP, phosphocreatine, and intramuscular glycogen regulate the rate of muscular fatigue for exercise tasks lasting between 1 and 30 min. Conversely, increases in core body temperature appear to contribute greatly to the onset and progression of muscular fatigue in exercise tasks lasting between 20 and 120 min. Likewise, a decline in intramuscular carbohydrate as an energy source contributes greatly to muscular fatigue in tasks lasting greater than 90 min (de Koning et al. 2011).

11.1.3 Pacing Strategies and RPE

As such, one marker of the integration of the psychological "exercise template" (St Clair Gibson et al. 2005) and physiologic feedback involving energy substrate evaluation is the overall rating of perceived exertion (RPE). An even pacing strategy, as measured by speed, power output, and/or RPE, is thought to provide an optimal performance level throughout the competitive event. Such a pace can forestall the onset of muscular fatigue by using stored information regarding previously performed pace and resulting metabolic and biomechanical feedback that occurred during a similar previously performed task. Therefore, the exercise duration and the pre-rehearsed "exercise template" that guides pace using in-task sensory feedback is continually updated with each subsequent repetition. The resulting pacing strategy for a competitive event is a composite of preset signals from a teleoanticipation sensory feedback system that is continually updated based on the effectiveness of the in-task competitive strategy (Hettinga et al. 2011).

Undifferentiated RPE (overall or whole-body) has been employed in many studies examining pacing strategies as linked to underlying teleoanticipation mechanisms. These studies used match–mismatch exercise paradigms or pharmacological interventions to selectively block afferent signals arising from active muscles and joints during exercise (Amann et al. 2009; Roelands et al. 2009, 2013). Match–mismatch exercise paradigms involve participants that are informed they would perform a predetermined number of trials at the same intensity and duration (Baden et al. 2005) or distance (Albertus et al. 2005). In reality, the paradigm was designed to deceive the participants into believing they exercised at different durations or distances. This was done by masking the participants to correct visual or verbal feedback regarding the performance.

Baden et al. (2005) had participants complete three trials at the same intensity and duration, in random order: (1) a 20 min run, (2) a 10 min run, but then with 1 min remaining subjects were instructed to run another 10 min (20 min total), and (3) a run of an unspecified duration, but subjects were instructed to stop at 20 min. Albertus et al. (2005) had participants complete four separate 20-km exercise time trials. However, the distance feedback was presented as: (1) correct, (2) under-reported, (3) over-reported, and (4) randomly between 25 and 250 m for each 1-km split. Regardless of the distance feedback, participants completed the 20-km cycle ergometer time trial in similar times. Albertus et al. (2005) suggested that factors other than distance feedback may have been responsible for pacing strategies among the four trials. They further speculated that the anticipatory response, indicative of the predetermined "exercise template," may have been more critical in setting the pace than the performance feedback during exercise. The support for teleoanticipation was evident in the last kilometer of each time trial in which the participants were still capable of increasing the power output suggesting that the participants utilized a "hardwired" motor-sensory mechanism to prevent the development of premature fatigue and subsequent loss of power or speed.

The results reported by Baden et al. (2005) and Albertus et al. (2005) suggest that perceived exertion is influenced by physiological changes during exercise. Therefore, this theory holds that prior to the exercise performance the anticipated intensity of the perceived level of exertion that will be experienced in-task is preset but subsequently fine-tuned on a moment-to-moment basis during exercise to ensure optimal performance. Lambert et al. (2005) summarizes the potential underlying mechanisms of teleoanticipation: "mechanoreceptor and metaboreceptor stimulation of the exercise reflex, perhaps initiated at the onset of exercise by mechanical feedback, and adjusted by chemoreceptor feedback according to the continuing energy demands of the muscle" (Lambert et al. 2005).

11.1.4 Just Noticeable Difference in Perceived Exertion

One method to optimize use of the "exercise template" containing the memory of preferred pacing strategies is to determine the just noticeable difference (JND). The JND reflects the acuity of human sensory processes. The concept of identifying a JND between isomorphic sensations of different intensities was explored by classic psychophysicists E. H. Weber and G. T. Fechner. These pioneers in experimental psychology sought to quantify the smallest detectable changes in physical stimuli that humans could perceive, i.e., the difference threshold, or JND (Noble and Robertson 1996; Wozniak 1999). Identifying the perceived exertion JND in an individual may improve the "exercise template" that guides pacing strategies, ensuring faster attainment of optimal performance. JND procedures identify physiological boundaries that define the intensity of an individual's perceived exertion above and below a given exercise intensity. These boundaries are indicative of an individual's perceptual acuity in the presence of changes in exercise intensity. Subsequent practice trials with feedback based on this JND range may allow an individual to more accurately self-regulate exercise intensity to achieve optimal performance strategies.

To measure the JND for a given type of sensation, a standard stimulus must first be determined. The standard stimulus represents a single point within the entire detectable range of stimuli. This point normally occurs somewhere between the stimulus threshold and terminal threshold, i.e., the lowest and highest perceived stimuli possible for that individual, respectively. The JND is then determined for that specific standard stimulus and is expressed as the smallest amount of change in the intensity of the stimulus that is required for that change to be perceived (Buckworth et al. 2013). The JND is determined for intensities that are both greater and less than the standard stimulus. It cannot be assumed that the JND above a standard stimulus is the same size as the JND below the same standard stimulus. In addition, it cannot be assumed that the JND determined for one standard stimulus is the same for other types of stimuli as perceived by any given individual.

11.1.5 Perceived Exertion JND Methodology

The JND for sensations such as hearing or touch can be measured easily using classic methods that involve multiple presentations of stimuli that are compared to the standard stimulus, each referred to as a comparative stimulus. For example, using the Method of Constant Stimuli, after presentation of the standard stimulus, a number of comparative stimuli are presented in random order and each one is presented multiple times. For each comparative stimulus the subject is asked whether it is similar to or different from the standard. The JND is calculated as the comparative stimulus that is perceived as being different from the standard 50 % of the time. The logic underlying their measurement proposes that the JND is not an absolute; rather it falls within a certain perceptual response range at any given time (Noble and Robertson 1996).

Classic psychophysical methods are not appropriate for determination of a perceived exertion JND. For senses such as hearing or touch, multiple comparative stimulus intensities can be presented without any residual or cumulative consequences that may change the individual's overall perception of that sense. However, multiple presentations of various intensities of exercise may increase RPE, lead to fatigue, and hinder the ability to measure the perceived exertion JND (Haile et al. 2013). Therefore, methods unique to the determination of the perceived exertion JND have been designed. A recent investigation determined a perceived exertion JND for cycle ergometer exercise in young adults (Haile et al. 2013). The individual PO corresponding to an OMNI RPE of 5 derived from an estimation trial served as the SS. This RPE was chosen because it was previously shown to correspond to the lower limit of the perceptual zone encompassing the response-normalized RPE at the ventilatory threshold. The perceived exertion JND above (JND-A) and below (JND-B) the standard stimulus was determined using four separate 5-min cycle ergometer exercise bouts with rest periods between bouts. Exercise bouts 1 and 3 presented the standard PO. In exercise bouts 2 and 4, subjects were allowed to adjust the PO to produce the smallest detectable change in perceived exertion, either greater than the standard (for JND-A) or less than the standard (for JND-B). VO_2 was measured during each bout. The change in VO_2 between each JND bout and the previously presented standard stimulus bout was used as the aerobic metabolic analog to define the perceived exertion JND. In addition, physical units (i.e., PO) describing the JND were also determined (Haile et al. 2013).

11.1.6 Evidence for Perceptual Acuity

Perceptual acuity varies greatly between individuals for all human senses, including perceived exertion. For cycle ergometer exercise in young adults, the average perceived exertion JND-A was 5.90 $\%VO_2$peak with a standard deviation of 4.09 $\%VO_2$peak. This corresponded to an average change in PO required for detection of 8.4 W. The average JND-B was 8.33 $\%VO_2$peak with a standard deviation of 6.01 $\%VO_2$peak, corresponding to an average change required for detection of 15.6 W (Haile et al. 2013). This indicates that some individuals had comparatively fine perceptual acuity and sensed very small changes in the exertional level associated with changes in exercise intensity. Other individuals required more substantial changes in intensity before a noticeable difference in perceived exertion was detected. In addition, JND-A was significantly smaller than JND-B indicating that, on average, perceptual acuity was more precise just above an OMNI RPE 5 compared to just below an OMNI RPE 5. This indicates that the perceived exertion JND may change with exercise intensity or target RPE's. Further research is needed to investigate the relation between the perceived exertion JND and exercise intensity, as well as the effects of fitness and PA levels. These potential differences in the perceived exertion JND, whether between exercise intensities in a single individual or between separate individuals of differing fitness or PA characteristics, may influence or even predict individual pacing strategy. Comparatively larger and smaller JND values may explain the appearance of positive and/or negative pacing strategies even when steady state pacing strategies may be considered the most optimal.

11.1.7 Perceived Exertion JND and Exercise Intensity Self-Regulation

A measurement of the perceived exertion JND can be used as an ideographic assessment of exercise intensity self-regulation error. A classic estimation–production paradigm analyzes prescription congruence by comparing physiological values that correspond to a specific RPE between estimation and production trials. This method does not take individual perceptual acuity into consideration. For example, VO_2 values could be found significantly different between trials by a statistical analysis. However, an individual with less perceptual acuity and hence a greater perceived exertion JND would not be able to perceive them as different.

Therefore, the addition of the JND method to examine prescription congruence between estimation and production protocols can improve the validity of exercise intensity self-regulation error assessment. The VO_2 values corresponding to JND-A and JND-B define a physiological range spanning an exercise intensity that corresponds to the target RPE for an exercise prescription. The VO_2 range analogous to the perceived exertion JND identifies the limits of perceptual acuity of an individual for that exercise intensity. By definition, when an individual produces the target RPE, if VO_2 values fall within the VO_2 range for that mode specific JND then that individual is accurately self-regulating exercise intensity. Exercise intensity self-regulation error would be defined as physiological values that fall outside of the VO_2 range corresponding to that perceptual JND when attempting to produce a prescribed target RPE.

The perceived exertion JND measures an individual's perceptual acuity during physical exercise. The JND identifies those with fine and broad perceptual acuity corresponding to the same relative exercise intensity. In addition, determination of the perceived exertion JND allows precise feedback regarding accuracy of intensity self-regulation during production protocols. This feedback can provide an individual with the knowledge of whether or not exercise intensity self-regulation error is present, i.e., if the individual's VO_2 is within or outside of the analogous JND range. In addition, this feedback would provide the individual with information about which direction to adjust exercise intensity to correct self-regulation error. This additional step in the traditional estimation–production paradigm may be crucial for some individuals to accurately self-regulate exercise intensity using target RPE's. Incorporating JND into exercise or fitness testing and training may provide a better-matched "exercise template" from which the individual may begin exercise and achieve an optimal pace to promote health-fitness gains. The improved "template" of the stored memory for the upcoming exercise may allow the individual's body to efficiently and effectively reach the metabolic requirements for optimal exercise performance.

11.1.8 Case Study

11.1.8.1 Client Information

A 30-year-old female school teacher comes to your exercise physiology laboratory for follow-up exercise testing. Previously she performed maximal and submaximal exercise testing under your supervision upon entering an exercise program as part of her employer's faculty health promotion program. The exercise testing involved both estimation and production procedures to test her ability to accurately self-regulate exercise intensity at a target RPE corresponding to the VT. The data from her production trial revealed that she seemed to have difficulty self-regulating exercise intensity. In this instance, she was demonstrating a significant amount of exercise intensity self-regulation error, as indicated by VO_2 values quite dissimilar to that corresponding to the VT achieved during a maximal exercise test. Further examination of her ability to self-regulate exercise intensity is recommended to better allow her to achieve her goals for the exercise conditioning.

11.1.8.2 Assessments, Results, and Analysis

Administer the perceived exertion JND exercise testing procedures with metabolic measurement to investigate the individual's perceptual acuity using the VO_2 analog to the perceived exertion JND.

Using the perceived exertion JND procedures:

1. Determine JND-A.
2. Determine JND-B.
3. Calculate the JND range.

Using the production trial:

1. Test the individual's ability to self-regulate exercise intensity at the VT by comparing physiological values collected during the production protocol to the upper and lower limits of the JND range, i.e., use JND-A and JND-B where appropriate.
2. At regular intervals during the production protocol, determine if the individual's physiological response (VO_2) is inside or outside of the JND range and provide appropriate feedback in order to improve exercise intensity self-regulation.

11.2 Methods

11.2.1 *Cycle Ergometer Procedures*

11.2.1.1 Equipment

1. Adult OMNI-Cycle RPE Scale (Fig. 2.4)
2. Cycle ergometer
3. Metronome
4. HR monitor
5. Respiratory-metabolic measurement system

11.2.1.2 Pre-exercise Procedures

1. Measure height (cm) and weight (kg) of subject.
2. Read the standard instructions for the Adult OMNI-Cycle RPE Scale for RPE-L to the subject (Appendix B.4). Perform the memory anchoring procedure as described in Chap. 5.

11.2.1.3 Graded Exercise Test

1. Position the HR monitor and respiratory-metabolic mouthpiece (with head support unit and nose clip if applicable) on the subject.

2. Set the proper seat height on the cycle ergometer according to leg length. When the foot is flat on the right pedal and the pedal is in the extreme down position, the right knee should be in approximately 5 degrees of flexion.
3. Load-incremented protocol for electronically braked and friction-braked cycle ergometers:

   (a) Instruct the subject to maintain a 50 rev · min^{-1} pedal cadence. Set the metronome to 100 beats · min^{-1} so each downward movement of each foot is synchronized with a beat of the metronome. The subject may also use the digital monitor on the cycle control panel to regulate pedal cadence.
   (b) For electronically braked cycle ergometers (e.g., Lode), begin stage 1 at 50 W then increase the resistance 25 W every 2 min.
   (c) For friction-braked cycle ergometers (e.g., Monark), begin stage 1 at 1 kg resistance then increase the resistance 0.5 kg every 2 min.
   (d) When the subject cannot maintain the pedal cadence for 10 consecutive seconds owing to fatigue, terminate the exercise test.
   (e) Instruct the subject to estimate RPE starting at 1:30 of each exercise stage. Because the position of the respiratory-metabolic mouth piece inhibits a verbal response, instruct the subject to point to the numbers on the RPE scale, which should be conveniently positioned within the subject's arm reach. State aloud the numerical ratings for each momentary assessment to which the subject pointed and request a confirmatory nod that the number stated was correct. If incorrect, allow the subject to point to the appropriate rating on the RPE scale once more. Ask the subject to hold his or her finger on the appropriate number on the scale for approximately 1 s.
   (f) Record HR (beats · min^{-1}) at 1:55 of each exercise stage.
   (g) Record the final 15-s VO_2 (l · min^{-1}) for each exercise stage.
   (h) Record HRpeak as the highest HR value recorded during the final exercise stage or immediately post-exercise.
   (i) Record VO_2peak as the highest 15-s VO_2 value recorded at the end of the test.
   (j) Determine the VO_2 (l · min^{-1}) and %VO_2peak associated with the VT using the respiratory-metabolic measurement system automatic VT calculator. The perceived exertion JND should be calculated for RPE-VT.
   (k) Record the PO (W) or brake resistance (kg) of the exercise test stage during which the VT occurred. For Monark cycle ergometers with a 6-m flywheel, multiply the brake resistance by 50 to convert the value from kg to W. This PO will serve as the standard stimulus during the JND trial.

11.2.1.4 Estimation Protocol: Data Organization and Analysis

1. In a Microsoft Excel spreadsheet, label columns of data for the following variables: Exercise Stage, VO_2 (l · min^{-1}), OMNI RPE-L.

2. If the respiratory-metabolic measurement system does not automatically calculate VT or if instruction on manual calculation and visual identification of the VT is desired, refer to Appendix D for detailed explanation of the following:
   (a) Calculation of $V_E \cdot VO_2^{-1}$ and $V_E \cdot VCO_2^{-1}$.
   (b) Plot of $V_E \cdot VO_2^{-1}$ and $V_E \cdot VCO_2^{-1}$ for visual identification of the VT using the ventilatory equivalent method.
   (c) Adjustment of automatic VT calculation using the appropriate application on the automated respiratory-metabolic measurement system.
3. Plot of VO_2 and OMNI RPE-L for determination of OMNI RPE-VT.
   (a) Click on the **INSERT** tab and in the **CHARTS** section click on **SCATTER**. Select the first available chart option. A blank or example scatter plot will appear on your screen.
   (b) Click on the **SELECT DATA** tab. Remove any entries found in the **LEGEND ENTRIES** text box then click **ADD**. Under **SERIES NAME**, enter VO_2 and RPE-L. Then click on the icon to the right of the **SERIES X VALUES** text box and highlight the VO_2 values. After the values are highlighted click the icon on the box that appeared. Then click on the icon to the right of the **SERIES Y VALUES** text box and highlight the RPE-L values. After the values are highlighted click the icon on the box that appeared. Click **OK** on the next two screens.
   (c) You should now have a scatter plot with RPE-L on the *y*-axis and VO_2 on the *x*-axis. Create a title for the plot and enter the appropriate axis labels and units of measure.
   (d) To determine RPE-VT, click on one of the data points to highlight the entire data series. Right click on one of the data points and a menu will appear. Click **ADD TRENDLINE** and the **FORMAT TRENDLINE** menu will appear. Select **LINEAR** and **DISPLAY EQUATION ON CHART** then click **CLOSE**. The trendline and equation will be displayed on the chart.
   (e) Use this linear equation to calculate RPE-VT. Use VO_2 ($l \cdot min^{-1}$) corresponding to the VT as the "x" value in the equation and solve for "y." The calculated "y" value, once rounded to the nearest whole integer, is the RPE-VT.
4. An example of these procedures with a screenshot depicting each step as performed using Microsoft Excel 2013 can be found in Appendix D.

11.2.1.5 JND Protocol

1. Position the HR monitor and respiratory-metabolic mouthpiece (with head support unit and nose clip if applicable) on the subject.
2. Set the proper seat height on the cycle ergometer according to leg length. When the foot is flat on the right pedal and the pedal is in the extreme down position, the right knee should be in approximately 5 degrees of flexion.

3. Instruct the subject to maintain a 50 rev · min^{-1} pedal cadence. Set the metronome to 100 beats · min^{-1} so each downward movement of each foot is synchronized with a beat of the metronome. The subject may also use the digital monitor on the cycle control panel to regulate pedal cadence.
4. Prior to bout 1 of the standard stimulus, instruct the subject to perform a 2-min warm-up.
   (a) For electronically braked cycle ergometers (e.g., Lode), set the PO at 25 W.
   (b) For friction-braked cycle ergometers (e.g., Monark), set the brake resistance at 0.5 kg.
   (c) During the second minute of the warm-up, gradually ramp the exercise workload (W or kg) up to the predetermined SS that was determined as the workload for the graded exercise test stage during which the VT occurred.
5. Standard stimulus bout 1
   (a) Instruct the subject to perform the standard intensity for 5 min.
   (b) Record the final 30-s average VO_2 (l · min^{-1}).
   (c) The subject should rest for 5 min.
6. Prior to the JND-A bout, instruct the subject to perform the same 2-min warm-up as performed prior to standard stimulus bout 1.
7. JND-A bout (5 min)
   (a) Instruct the subject to perform the standard intensity for 2 min.
   (b) During minutes 3 and 4, instruct the subject to adjust the resistance so the level of exertion is just noticeably above that of the standard. Remind the subject every 30 s that he/she may adjust the resistance as necessary to attain the JND. Do not allow further adjustment during minute 5.
   (c) During minute 5, instruct the subject to continue pedaling at the self-selected intensity.
   (d) Record the final 30-s average VO_2 (l · min^{-1}).
   (e) Record the final PO.
   (f) Then allow the subject to rest for 5 min.
8. Prior to standard stimulus bout 2, instruct the subject to perform the same 2-min warm-up as performed prior to standard stimulus bout 1.
9. Standard stimulus bout 2
   (a) Instruct the subject to perform the standard intensity for 5 min.
   (b) Record the final 30-s average VO_2 (l · min^{-1}).
   (c) Then allow the subject to rest for 5 min.
10. Prior to the JND-B bout, instruct the subject to perform the same 2-min warm-up as performed prior to standard stimulus bout 1.
11. JND-B bout (5 min)
    (a) Instruct the subject to perform the standard intensity for 2 min.

(b) During minutes 3 and 4, instruct the subject to adjust the resistance so the level of exertion is just noticeably below that of the standard. Remind the subject every 30 s that he/she may adjust the resistance as necessary to attain the JND. Do not allow further adjustment during minute 5.
(c) During minute 5, instruct the subject to continue pedaling at the self-selected intensity.
(d) Record the final 30-s average VO_2 ($l \cdot min^{-1}$).
(e) Record the final PO.

12. Instruct the subject to perform a 2-min cool-down.
    (a) For electronically braked cycle ergometers (e.g., Lode), set the PO at 25 W.
    (b) For friction-braked cycle ergometers (e.g., Monark), set the brake resistance at 0.5 kg.

11.2.1.6 JND Protocol: Data Organization and Analysis

1. JND-A Calculations
    (a) Calculate JND-A VO_2 ($l \cdot min^{-1}$) by subtracting the final 30-s average VO_2 of standard stimulus bout 1 from the final 30-s average VO_2 of the JND-A bout.
    (b) Calculate JND-A PO (W) for electronically braked cycle ergometers by subtracting the standard stimulus PO from the final PO of the JND-A bout.
    (c) Calculate JND-A PO for friction-loaded cycle ergometers by subtracting the standard stimulus brake resistance from the final resistance of the JND-A bout. For Monark cycle ergometers with a 6-m flywheel, multiply by 50 to convert the value from kg of resistance to W.
2. JND-B Calculations
    (a) Calculate JND-B VO_2 ($l \cdot min^{-1}$) by subtracting the final 30-s average VO_2 of the JND-B bout from the final 30-s average VO_2 of standard stimulus bout 2.
    (b) Calculate JND-B PO (W) for electronically braked cycle ergometers by subtracting the final PO from the JND-B bout from the standard stimulus PO.
    (c) Calculate JND-B PO for friction-loaded cycle ergometers by subtracting the final brake resistance from the JND-B bout from the standard stimulus brake resistance. For Monark cycle ergometers with a 6-m flywheel, multiply by 50 to convert the value from kg of resistance to W.
3. JND Range Calculations
    (a) To calculate the upper limit of the VO_2 Range for the JND, add JND-A VO_2 to the VO_2 corresponding to the VT.
    (b) To calculate the lower limit of the VO_2 Range for the JND, subtract JND-B VO_2 from the VO_2 corresponding to the VT.
    (c) To calculate the upper limit of the PO Range for the JND, add JND-A PO to the PO corresponding to the VT.
    (d) To calculate the lower limit of the PO Range for the JND, subtract JND-B PO from the PO corresponding to the VT.

11.2.1.7 Production Protocol

1. Position the HR monitor and respiratory-metabolic mouthpiece (with head support unit and nose clip if applicable) on the subject.
2. Set the proper seat height on the cycle ergometer according to leg length. When the foot is flat on the right pedal and the pedal is in the extreme down position, the right knee should be in approximately 5 degrees of flexion.
3. Instruct the subject to maintain a 50 $rev \cdot min^{-1}$ pedal cadence. Set the metronome to 100 $beats \cdot min^{-1}$ so each downward movement of each foot is synchronized with a beat of the metronome. The subject may also use the digital monitor on the cycle control panel to regulate pedal cadence.
4. Instruct the subject to perform a 2-min warm-up.

    (a) For electronically braked cycle ergometers (e.g., Lode), set the power output at 25 W.
    (b) For friction-braked cycle ergometers (e.g., Monark), set the brake resistance at 0.5 kg.

5. Following the warm-up, instruct the subject to exercise for 10 min at the predetermined target RPE-VT by adjusting the PO until the desired intensity is met.
6. Record HR ($beats \cdot min^{-1}$) every 2 min.
7. Record the final 15-s VO_2 ($l \cdot min^{-1}$) for each 1-min segment of exercise.
8. Following the 10-min exercise bout, instruct the subject to perform a 2-min cool-down period.

    (a) For electronically braked cycle ergometers (e.g., Lode), set the PO at 25 W.
    (b) For friction-braked cycle ergometers (e.g., Monark), set the brake resistance at 0.5 kg.

9. It may be necessary to have the subject perform a second production trial during which feedback is given regarding the accuracy of exercise intensity self-regulation. Following each 1-min segment of this additional exercise trial, the VO_2 associated with that time-point can be used immediately to determine whether the subject is within, above, or below the JND Range. Feedback can then be provided to the subject informing him/her that the current exercise intensity being self-regulated is correct, too high, or too low. Then, the effect of feedback on the frequency of exercise intensity self-regulation error can then be evaluated. If error in intensity self-regulation is still present, further practice and feedback is necessary.

11.2.1.8 Production Protocol: Data Organization and Analysis

1. In a Microsoft Excel spreadsheet, label columns of data for the following variables: Exercise Time (minutes), VO_2 ($l \cdot min^{-1}$).
2. Count the number of minutes in which the subject's VO_2 was within the JND VO_2 Range. Divide that number by 10 then multiple by 100 to determine the percent of time the subject exercised within the JND VO_2 Range.

3. Count the number of minutes in which the subject's VO_2 was below the JND VO_2 Range. Divide that number by 10 then multiple by 100 to determine the percent of time the subject exercised below the JND VO_2 Range.
4. Count the number of minutes in which the subject's VO_2 was higher than the JND VO_2 Range. Divide that number by 10 then multiple by 100 to determine the percent of time the subject exercised above the JND VO_2 Range.

11.3 Laboratory Discussion Questions

1. How does pacing strategy influence an athlete's race performance?
2. How is the psychophysiological feedback mechanism that influences performance pace influenced by mechanoreceptors, metaboreceptors, and chemoreceptors referenced by Lambert et al. (2005)?
3. Describe how an individual who is demonstrating intensity self-regulation error would benefit from an orientation trial that included the JND exercise procedures. Your description should be based on responses derived from an estimation protocol and a production protocol.
4. In a previous investigation that studied the perceived exertion JND for an exercise intensity corresponding to an OMNI RPE of 5, JND-A was found to be significantly smaller than JND-B. This indicated a more sensitive perceptual acuity at an intensity just above an OMNI RPE 5 than for an intensity just below an OMNI RPE 5. What is the meaning of these data for intensity self-regulation to produce a target RPE?
5. What psychophysiological mechanisms might explain why the perceived exertion JND differs between very low exercise intensity and an exercise intensity corresponding to the ventilatory threshold?
6. What mechanisms explain why the perceived exertion JND might be different between individuals of low fitness compared with elite runners?
7. Why would the perceived exertion JND for a specific absolute exercise intensity change after an intense, 6-month training program?

References

Abbiss CR, Laursen PB. Describing and understanding pacing strategies during athletic competition. Sports Med. 2008;38:239–52.

Albertus Y, Tucker R, St Clair Gibson A, Lambert EV, Hampson DB, Noakes TD. Effect of distance feedback on pacing strategy and perceived exertion during cycling. Med Sci Sports Exerc. 2005;37:461–8.

Amann M, Proctor LT, Sebranek JJ, Pegelow DF, Dempsey JA. Opioid-mediated muscle afferents inhibit central motor drive and limit peripheral muscle fatigue development in humans. J Physiol. 2009;587:271–83.

Baden DA, McLean TL, Tucker R, Noakes TD, St Clair Gibson A. Effect of anticipation during unknown or unexpected exercise duration on rating of perceived exertion, affect, and physiological function. Br J Sports Med. 2005;39:742–6.

Buckworth J, Dishman R, O'Connor P, Tomporowski P. Exercise psychology. 2nd ed. Champaign, IL: Human Kinetics; 2013.

de Koning JJ, Foster C, Bakkum A, Kloppenburg S, Thiel C, Joseph T, Cohen J, Porcari JP. Regulation of pacing strategy during athletic competition. PLoS One. 2011;6:e15863.

Haile L, Robertson RJ, Nagle EF, Krause MP, Gallagher Jr M, Ledezma CM, Wisniewski KS, Shafer AB, Goss FL. Just noticeable difference in perception of physical exertion during cycle exercise in young adult men and women. Eur J Appl Physiol. 2013;113:877–85.

Hettinga FJ, de Koning JJ, Schmidt LJ, Wind NA, Macintosh BR, Foster C. Optimal pacing strategy: from theoretical modeling to reality in 1500-m speed skating. Br J Sports Med. 2011;45:30–5.

Lambert EV, St Clair Gibson A, Noakes TD. Complex systems model of fatigue: integrative homeostatic control of peripheral physiological systems during exercise in humans. Br J Sports Med. 2005;39:52–62.

Noble BJ, Robertson RJ. Perceived exertion. Champaign, IL: Human Kinetics; 1996.

Roelands B, Goekint M, Buyse L, Pauwels F, De Schutter G, Piacentini F, Hasegawa H, Watson P, Meeusen R. Time trial performance in normal and high ambient temperature: is there a role for 5-HT? Eur J Appl Physiol. 2009;107:119–26.

Roelands B, de Koning J, Foster C, Hettinga F, Meeusen R. Neurophysical determinants of theoretical concepts and mechanisms involved in pacing. Sports Med. 2013;43:301–11.

St Clair Gibson A, Goedecke JH, Harley YX, Myers LJ, Lambert MI, Noakes TD, Lambert EV. Metabolic setpoint control mechanisms in different physiological systems at rest and during exercise. J Theor Biol. 2005;236:60–72.

Ulmer HV. Concept of an extracellular regulation of muscular metabolic rate during heavy exercise in humans by psychophysiological feedback. Experientia. 1996;52:416–20.

Wozniak RH. Classics in psychology, 1855–1914: historical essays. Bristol, UK: Thoemmes; 1999.

Chapter 12
Self-Selected Versus Imposed Exercise Intensities

The traditional method of prescribing the intensity of exercise is based on the scientific evidence that has shaped national PA guidelines regarding the overload training stimulus necessary to elicit health-fitness benefits. In this prescriptive paradigm, the health-fitness professional uses a GXT to determine target HR(s) or RPE(s) corresponding to a specific physiological threshold, such as the VT, or a range based on %VO_2max. The individual is instructed to self-regulate exercise at the prescribed intensity. This procedure can be termed *imposed* exercise because the individual does not choose the intensity. This paradigm ignores an individual's exercise intensity preference and may result in negative emotions that could decrease adherence. Allowing individuals to choose their own exercise intensity (i.e., perform self-selected exercise) has the potential to improve PA participation. In addition, research has shown that many individuals will choose to exercise at an intensity near the VT. Self-selected exercise may be an important link in the chain between the adoption and maintenance of regular PA that elicits both psychological and physiological benefits.

12.1 Background

12.1.1 Traditional Imposed Exercise Intensity Prescription

It is common practice within the health-fitness domain to employ exercise prescription procedures that are based on the scientific evidence approved by national organizations such as the American College of Sports Medicine (ACSM) and the Centers for Disease Control and Prevention (CDC). PA guidelines have been proposed that define the intensity, duration, and frequency of an exercise program designed to elicit health-fitness benefits (ACSM 2013). Traditional exercise programs based on these guidelines set specific intensity prescriptions that target physiological ranges,

L. Haile et al., *Perceived Exertion Laboratory Manual*,
DOI 10.1007/978-1-4939-1917-8_12

often expressed as a %VO_2max, and thresholds such as the VT. It has been suggested that a target RPE can be used to self-regulate exercise intensity to produce optimal health-fitness training outcomes based on physiological criteria. Various methods have been employed to confirm the validity of exercise intensity self-regulation using target RPE's, such as estimation–production paradigms for the assessment of prescription congruence and intensity discrimination.

However, prescribing exercise intensity based on guidelines that impose the intensity to be performed ignores an individual's preference for and tolerance of certain exercise intensities. It has been shown that increases in exercise intensity are associated with decreases in affective responses (AR) (Ekkekakis and Petruzzello 1999). This decrease in the pleasure derived from exercise can lower adherence to the prescribed exercise program (Cox et al. 2003; Lee et al. 1996; Perri et al. 2002). Based on these results, many individuals may not benefit from an initial exercise prescription where an imposed exercise intensity is employed. The goal of an exercise program should not only be health-fitness benefits, but should also include the development of one's willingness to participate in the activity. Therefore, one of the initial goals of an exercise prescription should be to promote program adherence. The adoption and maintenance of regular exercise will facilitate the ultimate goal of attaining the desired health-fitness benefits.

12.1.2 Imposed Versus Self-Selected Exercise Intensity Prescription

Prescribed exercise intensity can also be referred to as *imposed exercise intensity*. In this prescriptive scenario, the individual does not have a choice regarding the exercise intensity that he/she will perform. The imposition of exercise intensity can be seen by some individuals as highly controlling, negatively influencing AR and exercise adherence. Based on the theory of self-determination, engaging in an externally controlled behavior, such as performing imposed exercise intensity, decreases the intrinsic motivation to continue that behavior. Giving an individual the freedom to self-select the exercise intensity that will be performed may increase motivation for continued participation (Ryan and Deci 2000; Lind et al. 2008). Increased control over exercise intensity can lead to a greater enjoyment associated with exercise (Wankel 1993). This is important because of the positive link between exercise enjoyment and program adherence (Ryan and Deci 2000; Ryan et al. 1997; Caserta and Gillett 1998).

It has been proposed that a causal chain exists, linking (a) the intensity of PA (not only its level but whether it is self-selected or imposed), (b) AR to exercise (e.g., measured using Feeling Scale ratings), and (c) regular PA participation and/or adherence to exercise programming. Ekkekakis and Lind (2006) summarized the comparison between SS and imposed exercise intensity. They stated that during self-selected exercise, a sense of control is maintained such that the individual can avoid physical discomfort and fatigue. Therefore, self-selected exercise intensity may seem like a somewhat innocuous concept, negative emotions (e.g., social physique anxiety) may not arise, and a positive AR results. In contrast, when exercise intensity is imposed, personal control is taken away and overt signs of fatigue and discomfort may occur. The imposition of exercise intensity may be seen as posing a potential evaluative threat where negative emotions may arise, leading to decreased AR (Ekkekakis and Lind 2006). Studies have shown that imposed exercise intensities elicit a more negative (or comparatively less positive) AR when compared to self-selected exercise intensities (Ekkekakis and Lind 2006; Lind et al. 2008; Parfitt et al. 2000; Parfitt et al. 2006; Rose and Parfitt 2007).

It has been shown that for most individuals, self-selected exercise intensities are adequate to provide an overload training stimulus to promote gains in cardiorespiratory fitness while simultaneously providing a level of effort that results in a positive AR. Previous studies have found that, on average, self-selected exercise intensity is often similar to the VT or LT. Although many subjects chose self-selected intensity near the VT or LT, relatively few exceeded that level (Dishman et al. 1994; Ekkekakis and Lind 2006; Lind et al. 2005; Lind et al. 2008). Therefore, a prescription that employs self-selected exercise intensity may be beneficial for overall health-fitness without negatively affecting mood state or emotions.

12.1.3 RPE During Self-Selected Exercise

RPE has been used to quantify the perception of exertional intensity during self-selected exercise. In a 20-min cycle ergometer exercise bout, both high- and low-active men estimated similar Borg Scale RPE's at 5-min intervals. RPE increased from an average of 10.5 at minute 5 to an average of 14.2 at minute 20 (Dishman et al. 1994). In a 20-min bout of treadmill exercise in which speed was adjusted to produce self-selected intensity, previously sedentary middle-aged women estimated Borg RPE's averaging approximately 11 at minute 5 and increasing to an average of almost 14 at minute 20 (Lind et al. 2005). Using the OMNI Scale during a 20-min bout of cycle ergometer exercise, college-aged males estimated RPE's averaging 2 at minute 5 and increasing to an average of 5 at minute 20 (Haile et al. 2013). In each instance, subjects were allowed to adjust intensity every 5 min. In general, subjects increased intensity throughout the 20-min exercise bouts and, subsequently, RPE increased as well. On average, RPE's measured during the later stages of the exercise bouts were within the perceived exertion range that encompasses the VT in most individuals, i.e., from 11 to 16 using the Borg Scale and from 5 to 7 using the OMNI Scale.

12.1.4 RPE During Self-Selected Versus Imposed Exercise Intensity

Parfitt and colleagues (2000) conducted an investigation in which RPE was compared between self-selected and imposed treadmill exercise in young adult males and females. Imposed exercise intensity was set at 65 % of VO_2max and subjects self-selected exercise intensity at 71 % VO_2max. Even though exercise intensity was significantly different between trials, RPE's were similar. The exertion associated with the harder work rate during self-selected exercise intensity was perceived to be the same as for the lower, prescribed intensity. These findings indicate a potential positive perception of the self-selected exercise or conversely a negative perception of the imposed exercise (Parfitt et al. 2000).

The similarity in perceptual responses between self-selected and imposed exercise is important to consider when these formats are used for prescriptive purposes. If an individual feels that the level of exertion is lower when self-selected exercise is performed or higher when imposed exercise is performed, the individual may be underestimating or overestimating exertional intensity as it relates to physiological values. However, regardless of these subtle changes in the RPE response, an important aspect of self-selected and imposed exercise to consider is the individual's adherence to the exercise program. Some individuals may have better adherence during an initial exercise program involving self-selected exercise, yet others may prefer that the exercise professional sets a specific intensity based on established PA guidelines. Therefore, it is important to measure perceptual and psychosocial variables (i.e., RPE, pain, affect, enjoyment) during both self-selected and imposed exercise intensities before determining optimal exercise programming, especially in previously sedentary individuals.

12.1.5 *Case Study*

12.1.5.1 Client Information

A 44-year-old female enters your physical fitness program. She explains that in the past she exercised at home following her own pace but has discontinued the program. Now she feels that advice of a professional is needed in order to restart her exercise program. Recently, she has been attempting to get back into shape using the advice of a personal trainer at a commercial facility, but she complains that the trainer often pushed her too hard. She tells you that the trainer was a nice person, but she could not enjoy exercising when the trainer kept telling her to exercise at a higher intensity than she preferred.

12.1.5.2 Assessments, Results and Analysis

Have the subject perform a load-incremented estimation trial to determine the VT. Then, have the subject perform two submaximal exercise trials: one at self-selected exercise intensity and one at imposed exercise intensity. Measure VO_2 and RPE during each trial to determine the mode of exercise intensity most suitable for the subject based on perceived exertion responses and personal preference.

Using the estimation trial, determine VO_2, HR and RPE at the VT.
Using the imposed exercise intensity trial, determine the average VO_2, HR and RPE.
Using the self-selected exercise intensity trial, determine average VO_2, HR, and RPE.

12.2 Methods

12.2.1 Treadmill Procedures

12.2.1.1 Equipment

1. Adult OMNI-Walk/Run RPE Scale (Fig. A.2)
2. Treadmill
3. HR monitor
4. Respiratory-metabolic measurement system

12.2.1.2 Pre-exercise Procedures

1. Measure height (cm) and weight (kg) of subject.
2. Perform the memory anchoring procedure as described in Chap. 5. Read the standard instructions to the subject for the Adult OMNI-Walk/Run RPE Scale to measure RPE-O (Appendix B.1). If measurement of differentiated RPE (RPE-L and RPE-C) is desired, read the standard instructions for the Adult OMNI-Walk/Run RPE Scale for undifferentiated and differentiated RPE (Appendix B.2).

12.2.1.3 Graded Exercise Test

1. Position the HR monitor and respiratory-metabolic mouthpiece (with head support unit and nose clip if applicable) on the subject.
2. Instruct the subject to step onto the treadmill and review exercise termination procedures: When the subject cannot continue exercise due to exhaustion or discomfort, he/she should grasp the treadmill hand rails, at which time the test administrator will gradually slow the treadmill down for performance of a cool-down. The subject should be reminded not to step off the treadmill belt while it is still in motion.
3. Bruce Multistage Treadmill Test Protocol: this can be performed by manually adjusting treadmill speed and grade or using a program on a computer that is interfaced to the treadmill.

   (a) Begin the warm-up at 1.5 miles $\cdot$ h^{-1} and 0 % grade for 3 min.
   (b) Each exercise test stage will last for 3 min. The stages progress as follows:

Stage 1—1.7 miles $\cdot$ h^{-1} and 10 % grade
Stage 2—2.5 miles $\cdot$ h^{-1} and 12 % grade
Stage 3—3.4 miles $\cdot$ h^{-1} and 14 % grade
Stage 4—4.2 miles $\cdot$ h^{-1} and 16 % grade
Stage 5—5.0 miles $\cdot$ h^{-1} and 18 % grade
Stage 6—5.5 miles $\cdot$ h^{-1} and 20 % grade
Stage 7—6.0 miles $\cdot$ h^{-1} and 22 % grade
Stage 8—6.5 miles $\cdot$ h^{-1} and 24 % grade

(c) When the subject cannot continue any longer owing to exhaustion, terminate the exercise test by initiating the cool-down period at 1.5 miles $\cdot$ h^{-1} and 0 % grade. The cool-down should be 5 min in duration.
(d) Ask the subject to estimate RPE starting at 2:30 of each exercise stage using the OMNI Scale (RPE-O). Because the position of the respiratory-metabolic mouth piece inhibits a verbal response, instruct the subject to point to the numbers on the RPE scale, which should be conveniently positioned within the subject's arm reach. State aloud the numerical ratings for each momentary assessment to which the subject pointed and request a confirmatory nod that the number stated was correct. If incorrect, allow the subject to point to the appropriate rating on the RPE scale once more. Ask the subject to hold his or her finger on the appropriate number on the scale for approximately 1 s.
(e) Record HR (b $\cdot$ min^{-1}) at 2:55 of each exercise stage.
(f) Record the final 15-s VO_2 (ml $\cdot$ kg^{-1} $\cdot$ min^{-1}) for each exercise stage.
(g) Record HRmax as the highest HR value recorded during the final exercise stage or immediately post-exercise.
(h) Record VO_2max as the highest 15-s VO_2 value recorded at the end of the test.
(i) Determine the VO_2 (ml $\cdot$ kg^{-1} $\cdot$ min^{-1}) and %VO_2max associated with the VT using the respiratory-metabolic measurement system automatic VT calculator.

12.2.1.4 Estimation Protocol: Data Organization and Analysis

1. In a Microsoft Excel spreadsheet, label columns of data for the following variables: Exercise Stage, VO_2 (ml $\cdot$ kg^{-1} $\cdot$ min^{-1}), OMNI RPE-O, HR (b $\cdot$ min^{-1}). Include columns for OMNI RPE-L and OMNI RPE-C if applicable.
2. If the respiratory-metabolic measurement system does not automatically calculate VT or if instruction on manual calculation and visual identification of the VT is desired, refer to Appendix D for a detailed explanation for the following:

   (a) Calculation of $V_E \cdot VO_2^{-1}$ and $V_E \cdot VCO_2^{-1}$.
   (b) Plot of $V_E \cdot VO_2^{-1}$ and $V_E \cdot VCO_2^{-1}$ for visual identification of the VT using the ventilatory equivalent method.
   (c) Adjustment of automatic VT calculation using a respiratory-metabolic measurement system.

3. Plot of VO_2 as a function of OMNI RPE-O for determination of OMNI RPE-VT.
   (a) Click on the **INSERT** tab and in the **CHARTS** section click on **SCATTER**. Select the first available chart option. A blank or example scatter plot will appear on your screen.
   (b) Click on the **SELECT DATA** tab. Remove any entries found in the **LEGEND ENTRIES** text box then click **ADD**. Under **SERIES NAME**, enter VO_2 and OMNI RPE-O. Then click on the icon to the right of the **SERIES X VALUES** text box and highlight the VO_2 values. After the values are highlighted click the icon on the box that appeared. Then click on the icon to the right of the **SERIES Y VALUES** text box and highlight the OMNI RPE-O values. After the values are highlighted click the icon on the box that appeared. Click **OK** on the next two screens.
   (c) You should now have a scatter plot with OMNI RPE-O on the y-axis and VO_2 on the x-axis. Create a title for the plot and enter the appropriate axis labels and units of measure.
   (d) To determine OMNI RPE-VT, click on one of the data points to highlight the entire data series. Right click on one of the data points and a menu will appear. Click **ADD TRENDLINE** and the **FORMAT TRENDLINE** menu will appear. Select **LINEAR** and **DISPLAY EQUATION ON CHART** then click **CLOSE**. The trendline and equation will be displayed on the chart.
   (e) Use this linear equation to calculate RPE-VT. Use the previously determined VO_2 ($ml \cdot kg^{-1} \cdot min^{-1}$) corresponding to the VT as the "x" value in the equation and solve for "y." The calculated "y" value, once rounded to the nearest whole integer, is the OMNI RPE-VT.
4. Repeat the above steps for VO_2 and HR to determine HR-VT.
5. An example of these procedures with a screenshot depicting each step as performed using Microsoft Excel 2013 can be found in Appendix D.

12.2.1.5 Imposed Exercise Intensity Trial

1. Position the HR monitor and respiratory-metabolic mouthpiece (with head support unit and nose clip if applicable) on the subject.
2. Instruct the subject to step onto the treadmill and review exercise termination procedures: When the subject cannot continue exercise due to exhaustion or discomfort, he/she should grasp the treadmill hand rails, at which time the test administrator will slow down or stop the treadmill. The subject should be reminded not to step off the treadmill belt while it is still in motion.
3. Start the treadmill at 3 $miles \cdot h^{-1}$ and 0 % grade for a 2-min warm-up.
4. Following the warm-up, instruct the subject that he/she will exercise for 10 min at the predetermined target VO_2 corresponding to the VT, which will be achieved by adjusting the treadmill speed.
   (a) Increase the speed of the treadmill to 5 $miles \cdot h^{-1}$ immediately following the warm-up. Then monitor every 15-s average VO_2 ($ml \cdot kg \cdot min^{-1}$) on the computer display.

(b) If VO_2 remains below the VT after 1 min, increase the treadmill speed by 0.5 miles · h^{-1}. Continue to increase speed by 0.5 miles · h^{-1} every 30 s until VO_2 is within 2–3 ml · kg^{-1} · min^{-1} of VO_2 corresponding to the VT.
(c) If VO_2 increases to a level above the VT, make adjustments by decreasing treadmill speed in increments of 0.1 miles · h^{-1} every 30 s to fine-tune exercise intensity.

5. Record treadmill speed every minute.
6. Record HR (b · min^{-1}) every 2 min.
7. Record the final 15-s VO_2 in ml · kg^{-1} · min^{-1} for each 2-min segment of exercise.
8. Instruct the subject to estimate RPE starting at 1:30 of each 2-min segment of exercise using the OMNI Scale (RPE-O).
9. Following the 10-min exercise, the treadmill should be set to 3 miles · h^{-1} and 0 % grade for a 2-min cool-down period.

12.2.1.6 Self-Selected Exercise Intensity Trial

1. Position the HR monitor and respiratory-metabolic mouthpiece (with head support unit and nose clip if applicable) on the subject.
2. Instruct the subject to step onto the treadmill and review exercise termination procedures: When the subject cannot continue exercise due to exhaustion or discomfort, he/she should grasp the treadmill hand rails, at which time the test administrator will slow down or stop the treadmill. The subject should be reminded not to step off the treadmill belt while it is still in motion.
3. Read the following instructions to the subject for self-selected exercise (Dishman et al. 1994; Parfitt et al. 2000): "You will be allowed to select an intensity you prefer to perform on the treadmill. This should be an intensity that you would choose for a 10-min workout if you were participating in a fitness program. The intensity should be high enough that you would get a good workout, but not so high that you would not prefer to exercise at that intensity daily or at least every other day. It should be an intensity that is appropriate for you."
4. Start the treadmill at 3 miles · h^{-1} and 0 % grade for a 2-min warm-up.
5. Following the warm-up, instruct the subject to exercise for 10 min at a SS exercise intensity by adjusting treadmill speed whenever desired.
6. Record treadmill speed every minute.
7. Record HR (b · min^{-1}) every 2 min.
8. Record the final 15-s VO_2 in ml · kg^{-1} · min^{-1} for each 2-min segment of exercise.
9. Ask the subject to estimate RPE starting at 1:30 of each 2-min segment of exercise using the OMNI Scale (RPE-O).
10. Following the 10-min exercise, the treadmill should be set to 3 miles · h^{-1} and 0 % grade for a 2-min cool-down period.

12.2.1.7 Imposed and Self-Selected Exercise Intensity Trials: Data Organization and Analysis

1. For the imposed exercise intensity trial:
   (a) Calculate the average treadmill speed (miles · h^{-1}) using the data from each minute of exercise.
   (b) Calculate the average HR (b · min^{-1}) using each 2-min HR response.
   (c) Calculate the average OMNI RPE-O using each 2-min RPE response.
   (d) Calculate the average VO_2 (ml · kg^{-1} · min^{-1}) using each 2-min VO_2.
2. For the self-selected exercise intensity trial:
   (a) Calculate the average treadmill speed (miles · h^{-1}) using the data from each minute of exercise.
   (b) Calculate the average HR (b · min^{-1}) using each 2-min HR response.
   (c) Calculate the average OMNI RPE-O using each 2-min RPE response.
   (d) Calculate the average VO_2 (ml · kg^{-1} · min^{-1}) using each 2-min VO_2.

12.2.2 Cycle Ergometer Procedures

12.2.2.1 Equipment

1. Adult OMNI-Cycle RPE Scale (Fig. 2.4)
2. Cycle ergometer
3. Metronome
4. HR monitor
5. Respiratory-metabolic measurement system

12.2.2.2 Pre-exercise Procedures

1. Measure height (cm) and weight (kg) of subject.
2. Read the standard instructions for the Adult OMNI-Cycle RPE Scale for RPE-L to the subject (Appendix B.4). If the measurement of undifferentiated (RPE-O) and differentiated RPE for chest/breathing (RPE-C) is desired, read the standard instructions for the Adult OMNI-Cycle RPE Scale for undifferentiated and differentiated RPE (Appendix B.5). Perform the memory anchoring procedure as described in Chap. 5.

12.2.2.3 Graded Exercise Test

1. Position the HR monitor and respiratory-metabolic mouthpiece (with head support unit and nose clip if applicable) on the subject.

2. Set the proper seat height on the cycle ergometer according to leg length. When the foot is flat on the right pedal and the pedal is in the extreme down position, the right knee should be in approximately 5 degrees of flexion.
3. Load-incremented protocol for electronically braked and friction-braked cycle ergometers:

   (a) Instruct the subject to maintain a 50 $rev \cdot min^{-1}$ pedal cadence. Set the metronome to 100 $beats \cdot min^{-1}$ so each downward movement of each foot is synchronized with a beat of the metronome. The subject may also use the digital monitor on the cycle control panel to regulate pedal cadence.
   (b) For electronically braked cycle ergometers (e.g., Lode), begin stage 1 at 50 W then increase the resistance 25 W every 2 min.
   (c) For friction-braked cycle ergometers (e.g., Monark), begin stage 1 at 1 kg resistance then increase the resistance 0.5 kg every 2 min.
   (d) When the subject cannot maintain the pedal cadence for 10 consecutive seconds owing to fatigue, terminate the exercise test.
   (e) Instruct the subject to estimate RPE starting at 1:30 of each exercise stage using the OMNI Scale (RPE-L). Because the position of the respiratory-metabolic mouth piece inhibits a verbal response, instruct the subject to point to the numbers on the RPE scale, which should be conveniently positioned within the subject's arm reach. State aloud the numerical ratings for each momentary assessment to which the subject pointed and request a confirmatory nod that the number stated was correct. If incorrect, allow the subject to point to the appropriate rating on the RPE scale once more. Ask the subject to hold his or her finger on the appropriate number on the scale for approximately 1 s.
   (f) Record HR ($b \cdot min^{-1}$) at 1:55 of each exercise stage.
   (g) Record the final 15-s VO_2 in $l \cdot min^{-1}$ for each exercise stage.
   (h) Record HRpeak as the highest HR value recorded during the final exercise stage or immediately post-exercise.
   (i) Record VO_2peak as the highest 15-s VO_2 value recorded at the end of the test.
   (j) Determine the VO_2 ($l \cdot min^{-1}$) and %VO_2peak associated with the VT using the respiratory-metabolic measurement system automatic VT calculator.

12.2.2.4 Estimation Protocol: Data Organization and Analysis

1. In a Microsoft Excel spreadsheet, label columns of data for the following variables: Exercise Stage, VO_2 ($l \cdot min^{-1}$), OMNI RPE-L, HR ($b \cdot min^{-1}$). Include columns for OMNI RPE-O and OMNI RPE-C if applicable.
2. If the respiratory-metabolic measurement system does not automatically calculate VT or if instruction on manual calculation and visual identification of the VT is desired, refer to Appendix D for detailed instructions for the following:

   (a) Calculation of $V_E \cdot VO_2^{-1}$ and $V_E \cdot VCO_2^{-1}$.

(b) Plot of $V_E \cdot VO_2^{-1}$ and $V_E \cdot VCO_2^{-1}$ for visual identification of the VT using the ventilatory equivalent method.
(c) Adjustment of automatic VT calculation using a computer application available for most respiratory-metabolic measurement systems.

3. Plot of VO_2 and OMNI RPE-L for determination of OMNI RPE-VT.

(a) Click on the **INSERT** tab and in the **CHARTS** section click on **SCATTER**. Select the first available chart option. A blank or example scatter plot will appear on your screen.
(b) Click on the **SELECT DATA** tab. Remove any entries found in the **LEGEND ENTRIES** text box then click **ADD**. Under **SERIES NAME**, enter VO_2 and OMNI RPE-L. Then click on the icon to the right of the **SERIES X VALUES** text box and highlight the VO_2 values. After the values are highlighted click the icon on the box that appeared. Then click on the icon to the right of the **SERIES Y VALUES** text box and highlight the OMNI RPE-L values. After the values are highlighted click the icon on the box that appeared. Click **OK** on the next two screens.
(c) You should now have a scatter plot with OMNI RPE-L on the *y*-axis and VO_2 on the *x*-axis. Create a title for the plot and enter the appropriate axis labels and units of measure.
(d) To determine OMNI RPE-VT, click on one of the data points to highlight the entire data series. Right click on one of the data points and a menu will appear. Click **ADD TRENDLINE** and the **FORMAT TRENDLINE** menu will appear. Select **LINEAR** and **DISPLAY EQUATION ON CHART** then click **CLOSE**. The trendline and equation will be displayed on the chart.
(e) Use this linear equation to calculate RPE-VT. Use the previously determined VO_2 ($l \cdot min^{-1}$) corresponding to the VT as the "x" value in the equation and solve for "y." The calculated "y" value, once rounded to the nearest whole integer, is the OMNI RPE-VT.

4. Repeat the above steps for VO_2 and HR to determine HR-VT.
5. An example of these procedures with a screenshot depicting each step as performed using Microsoft Excel 2013 can be found in Appendix D.

12.2.2.5 Imposed Exercise Intensity Trial

1. Position the HR monitor and respiratory-metabolic mouthpiece (with head support unit and nose clip if applicable) on the subject.
2. Set the proper seat height on the cycle ergometer according to leg length. When the foot is flat on the right pedal and the pedal is in the extreme down position, the right knee should be in approximately 5 degrees of flexion.
3. Instruct the subject to maintain a 50 $rev \cdot min^{-1}$ pedal cadence. Set the metronome to 100 $beats \cdot min^{-1}$ so each downward movement of each foot is synchronized with a beat of the metronome. The subject may also use the digital monitor on the cycle control panel to regulate pedal cadence.

4. Instruct the subject to perform a 2-min warm-up.
   (a) For electronically braked cycle ergometers (e.g., Lode) set the PO at 25 W.
   (b) For friction-braked cycle ergometers (e.g., Monark), set the brake resistance at 0.5 kg.
5. Following the warm-up, inform the subject that he/she will exercise for 10 min at the predetermined target VO_2 corresponding to the VT which will be achieved by adjusting the cycle brake resistance.
6. For electronically braked cycle ergometers (e.g., Lode)
   (a) Increase the cycle brake resistance to 75 W immediately following the warm-up. Then monitor every 15-s average VO_2 ($l \cdot min^{-1}$) on the computer display.
   (b) If VO_2 remains below the VT after 1 min, increase the cycle brake resistance by 25 W. Continue to increase the brake resistance by 25 W every 30 s until VO_2 is within 0.2–0.3 $l \cdot min^{-1}$ of the VO_2 corresponding to the VT.
   (c) If VO_2 increases to a level above the VT, make further adjustments by decreasing the brake resistance in increments of 5–10 W to fine-tune exercise intensity.
7. For friction-braked cycle ergometers (e.g., Monark)
   (a) Increase the cycle brake resistance to 1.5 kg immediately following the warm-up. Then monitor every 15-s average VO_2 ($l \cdot min^{-1}$) on the computer display.
   (b) If VO_2 remains below the VT after 1 min, increase the cycle brake resistance by 0.5 kg. Continue to increase the brake resistance by 0.5 kg every 30 s until VO_2 is within 0.2–0.3 $l \cdot min^{-1}$ of the VO_2 corresponding to the VT.
   (c) If VO_2 increases to a level above the VT, make further adjustments by decreasing the brake resistance in increments of 0.1 kg to fine-tune exercise intensity.
8. Record cycle brake resistance (W or kg) every minute.
9. Record HR ($b \cdot min^{-1}$) every 2 min.
10. Record the final 15-s VO_2 in $l \cdot min^{-1}$ for each 2-min segment of exercise.
11. Ask the subject to estimate RPE-L starting at 1:30 of each 2-min segment of exercise using the OMNI Scale.
12. Following the 10-min exercise, instruct the subject to perform a 2-min cool-down.
    (a) For electronically braked cycle ergometers (e.g., Lode) set the PO at 25 W.
    (b) For friction-braked cycle ergometers (e.g., Monark), set the brake resistance at 0.5 kg.

12.2.2.6 Self-Selected Exercise Intensity Trial

1. Position the HR monitor and respiratory-metabolic mouthpiece (with head support unit and nose clip if applicable) on the subject.

2. Set the proper seat height on the cycle ergometer according to leg length. When the foot is flat on the right pedal and the pedal is in the extreme down position, the right knee should be in approximately 5 degrees of flexion.
3. Instruct the subject to maintain a 50 rev · min^{-1} pedal cadence. Set the metronome to 100 beats · min^{-1} so each downward movement of each foot is synchronized with a beat of the metronome. The subject may also use the digital monitor on the cycle control panel to regulate pedal cadence.
4. Read the following instructions for self-selected exercise to the subject (Dishman et al. 1994; Parfitt et al. 2000): "You will be allowed to select an intensity you prefer to perform on the cycle. This should be an intensity that you would choose for a 10-min workout if you were participating in a fitness program. The intensity should be high enough that you would get a good workout, but not so high that you would not prefer to exercise at that intensity daily or at least every other day. It should be an intensity that is appropriate for you."
5. Instruct the subject to perform a 2-min warm-up.
   (a) For electronically braked cycle ergometers (e.g., Lode) set the PO at 25 W.
   (b) For friction-braked cycle ergometers (e.g., Monark), set the brake resistance at 0.5 kg.
6. Following the warm-up, instruct the subject to exercise for 10 min at a SS exercise intensity by adjusting the cycle brake resistance whenever desired.
7. Record cycle brake resistance every minute.
8. Record HR (b · min^{-1}) every 2 min.
9. Record the final 15-s VO_2 in l · min^{-1} for each 2-min exercise segment.
10. Ask the subject to estimate RPE-L starting at 1:30 of each 2-min segment of exercise using the OMNI Scale.
11. Following the 10-min exercise, instruct the subject to perform a 2-min cool-down.
    (a) For electronically braked cycle ergometers (e.g., Lode) set the PO at 25 W.
    (b) For friction-braked cycle ergometers (e.g., Monark), set the brake resistance at 0.5 kg.

12.2.2.7 Imposed and Self-Selected Exercise Intensity Trials: Data Organization and Analysis

1. For the imposed exercise intensity trial:
   (a) Calculate the average cycle brake resistance using the data from each minute of exercise.
   (b) Calculate the average HR (b · min^{-1}) using each 2-min HR response.
   (c) Calculate the average OMNI RPE-L using each 2-min RPE response.
   (d) Calculate the average VO_2 (l · min^{-1}) using each 2-min VO_2.

2. For the self-selected exercise intensity trial:
   (a) Calculate the average cycle brake resistance using the data from each minute of exercise.
   (b) Calculate the average HR ($b \cdot min^{-1}$) using each 2-min HR response.
   (c) Calculate the average OMNI RPE-L using each 2-min RPE response.
   (d) Calculate the average VO_2 ($l \cdot min^{-1}$) using each 2-min VO_2.

12.3 Laboratory Discussion Questions

1. Did the imposed trial produce a target intensity similar to the VT? Did the subject exhibit RPE responses similar to those corresponding to the VT as calculated using responses to the estimation trial?
2. Were physiological variables (i.e., VO_2, HR, V_E) similar or dissimilar between the self-selected and imposed exercise trials? Which mode of identifying exercise intensity, self-selected or imposed, would you recommend based on the potential for physiological and health-fitness benefits? Why?
3. Were RPE responses similar or dissimilar between the self-selected and imposed exercise trials? Which mode of identifying exercise intensity, self-selected or imposed, would you recommend based on the potential to promote maximum exercise program adherence? Why?
4. Was there a perceptual-physiological link between RPE responses and physiological responses to self-selected and imposed exercise intensities as predicted by the Effort Continua Model? Would you use a target RPE or target physiological response to set the initial intensity for a sedentary individual's exercise program?
5. How would you progress the intensity of an individual's exercise program if it was based on (a) an imposed protocol or (b) a self-selected protocol?

References

American College of Sports Medicine. ACSM's guidelines for exercise testing and prescription. 9th ed. Philadelphia, PA: Lippincott Williams & Wilkins; 2013.

Caserta MS, Gillett PA. Older women's feelings about exercise and their adherence to an aerobic regimen over time. Gerontologist. 1998;38:602–9.

Cox KL, Burke V, Gorely TJ, Beilin LJ, Puddey IB. Controlled comparison of retention and adherence in home- vs center-initiated exercise interventions in women ages 40–65 years: the S.W.E.A.T. study (sedentary women exercise adherence trial). Prev Med. 2003;36:17–29.

Dishman RK, Farquhar RP, Curetone KG. Responses to preferred intensities of exertion in men differing in activity levels. Med Sci Sports Exerc. 1994;26:783–90.

Ekkekakis P, Petruzzello SJ. Acute aerobic exercise and affect: current status, problems and prospects regarding dose–response. Sports Med. 1999;28:337–74.

Ekkekakis P, Lind E. Exercise does not feel the same when you are overweight: the impact of self-selected and imposed intensity on affect and exertion. Int J Obes. 2006;30:652–60.

Haile L, Goss FL, Robertson RJ, Andreacci JL, Gallagher Jr M, Nagle EF. Session perceived exertion and affective responses to self-selected and imposed cycle exercise of the same intensity in young men. Eur J Appl Physiol. 2013;116:1755–65.

Lee JY, Jensen BE, Oberman A, Fletcher GF, Fletcher BJ, Raczynski JM. Adherence in the training levels comparison trial. Med Sci Sports Exerc. 1996;28:47–52.

Lind E, Joens-Matre RR, Ekkekakis P. What intensity of physical activity do previously sedentary middle-aged women select? Evidence of a coherent pattern from physiological, perceptual, and affective markers. Prev Med. 2005;40:407–19.

Lind E, Ekkekakis P, Vazou S. The affective impact of exercise intensity that slightly exceeds the preferred level: 'pain' for no additional 'gain'. J Health Psychol. 2008;13:464–8.

Parfitt G, Rose EA, Markland D. The effect of prescribed and preferred intensity exercise on psychological affect and the influence of baseline measures of affect. J Health Psychol. 2000;5: 231–40.

Parfitt G, Rose EA, Burgess WM. The psychological and physiological responses of sedentary individuals to prescribed and preferred intensity exercise. Br J Health Psychol. 2006;11: 39–53.

Perri MG, Anton SD, Durning PE, Ketterson TU, Sydeman SJ, Berlant NE, Kanasky Jr WF, Newton Jr RL, Limacher MC, Martin AD. Adherence to exercise prescriptions: effects of prescribing moderate versus higher levels of intensity and frequency. Health Psychol. 2002;21: 452–8.

Rose EA, Parfitt G. A quantitative analysis and qualitative explanation of the individual differences in affective responses to prescribed and self-selected exercise intensities. J Sport Exerc Psychol. 2007;29:335–54.

Ryan RM, Deci EL. Self-determination theory and the facilitation of intrinsic motivation, social development, and well-being. Am Psychol. 2000;55:68–78.

Ryan RM, Frederick CM, Lepes D, Rubio N, Sheldon KM. Intrinsic motivation and exercise adherence. Int J Sports Psychol. 1997;28:335–54.

Wankel LM. The importance of enjoyment to adherence and psychological benefits from physical activity. Int J Sports Psychol. 1993;24:151–69.

Chapter 13
Predicted, Momentary and Session RPE

The common method of rating perceived exertion is for individuals to estimate *momentary RPE*, also referred to as in-task or on-stimulus RPE. This method is used when individuals are asked to estimate RPE during any type of acute exercise or PA performance, including warm-up and recovery periods. Momentary RPE has many uses as noted in previous chapters, but does not provide information about an individual's perception of physical exertion related to an exercise bout to be performed in the near future (*predicted* RPE) or that has already been performed in the recent past (*session* RPE). Therefore, the assessment of off-stimulus RPE values may provide additional information related to past performance and future participation of exercise and physical activity. A mismatch between predicted and/or session RPE values and momentary RPE for a given bout of exercise may help to identify an individual at risk of drop-out from an exercise program, especially if the predicted or session RPE values are greater than momentary RPE. In addition, a mismatch between session RPE and momentary RPE may affect an individual's ability to properly report exercise intensity. The primary purpose of this laboratory experiment is to measure and compare an individual's predicted, momentary and session RPE responses to exercise.

13.1 Background

13.1.1 Predicted RPE

An individuals' prediction of the level of perceived exertion that they expect to experience in a future bout of exercise (i.e., predicted RPE) may help to explain their motivation to perform that exercise. *Predicted RPE* can be defined as a global estimate of the average RPE that is expected for an entire bout of upcoming exercise or PA. It is rated prior to performance of that activity. Poulton and colleagues (2002)

L. Haile et al., *Perceived Exertion Laboratory Manual*,
DOI 10.1007/978-1-4939-1917-8_13

studied predicted physical discomfort. Although the investigation did not measure predicted RPE directly, the definition of perceived exertion includes the construct of physical discomfort. Significant relations were found between patterns of the over-prediction of physical discomfort and negative self-reported health status, negative attitudes about exercise, high body mass index, and poor CRF. In addition, the greater the mismatch between predicted and actual physical discomfort, the lower the physical activity level (Poulton et al. 2002). Therefore, measuring predicted RPE prior to a bout of exercise may help determine whether an individual is at risk of low adherence to an exercise program, identified by the overprediction (i.e., higher value) of the expected RPE.

It is of interest to determine whether predicted RPE matches momentary RPE for a given type of exercise programming. A significant difference between predicted RPE and the mean of momentary RPE's for a single exercise session indicates a perceptual mismatch potentially leading to negative attitudes regarding adoption and maintenance of PA participation. Few investigations have studied predicted RPE. Young adult men and women overpredicted RPE prior to load-incremented cycle ergometer exercise (Haile et al. 2008; Hunt et al. 2007). However, the relation of this mismatch with attitudes towards exercise, PA level or adherence to exercise was not investigated. In addition, a load-incremented exercise test is not a common exercise format that is included in steady state or interval prescriptions for health-related fitness. Kane and colleagues (2010) compared predicted RPE and momentary RPE in response to a shuttle run test called the Progressive Aerobic Cardiovascular Endurance Run (PACER). The subjects for the investigation were children who reported moderate experience with aerobic exercise. Predicted RPE exhibited a match with momentary RPE, which may have been a result of previous experience with similar aerobic activities resulting in an accurate expectation of the exertion to be experienced (Kane et al. 2010). Future research is necessary to determine if predicted RPE can be used to identify barriers to exercise adherence. This can be done by calculating the difference between predicted RPE and the mean of the sequentially determined momentary RPE of exercise bouts normally prescribed to improve CRF.

13.1.2 Session RPE

An individual's memory of the perceptual experience of a previously performed bout of exercise may affect the accurate reporting of the specific exercise intensity performed during that bout. It may also affect an individual's desire to perform similar exercise in the future. *Session RPE* can be defined as a post-exercise estimate of the global RPE for an entire bout of exercise or PA. It is rated following completion of that activity. Session RPE has been measured following a wide variety of exercise and sport modalities, including cycling (Foster et al. 1996, 2001; Green et al. 2007; Haile et al. 2008, 2013b; Herman et al. 2006; Hunt et al. 2007; Killen et al. 2013; Kilpatrick et al. 2009; Rodriguex-Marroyo et al. 2012), walking/running (Davis et al. 2012; Foster et al. 1996; Green et al. 2009; Haile et al. 2013a;

Minganti et al. 2011b; Seiler and Kjerland 2006), swimming (Wallace et al. 2009), diving (Minganti et al. 2011a), soccer (Alexiou and Coutts 2008; Algroy et al. 2011; Gomez-Piriz et al. 2011; Impellizzeri et al. 2004; Tessitore et al. 2011), basketball (Foster et al. 2001; Moreira et al. 2012), speed-skating (Foster et al. 1996), rugby (McLean et al. 2010), judo (Viveiros et al. 2011), taekwondo (Haddad et al. 2011), futsal (Milanez et al. 2011), teamgym (Minganti et al. 2010), sprint kayaking (Borges et al. 2014), field-based speed training (Lockie et al. 2011), and resistance exercise (Bacon et al. 2012; Charro et al. 2010; Day et al. 2004; Egan 2003; Egan et al. 2006; McGuigan et al. 2004, 2008; Pritchett et al. 2009; Singh et al. 2007; Sweet et al. 2004). In most of these investigations, session RPE was rated between 5 and 30 min following the exercise bout.

Session RPE was originally developed to track overtraining in endurance athletes (Foster et al. 2001), but was recently proposed as a method to track the relative exercise intensity for individuals participating in PA intervention programs (Haile et al. 2013b). Whether an individual is previously sedentary and attempting to adopt a new exercise program or an elite athlete, session RPE can be a practical, affordable and non-invasive method to describe the relative intensity of physical activity and exercise performed within training or a general conditioning program. In addition, it may be useful for predicting injury or illness as a result of overtraining and the failure to achieve physiological benefits as a result of undertraining (Foster et al. 2001; Haile et al. 2013b).

13.1.3 Validity of Session RPE

Session RPE has been widely studied, especially in comparison to predicted RPE. However, few studies have investigated the validity of session RPE as a global value that accurately represents the mean of momentary RPE responses measured during exercise. It is known that session RPE generally changes in correspondence with variations in physiological variables measured during exercise, including HR, VO_2 and blood lactate concentrations (Herman et al. 2006; Rodriguex-Marroyo et al. 2012; Seiler and Kjerland 2006). In previous investigations, session RPE has exhibited an acceptable degree of concurrent validity because it changes in a predictable direction with well-known physiological exertional mediators. However, by definition, session RPE is the global estimate of the average of the sequentially estimated momentary RPE determined during a previous exercise bout. To exhibit construct validity, session RPE should be equal to the mean of momentary RPE responses. Such validity is achieved when scale anchoring procedures as presented during RPE scale orientation are understood by the individual and conform to Borg's Range Model. Given these measurement procedures, session RPE can be used to estimate the relative exercise intensity. Therefore, session RPE that is equal to the mean of the momentary RPE could be used to predict the physiological responses that describe the relative intensity of a previously performed exercise session. This information would be appropriate for use in PA logs as part of a behavioral intervention involving systematic exercise participation. However, if session RPE exhibits a mismatch compared with the mean of the momentary RPE, then the measure likely does not accurately represent the global exertion responses for the entire exercise session. In such an instance, the session RPE response may not be a useful index of the global exertional experience and as such the relative exercise intensity for the entire preceding exercise bout.

Green and colleagues (2009) measured session RPE 20 min following treadmill exercise bouts at 70 % of VO_2max. Whether the exercise was performed for 20, 30 or 40 min in separate bouts, session RPE was significantly greater than the mean of the momentary RPE. Kilpatrick and colleagues (2009) measured session RPE 15 min following 30 min of treadmill exercise. Subjects performed three separate bouts at an intensity of their choosing based on verbal instructions to attain light, moderate, or vigorous intensity. For all three bouts session RPE was significantly greater than mean momentary RPE. Haile and colleagues (2013b) measured session RPE 15 min following two separate 20 min bouts of cycle ergometer exercise. In the first bout, subjects performed self-selected exercise intensity based on what they felt was a "good workout." In the second bout, each subject's own previously determined self-selected intensity was then prescribed (i.e., imposed) on that subject. However, the subject was told the brake resistance was "selected by the investigators." In either case, whether the same intensity was self-selected or imposed, session RPE was significantly greater than the mean of the momentary RPE (Haile et al. 2013b). In all three studies, session RPE was not representative of the mean momentary RPE. However, session RPE was similar to momentary RPE estimated near the end of exercise. Therefore, it seems that the most recently experienced momentary RPE values may dominate the session RPE response. In some individuals, there may be a certain amount of perceptual memory fade in the minutes following exercise cessation. This time dependent decay may not allow the momentary RPE experienced near the beginning of exercise to contribute to the global response. It may be beneficial to ask subjects to rate session RPE sooner following exercise than the 15–20-min period that has been commonly used in many investigations.

13.1.4 Effect of Exercise Intensity on Session RPE

Another factor that has been shown to mediate the session RPE response is the effect of changes in exercise intensity during a single performance. This effect may be a potential reason why the session RPE has been rated consistently higher than the mean of the momentary RPE. Green and colleagues (2007) found that session RPE was higher following interval exercise in comparison to constant load cycle exercise when overall workload was the same between test protocols. In this experimental design, average workload was the same between interval and constant load exercise trials. However, the intensity of each exercise interval was higher than that at any time point in the constant load protocol, likely resulting in a greater session RPE for the interval protocol. Such a difference was expected because of the greater disruption in metabolic homeostasis experienced during interval exercise relative to constant workload exercise, regardless that average workload was the same (Green et al. 2007). In addition, two other investigations found that subjects rated session RPE greater than the mean momentary RPE following load-incremented cycle ergometer exercise (Haile et al. 2008; Hunt et al. 2007). This response was most likely influenced by the high exercise intensities performed closer to the end of the exercise session. However, these studies did not compare session RPE with momentary RPE values. In the study by Haile and colleagues (2013b), subjects were allowed to self-select exercise intensity every 5 min during a 20-min cycle ergometer exercise trial. On average, subjects increased intensity and momentary RPE values throughout the trial. Subsequently, session RPE was greater than the mean momentary RPE. However, the session RPE value was similar to the momentary RPE values rated near the end of exercise, where the highest momentary RPE values occurred (Haile et al. 2013b).

13.1.5 Segmented Session RPE

Based on this previous research, it is possible that the time segments of a previous exercise bout can differentially dominate the global session RPE response. Higher exertional perceptions that are linked to comparatively higher exercise intensity, either performed during interval exercise or near the end of load-incremented and self-selected exercise, may result in a higher session RPE response (Green et al. 2007; Haile et al. 2008, 2013b; Hunt et al. 2007). In addition, the in-task level of exertion that is experienced near the end of the exercise trial may intensify session RPE. This holds regardless of whether intensity was constant (Green et al. 2009), self-selected based on specific verbal prescriptions (Kilpatrick et al. 2009), or freely self-selected based on what the subject thought was a "good workout." Therefore, it may be necessary to separately measure session RPE for the different segments of the previous exercise bout. This is termed *segmented session RPE*. In the post-exercise period, the subject is asked to estimate the average RPE for a specific segment (time-period) of a bout of exercise or PA that was just completed. Segmented session RPE can take the place of or be measured in addition to a full session RPE.

Haile and colleagues (2013a) asked subjects to rate session RPE and segmented session RPE following 20 min of self-selected treadmill exercise. Segmented session RPE values were rated for the first and second halves (first and second 10-min time-periods) of exercise. Each segmented session RPE value was similar to the momentary RPE values rated during its respective time-period. In addition, the mean of the two segmented session RPE values was similar to the mean momentary RPE for the total exercise session. Allowing the subjects to rate session RPE for each segment of the previous exercise bout resulted in a response that was representative of the mean of momentary RPE values. In this investigation, session RPE was also similar to the mean momentary RPE. Therefore, simply asking the subjects to focus on the different segments of exercise prior to rating session RPE may result in a response that represents the mean momentary RPE (Haile et al. 2013a).

13.1.6 Case Study

13.1.6.1 Client Information

A 49-year-old male has agreed to participate in your intervention study during which he will be recording details about his exercise in a log over a 1-year period. For this intervention you are asking him to perform aerobic exercise most days of the week as per national PA guidelines. You are not requiring him to exercise in a specific fitness facility or use a specific type of ergometer because you do not want to restrict his activity participation. You encourage him to exercise wherever he prefers, be it at a local gym, at home, or outside. You instruct him to perform exercise at an intensity that he believes to be a good workout, but not to work out so hard that he would not be able to perform the exercise at least every other day. However, with such general exercise instructions and the potential for using various modalities, you want to employ a common method to record information about exercise intensity that the subject selected.

13.1.6.2 Assessments, Results and Analysis

Have the subject perform a load-incremented estimation trial to determine the VT. Then, after a brief description of the subsequent submaximal exercise trial, ask the subject to estimate their predicted RPE. Have the subject perform the submaximal exercise trial at an exercise intensity equivalent to the VT (i.e., an imposed intensity). Following exercise, ask the subject to estimate session and segmented session RPE's.

Using the estimation trial, determine VO_2 at the VT.
Prior to the submaximal exercise trial, measure predicted RPE.
During the submaximal exercise trial, measure momentary RPE.
Following the submaximal exercise trial, measure session and segmented session RPE.

13.2 Methods

13.2.1 Treadmill Procedures

13.2.1.1 Equipment

1. Adult OMNI-Walk/Run RPE Scale (Fig. A.2)
2. Treadmill
3. HR monitor
4. Respiratory-metabolic measurement system

13.2.1.2 Pre-exercise Procedures

1. Measure height (cm) and weight (kg) of subject.
2. Read the standard instructions for the Adult OMNI-Walk/Run RPE Scale for RPE-O (Appendix B.1) to the subject. If measurement of differentiated RPE (RPE-L and RPE-C) is desired, read the standard instructions for the Adult OMNI-Walk/Run RPE Scale for undifferentiated and differentiated RPE (Appendix B.2) to the subject. Perform the memory anchoring procedure as described in Chap. 5.

13.2.1.3 Graded Exercise Test

1. Position the HR monitor and respiratory-metabolic mouthpiece (with head support unit and nose clip if applicable) on the subject.
2. Instruct the subject to step onto the treadmill and review exercise termination procedures: When the subject cannot continue exercise due to exhaustion or discomfort, he/she should grasp the treadmill hand rails, at which time the test administrator will gradually slow the treadmill down for performance of a cool-down. The subject should be reminded not to step off the treadmill belt while it is still in motion.
3. Bruce Multistage Treadmill Test Protocol: this can be performed by manually adjusting treadmill speed and grade or using a program on a computer that is interfaced to the treadmill.

   (a) Begin the warm-up at 1.5 miles $\cdot$ h^{-1} and 0 % grade for 3 min.
   (b) Each exercise test stage will last for 3 min. The stages progress as follows:

      Stage 1—1.7 miles $\cdot$ h^{-1} and 10 % grade
      Stage 2—2.5 miles $\cdot$ h^{-1} and 12 % grade
      Stage 3—3.4 miles $\cdot$ h^{-1} and 14 % grade
      Stage 4—4.2 miles $\cdot$ h^{-1} and 16 % grade
      Stage 5—5.0 miles $\cdot$ h^{-1} and 18 % grade
      Stage 6—5.5 miles $\cdot$ h^{-1} and 20 % grade
      Stage 7—6.0 miles $\cdot$ h^{-1} and 22 % grade
      Stage 8—6.5 miles $\cdot$ h^{-1} and 24 % grade

   (c) When the subject cannot continue any longer, terminate the exercise test by initiating the cool-down period at 1.5 miles $\cdot$ h^{-1} and 0 % grade. The cool-down should be 5 min in duration.
   (d) Ask the subject to estimate RPE starting at 2:30 of each exercise stage using the OMNI Scale (RPE-O). Because the position of the respiratory-metabolic mouth piece inhibits a verbal response, instruct the subject to point to the numbers on the RPE scale, which should be conveniently positioned within the subject's arm reach. State aloud the numerical ratings for each momentary assessment to which the subject pointed and request a confirmatory nod

that the number stated was correct. If incorrect, allow the subject to point to the appropriate rating on the RPE scale once more. Ask the subject to hold his or her finger on the appropriate number on the scale for approximately 1 s.

(e) Record HR ($b \cdot min^{-1}$) at 2:55 of each exercise stage.
(f) Record the final 15-s VO_2 in $ml \cdot kg^{-1} \cdot min^{-1}$ for each exercise stage.
(g) Record HRmax as the highest HR value recorded during the final exercise stage or immediately post-exercise.
(h) Record VO_2max as the highest 15-s VO_2 value recorded during the test.
(i) Determine the VO_2 ($ml \cdot kg^{-1} \cdot min^{-1}$) and %$VO_2max$ associated with the VT using the respiratory-metabolic measurement system automatic VT calculator.
(j) If the respiratory-metabolic measurement system does not automatically calculate VT or if an explanation of the manual calculation and visual identification of the VT is desired, refer to Appendix D for detailed instructions.

13.2.1.4 Submaximal Exercise Intensity Trial

1. Position the HR monitor and respiratory-metabolic mouthpiece (with head support unit and nose clip if applicable) on the subject.
2. Instruct the subject to step onto the treadmill and review exercise termination procedures: When the subject cannot continue exercise due to exhaustion or discomfort, he/she should grasp the treadmill hand rails, at which time the test administrator will slow down or stop the treadmill. The subject should be reminded not to step off the treadmill belt while it is still in motion.
3. Predicted RPE

   (a) Prior to the warm-up, instruct the subject that he/she will exercise for 20 min at the predetermined target VO_2 corresponding to the VT, which will be achieved by adjusting the treadmill speed.
   (b) Instruct the subject to predict the global, or average, perceived exertion for the overall body (RPE-O) that will be experienced in the forthcoming submaximal exercise bout. Record predicted RPE-O.

4. Start the treadmill at 3 $miles \cdot h^{-1}$ and 0 % grade for a 2-min warm-up.
5. Submaximal exercise intensity at the VT

   (a) Increase the speed of the treadmill to 5 $miles \cdot h^{-1}$ immediately following the warm-up. Then monitor every 15-s average VO_2 ($ml \cdot kg \cdot min^{-1}$) on the computer display.
   (b) If VO_2 remains below the target VT value after 1 min, increase the treadmill speed by 0.5 $miles \cdot h^{-1}$. Continue to increase speed by 0.5 $miles \cdot h^{-1}$ every 30 s until VO_2 is within 2–3 $ml \cdot kg \cdot min^{-1}$ of the target VO_2 equivalent to the VT.
   (c) If VO_2 increases to a level above the VT, make further adjustments by decreasing treadmill speed in increments of 0.1 $miles \cdot h^{-1}$ to fine-tune exercise intensity.

6. Record treadmill speed every minute.
7. Record HR ($b \cdot min^{-1}$) every 2 min.
8. Record the final 15-s VO_2 in $ml \cdot kg^{-1} \cdot min^{-1}$ for each 2-min segment of exercise.
9. Ask the subject to estimate RPE-O starting at 1:30 of each 2-min segment of exercise using the OMNI Scale.
10. Following the 20-min exercise, the treadmill should be set to 3 $miles \cdot h^{-1}$ and 0 % grade for a 2-min cool-down period.
11. Provide the subject with 5 min of seated rest.
12. Instruct the subject to estimate session and segmented session RPE values in random order.

    (a) Session RPE: instruct the subject to estimate the global, or average, perceived exertion for the overall body (RPE-O) that was experienced during the preceding exercise bout. Record session RPE-O.
    (b) Segmented session RPE for the first half of the exercise bout: instruct the subject to estimate the average perceived exertion for the overall body (RPE-O) that was experienced during the first 10 min of the preceding exercise bout. Record the first segmented session RPE-O.
    (c) Segmented session RPE for the second half of the exercise bout: instruct the subject to estimate the average perceived exertion for the overall body (RPE-O) that was experienced during the second 10 min of the preceding exercise bout. Record the second segmented session RPE-O.

13.2.1.5 Submaximal Exercise Trial: Data Organization and Analysis

1. Calculate the average momentary OMNI RPE-O using each 2-min RPE measured during exercise.
2. Calculate the average momentary OMNI RPE-O for the first half of exercise using each 2-min RPE measured during the first 10 min of exercise.
3. Calculate the average momentary OMNI RPE-O for the second half of exercise using each 2-min RPE measured during the second 10 min of exercise.
4. Calculate the average segmented session OMNI RPE-O using the two separate segmented session RPE values for each half of the exercise bout.

13.2.2 Cycle Ergometer Procedures

13.2.2.1 Equipment

1. Adult OMNI-Cycle RPE Scale (Fig. 2.4)
2. Cycle ergometer
3. Metronome
4. HR monitor
5. Respiratory-metabolic measurement system

13.2.2.2 Pre-exercise Procedures

1. Measure height (cm) and weight (kg) of subject.
2. Read the standard instructions for the Adult OMNI-Cycle RPE Scale for RPE-L (Appendix B.4) to the subject. If the measurement of undifferentiated (RPE-O) and differentiated RPE for chest/breathing (RPE-C) is desired, read the standard instructions for the Adult OMNI-Cycle RPE Scale for undifferentiated and differentiated RPE (Appendix B.5). Perform the memory anchoring procedure as described in Chap. 5.

13.2.2.3 Graded Exercise Test

1. Position the HR monitor and respiratory-metabolic mouthpiece (with head support unit and nose clip if applicable) on the subject.
2. Set the proper seat height on the cycle ergometer according to leg length. When the foot is flat on the right pedal and the pedal is in the extreme down position, the right knee should be in approximately 5 degrees of flexion.
3. Load-incremented protocol for electronically braked and friction-braked cycle ergometers:
   (a) Instruct the subject to maintain a 50 $rev \cdot min^{-1}$ pedal cadence. Set the metronome to 100 $beats \cdot min^{-1}$ so each downward movement of each foot is synchronized with a beat of the metronome. The subject may also use the digital monitor on the cycle control panel to regulate pedal cadence.
   (b) For electronically braked cycle ergometers (e.g., Lode), begin stage 1 at 50 W then increase the resistance 25 W every 2 min.
   (c) For friction-braked cycle ergometers (e.g., Monark), begin stage 1 at 1 kg resistance then increase the resistance 0.5 kg every 2 min.
   (d) When the subject cannot maintain the pedal cadence for 10 consecutive seconds owing to fatigue, terminate the exercise test.
   (e) Ask the subject to estimate RPE-L starting at 1:30 of each exercise stage using the OMNI Scale. Because the position of the respiratory-metabolic mouth piece inhibits a verbal response, instruct the subject to point to the numbers on the RPE scale, which should be conveniently positioned within the subject's arm reach. State aloud the numerical ratings for each momentary assessment to which the subject pointed and request a confirmatory nod that the number stated was correct. If incorrect, allow the subject to point to the appropriate rating on the RPE scale once more. Ask the subject to hold his or her finger on the appropriate number on the scale for approximately 1 s.
   (f) Record HR ($b \cdot min^{-1}$) at 1:55 of each exercise stage.
   (g) Record the final 15-s VO_2 in $l \cdot min^{-1}$ for each exercise stage.
   (h) Record HRpeak as the highest HR value recorded during the final exercise stage or immediately post-exercise.

(i) Record VO_2peak as the highest 15-s VO_2 value recorded at the end of the test.
(j) Determine the VO_2 ($l \cdot min^{-1}$) and %VO_2peak associated with the VT using the respiratory-metabolic measurement system automatic VT calculator.
(k) If the respiratory-metabolic measurement system does not automatically calculate VT or if an explanation of manual calculation and visual identification of the VT is desired, refer to Appendix D for detailed instructions.

13.2.2.4 Submaximal Exercise Intensity Trial

1. Position the HR monitor and respiratory-metabolic mouthpiece (with head support unit and nose clip if applicable) on the subject.
2. Set the proper seat height on the cycle ergometer according to leg length. When the foot is flat on the right pedal and the pedal is in the extreme down position, the right knee should be in approximately 5 degrees of flexion.
3. Instruct the subject to maintain a 50 $rev \cdot min^{-1}$ pedal cadence. Set the metronome to 100 $beats \cdot min^{-1}$ so each downward movement of each foot is synchronized with a beat of the metronome. The subject may also use the digital monitor on the cycle control panel to regulate pedal cadence.
4. Predicted RPE

   (a) Prior to the warm-up, instruct the subject that he/she will exercise for 20 min at the predetermined target VO_2 corresponding to the VT, which will be achieved by adjusting the cycle brake resistance.
   (b) Instruct the subject to predict the global, or average, perceived exertion for the leg muscles (RPE-L) that will be experienced in the forthcoming submaximal exercise bout. Record predicted RPE-L.

5. Instruct the subject to perform a 2-min warm-up.

   (a) For electronically braked cycle ergometers (e.g., Lode) set the PO at 25 W.
   (b) For friction-braked cycle ergometers (e.g., Monark), set the brake resistance at 0.5 kg.

6. Submaximal exercise intensity at the VT for electronically braked cycle ergometers (e.g., Lode)

   (a) Increase the cycle brake resistance to 75 W immediately following the warm-up. Then monitor every 15-s average VO_2 ($l \cdot min^{-1}$) on the computer display.
   (b) If VO_2 remains below the VT after 1 min, increase the cycle brake resistance to by 25 W. Continue to increase the brake resistance by 25 W every 30 s until VO_2 is within 0.2–0.3 $l \cdot min^{-1}$ of the target VO_2 corresponding to the VT.
   (c) If VO_2 increases to a level above the VT, make further adjustments by decreasing the brake resistance in increments of 5–10 W to fine-tune exercise intensity.

7. Submaximal exercise intensity at the VT for friction-braked cycle ergometers (e.g., Monark)

   (a) Increase the cycle brake resistance to 1.5 kg immediately following the warm-up. Then monitor every 15-s average VO_2 ($l \cdot min^{-1}$) on the computer display.
   (b) If VO_2 remains below the VT after 1 min, increase the cycle brake resistance by 0.5 kg. Continue to increase the brake resistance by 0.5 kg every 30 s until VO_2 is within 0.2–0.3 $l \cdot min^{-1}$ of the target VO_2 corresponding to the VT.
   (c) If VO_2 increases to a level above the VT, make further adjustments by decreasing the brake resistance in increments of 0.1 kg to fine-tune exercise intensity.

8. Record cycle brake resistance (W or kg) every minute.
9. Record HR ($b \cdot min^{-1}$) every 2 min.
10. Record the final 15-s VO_2 in $l \cdot min^{-1}$ for each 2-min segment of exercise.
11. Ask the subject to estimate RPE starting at 1:30 of each 2-min segment of exercise using the OMNI Scale (RPE-L).
12. Following the 10-min exercise, instruct the subject to perform a 2-min cool-down.

    (a) For electronically braked cycle ergometers (e.g., Lode) set the PO at 25 W.
    (b) For friction-braked cycle ergometers (e.g., Monark), set the brake resistance at 0.5 kg.

13. Provide the subject with 5 min of seated rest.
14. Instruct the subject to estimate Session and Segmented Session RPE values in random order.

    (a) Session RPE: instruct the subject to estimate the global, or average, perceived exertion for the leg muscles (RPE-L) that was experienced during the preceding exercise bout. Record session RPE-L.
    (b) Segmented session RPE for the first half of exercise: instruct the subject to estimate the average perceived exertion for the leg muscles (RPE-L) that was experienced during the first 10 min of the preceding exercise bout. Record the first segmented session RPE-L.
    (c) Segmented session RPE for the second half of exercise: instruct the subject to estimate the average perceived exertion for the leg muscles (RPE-L) that was experienced during the second 10 min of the preceding exercise bout. Record the second segmented session RPE-L.

13.2.2.5 Submaximal Exercise Trial: Data Organization and Analysis

1. Calculate the average momentary OMNI RPE-L using each 2-min RPE measured during exercise.
2. Calculate the average momentary OMNI RPE-L for the first half of exercise using each 2-min RPE measured during the first 10 min of exercise.

3. Calculate the average momentary OMNI RPE-L for the second half of exercise using each 2-min RPE measured during the second 10 min of exercise.
4. Calculate the average segmented session OMNI RPE-L using the two separate segmented session RPE values.

13.3 Laboratory Discussion Questions

1. Were the predicted and momentary RPE responses similar or dissimilar values for your subject? Did the subject accurately predict the perceived exertion experienced during the exercise bout?
2. Were the session and segmented session values similar or dissimilar to the average of momentary RPE values for your subject? Did the subject accurately rate the perceived exertion experienced during the exercise bout based on memory? Using segmented session RPE, did the subject accurately rate the perceived exertion experienced during the first and second halves of the exercise bout?
3. Based on your results, discuss the validity of session and segmented session RPE for tracking perceived exertion experienced during previous exercise. Which RPE more accurately represented the mean of momentary RPE during the exercise bout, session RPE or the average of the two segmented session RPE values?
4. Describe the potential methodological difficulty with using momentary RPE values to estimate exercise intensity during a PA behavior intervention when PA is unsupervised.
5. What advantages does session RPE have over traditional methods of tracking PA intensity such as HR or ergometer setting?
6. Compare and contrast predicted RPE and session RPE. Use a match–mismatch paradigm in your explanation.
7. How might the perceived exertion assessed prior to exercise (predicted RPE) influence the perceived exertion response during exercise (momentary RPE)?
8. How could session RPE be used to track PA intensity during a PA program? What information could you derive from session RPE values that indicate the prescribed intensity was inappropriate, i.e., too low or too high?
9. How might you use segmented session RPE values to rate an interval exercise bout?

13.4 Laboratory Addendum

13.4.1 Segmented Session RPE for Resistance Exercise

Numerous investigations have measured session RPE following resistance exercise (Bacon et al. 2012; Charro et al. 2010; Day et al. 2004; Egan 2003; Egan et al. 2006; McGuigan et al. 2004, 2008; Pritchett et al. 2009; Singh et al. 2007; Sweet et al. 2004).

These studies have only employed a single measured of session RPE (i.e., one rating to represent an entire resistance exercise workout). In contrast, previous investigations have not measured segmented session RPE. Segmented session RPE may be useful to differentiate the exertion perceived between the various components of a resistance exercise workout, especially one including multiple sets of multiple resistance exercises. Such measurement would be similar to the use of segmented session RPE for aerobic exercise to differentiate between halves or thirds of an aerobic exercise bout. Different segmented session RPE values could be used for separate resistance exercises, muscle groups, or limbs. Finally, segmented session RPE could even be used to differentiate between sets of the same resistance exercise.

13.4.2 Exertional Recall

Self-report recall questionnaires are often used to assess an individual's current PA level prior to exercise testing or entering an exercise program. In addition, PA recall questionnaires can be used for research purposes to describe activity levels of a sample population. The common model for recall questionnaires uses the FITT principle, i.e., they ask an individual to recall information regarding the frequency, intensity, time (or duration), and type of PA performed. It is quite easy for individuals to describe the frequency of exercise as days per week of participation and the exercise duration in minutes per session for each type of exercise. However, it can be quite difficult to describe the intensity of PA across a wide range of activities, many of which involve various modes and are intermittent in nature such as occupational and recreational activities (Schafer et al. 2013). The metabolic equivalent task (MET) method is widely used to describe the intensity of PA. Using this method, 1 MET represents a resting value and activities are described as multiples of resting energy expenditure (e.g., an activity that is twice the energy expenditure of rest is described as 2 METs). Research has determined an average MET level for a wide variety of PA modalities that are indexed using a coding system in the compendium of physical activities (Ainsworth et al. 1993, 2000). Using the compendium, the MET level (i.e., estimated energy expenditure) assigned to each activity is predetermined and does not take into account individual differences in performance of that activity.

Schafer and colleagues (2013) have proposed and validated a new self-report recall questionnaire for the assessment of the intensity of recently performed exercise using RPE. This method is based on the strong positive relation between RPE and the relative aerobic metabolic rate (%VO_2max). Using this method, termed *exertional recall*, an individual can be asked to rate the perceived exertion experienced during exercise performed in the days, weeks, or months prior. Exertional recall can be considered a form of Session RPE. Previous investigations that have included RPE as part of a PA recall questionnaire reported that the achievement of health-fitness benefits is positively related to the level of perceived exertion experienced during previous exercise (Lee et al. 2003; Rudra et al. 2005, 2006).

In the validity experiment conducted by Schafer et al. (2013), correlation coefficients ranging from $r=0.47$ to $r=0.76$ were found for the relation between both VO_2 and HR responses that were measured during an intermittent walking and running exercise bout and recall RPE rated 1 week following the exercise performance. In addition, correlation coefficients ranging from $r=0.51$ to $r=0.87$ were found for the relation between momentary RPE responses (i.e., measured during actual treadmill exercise) and recall RPE. The difference between momentary RPE and recall RPE was less than 1 OMNI Scale unit. These results support the validity of an exertional recall questionnaire to rate both undifferentiated and differentiated RPEs 1 week following exercise performance (Schafer et al. 2013). This new method of measuring recall RPE may be a useful and accurate descriptor of the subjective intensity of previous exercise that could be included in PA questionnaires.

References

Ainsworth BE, Haskell WL, Leon AS, Jacobs Jr DR, Montoye HJ, Sallis JF, Paffenbarger Jr RS. Compendium of physical activities: classification of energy costs of human physical activities. Med Sci Sports Exerc. 1993;25:71–80.

Ainsworth BE, Haskell WL, Whitt MC, Irwin ML, Swartz AM, Strath SJ, O'Brien WL, Bassett Jr DR, Schmitz KH, Emplaincourt PO, Jacobs Jr DR, Leon AS. Compendium of physical activities: an update of activity codes and MET intensities. Med Sci Sports Exerc. 2000;32: S498–504.

Alexiou H, Coutts AJ. A comparison of methods used for quantifying internal training load in women soccer players. Int J Sports Physiol Perform. 2008;3:320–30.

Algroy EA, Hetlelid KJ, Seiler S, Pedersen JIS. Quantifying training intensity distribution in a group of Norwegian professional soccer players. Int J Sports Physiol Perform. 2011;6:70–81.

Bacon NT, Wingo JE, Richardson MT, Ryan GA, Pangallo TC, Bishop PA. Effect of two recovery methods on repeated closed-handed and open-handed weight-assisted pull-ups. J Strength Cond Res. 2012;26:1348–52.

Borges TO, Bullock N, Duff C, Coutts AJ. Methods for quantifying training in sprint kayak. J Strength Cond Res. 2014;28:474–82.

Charro MA, Aoki MS, Coutts AJ, Araujo RC, Bacurau RF. Hormonal, metabolic and perceptual responses to different resistance training systems. J Sports Med Phys Fitness. 2010;50:229–34.

Davis JK, Bishop PA, Zhang Y, Green JM, Casaru C, Orrick KD, Curtner-Smith M, Richardson MT, Schumacker RE. Fluid balance, thermal stress, and post exercise response in women's Islamic athletic clothing. Eur J Appl Physiol. 2012;112:725–34.

Day ML, McGuigan MR, Brice G, Foster C. Monitoring exercise intensity during resistance training using the session RPE scale. J Strength Cond Res. 2004;18:353–8.

Egan AD. Session rating of perceived exertion during high intensity and low intensity bouts of resistance exercise. J Undergrad Res. 2003;6:1–6.

Egan AD, Winchester JB, Foster C, McGuigan MR. Using session RPE to monitor different methods of resistance exercise. J Sports Sci Med. 2006;5:289–95.

Foster C, Daines E, Hector L, Snyder AC, Welsh R. Athletic performance in relation to training load. Wis Med J. 1996;95:370–4.

Foster C, Florhaug A, Franklin J, Gottschall L, Hrovatin LA, Parker S, Doleshal P, Dodge C. A new approach to monitoring exercise training. J Strength Cond Res. 2001;15:109–15.

Gomez-Piriz PT, Jiminez-Reyes P, Ruiz-Ruiz C. Relation between total body load and session rating of perceived exertion in professional soccer players. J Strength Cond Res. 2011;25: 2100–3.

Green JM, Yang Z, Laurent CM, Davis JK, Kerr K, Pritchett RC, Bishop PA. Session RPE following interval and constant-resistance cycling in hot and cool environments. Med Sci Sports Exerc. 2007;39:2051–7.
Green JM, McIntosh JR, Hornsby J, Timme L, Gover L, Mayes JL. Effect of exercise duration on session RPE at an individualized constant workload. Eur J Appl Physiol. 2009;107:501–7.
Haddad M, Chaouachi A, Castagna C, Wong DP, Behm DG, Chamari K. The construct validity of session RPE during an intensive camp in young male taekwondo athletes. Int J Sports Physiol Perform. 2011;6:252–63.
Haile L, Ledezma CM, Koch KA, Shouey LB, Aaron DJ, Goss FL, Robertson RJ. Predicted, actual and session muscle pain and perceived exertion during cycle exercise in young men. Med Sci Sports Exerc. 2008;40:S301.
Haile L, Gallagher M, Haile AM, Dixon CB, Goss FL, Robertson RJ. Session, segmented session, and acute RPE and affective responses to self-selected treadmill exercise. Med Sci Sports Exerc. 2013a;45:S167.
Haile L, Goss FL, Robertson RJ, Andreacci JL, Gallagher Jr M, Nagle EF. Session perceived exertion and affective responses to self-selected and imposed cycle exercise of the same intensity ion young men. Eur J Appl Physiol. 2013b;116:1755–65.
Herman L, Foster C, Maher MA, Mikat RP, Porcari JP. Validity and reliability of the session RPE method for monitoring exercise training intensity. S Afr J Sports Med. 2006;18:14–7.
Hunt SE, DiAlesandro A, Lambright G, Williams D, Aaron D, Goss F, Robertson R. Predicted and actual leg pain and perceived exertion during cycle exercise in young women. Med Sci Sports Exerc. 2007;39:S485.
Impellizzeri FM, Rampinini E, Coutts AJ, Sassi A, Marcora SM. Use of RPE-based training load in soccer. Med Sci Sports Exerc. 2004;36:1042–7.
Kane I, Robertson RJ, Fertman CI, McConnaha WR, Nagle EF, Rabin BS, Rubinstein EN. Predicted and actual exercise discomfort in middle school children. Med Sci Sports Exerc. 2010;42: 1013–21.
Killen LG, Green JM, O'Neal EK, McIntosh JR, Hornsby J, Coates TE. Effects of caffeine on session ratings of perceived exertion. Eur J Appl Physiol. 2013;113:721–7.
Kilpatrick MW, Robertson RJ, Powers JM, Mears JL, Ferrer NF. Comparisons of RPE before, during, and after self-regulated aerobic exercise. Med Sci Sports Exerc. 2009;41:681–6.
Lee IM, Sesso HD, Oguma Y, Paffenbarger Jr RS. Relative intensity of physical activity and risk of coronary heart disease. Circulation. 2003;107:1110–6.
Lockie RG, Murphy AJ, Scott BR, Janse de Jonge XA. Quantifying session ratings of perceived exertion for field-based speed training methods in team sport athletes. J Strength Cond Res. 2011;26:2721–8.
McGuigan MR, Egan AD, Foster C. Salivary cortisol responses and perceived exertion during high intensity and low intensity bouts of resistance exercise. J Sports Sci Med. 2004;3:8–15.
McGuigan MR, Dayel AA, Tod D, Foster C, Newton RU, Pettigrew S. Use of session rating of perceived exertion for monitoring resistance exercise in children who are overweight or obese. Pediatr Exerc Sci. 2008;20:333–41.
McLean BD, Coutts AJ, Kelly V, McGuigan MR, Cormack SJ. Neuromuscular, endocrine, and perceptual fatigue responses during different length between-match microcycles in professional rugby league players. Int J Sports Physiol Perform. 2010;5:367–83.
Milanez VF, Pedro RE, Moreira A, Boullosa DA, Salle-Neto F, Nakamura FY. The role of aerobic fitness on session rating of perceived exertion in futsal players. Int J Sports Physiol Perform. 2011;6:358–66.
Minganti C, Capranica L, Meeusen R, Amici S, Piacentini MF. The validity of session-rating of perceived exertion method for quantifying training load in teamgym. J Strength Cond Res. 2010;24:3063–8.
Minganti C, Capranica L, Meeusen R, Piacentini MF. The use of session-RPE method for quantifying training load in diving. Int J Sports Physiol Perform. 2011a;6:408–18.
Minganti C, Ferragina A, Demarie S, Verticchio N, Meeusen R, Piacentini MF. The use of session RPE for interval training in master endurance athletes: should rest be included? J Sports Med Phys Fitness. 2011b;51:547–54.

Moreira A, McGuigan MR, Arruda AFS, Freitas CG, Aoki MS. Monitoring internal load parameters during simulated and official basketball matches. J Strength Cond Res. 2012;26:861–6.
Poulton R, Trevena J, Reeder AI, Richard R. Physical health correlates of overprediction of physical discomfort during exercise. Behav Res Ther. 2002;40(4):401–14.
Pritchett RC, Green JM, Wickwire PJ, Pritchett KL, Kovacs MS. Acute and session RPE responses during resistance training: bouts to failure at 60 % and 90 % of 1RM. S Afr J Sports Med. 2009;21:23–6.
Rodriguex-Marroyo JA, Villa G, Garcia-Lopez J, Foster C. Comparison of heart rate and rating of perceived exertion methods of defining exercise load in cyclists. J Strength Cond Res. 2012; 26:2249–57.
Rudra CB, Williams MA, Lee IM, Miller RS, Sorensen TK. Perceived exertion during prepregnancy physical activity and preeclampsia risk. Med Sci Sports Exerc. 2005;37:1836–41.
Rudra CB, Williams MA, Lee IM, Miller RS, Sorensen TK. Perceived exertion in physical activity and risk of gestational diabetes mellitus. Epidemiology. 2006;17:31–7.
Schafer MA, Robertson RJ, Thakkada SJ, Gallagher Jr M, Hunt SE, Goss FL, Aaron DJ. Validation of the OMNI RPE seven day exertional recall questionnaire. Res Q Exerc Sport. 2013;84: 363–72.
Seiler KS, Kjerland GO. Quantifying training intensity distribution in elite endurance athletes: is there evidence for an "optimal" distribution? Scand J Med Sci Sports. 2006;16:49–56.
Singh F, Foster C, Tod D, McGuigan MR. Monitoring different types of resistance training using session rating of perceived exertion. Int J Sports Physiol Perform. 2007;2:34–45.
Sweet TW, Foster C, McGuigan MR, Brice G. Quantitation of resistance training using the session rating of perceived exertion method. J Strength Cond Res. 2004;18:796–802.
Tessitore A, Perroni F, Cortis C, Meeusen R, Lupo C, Capranica L. Coordination of soccer players during preseason training. J Strength Cond Res. 2011;25:3059–69.
Viveiros L, Costa EC, Moreira A, Nakamura FY, Aoki MS. Training load monitoring in judo: comparison between the training load intensity planned by the coach and the intensity experienced by the athlete. Rev Bras Med Esporte. 2011;17:266–9.
Wallace LK, Slattery KM, Coutts AJ. The ecological validity and application of the session-RPE method for quantifying training loads in swimming. J Strength Cond Res. 2009;23:33–8.

Part IV
Applied Perceptual and Psychosocial Research

Chapter 14
Application of Perceptual Models to the Measurement of Pain and Affective Responses to Exercise

Thus far, this laboratory manual has presented the conceptual models, background information, previous literature, and current methodologies for the measurement of perceived exertion responses to exercise. The application of perceptual responses to exercise assessment, prescription and program monitoring has been discussed. The study and development of the perceived exertion knowledge base, however, has expanded over the years to include other perceptual and psychosocial constructs, i.e., naturally occurring muscle pain, affect, and enjoyment. It has been argued that, in addition to an individual's perception of physical exertion, variables such as pain, affect, and enjoyment may play an important role in determining the level of regular PA participation. Part 4 of this manual is titled Applied Perceptual and Psychosocial Research. This, the first chapter in Part 4, presents a series of *power reviews*, or brief summaries of the literature, concerning the measurement of naturally occurring muscle pain, affect, and enjoyment during exercise. Each section of this chapter can be linked retroactively to specific content presented previously regarding perceived exertion. Then, the remaining chapters of Part 4 present more extensive literature reviews for topics that are of growing interest concerning perceptual and psychosocial responses to exercise. These topics include the effects of caffeine supplementation, acute carbohydrate feeding, and music on perceptual, affective, and physiological responses to exercise.

14.1 Application of Perceived Exertion Scaling Procedures to Pain and Affect

See Chap. 5. Perceived Exertion Scaling Procedures.

L. Haile et al., *Perceived Exertion Laboratory Manual*,
DOI 10.1007/978-1-4939-1917-8_14

14.1.1 Comment on Anchoring for Exercise-Induced Pain

Anchoring procedures are not as extensive for measurement of naturally occurring muscle pain during exercise compared to those required for perceived exertion metrics. The memory anchoring procedures can be quite similar for exercise-induced pain as described above for perceived exertion. Prior to exercise, while presenting a standardized instructional set, the individual is asked to think about the pain experienced in the active muscles during previous exercise or physical activity. Then, the individual is asked to remember times when levels of muscle pain equal to the low and high anchor points on the scale were experienced. During subsequent bouts of exercise, the individual is asked to rate muscle pain levels based on the memory of muscle pain at the low and high anchor points.

However, an exercise anchoring procedure cannot be used in conjunction with a pain scale in the same manner that it is used with a perceived exertion scale. The psychophysical concept underlying perceived exertion scale anchoring is based on the predictions of Borg's Range Model. The basic tenets of this model assume that as individuals undertake exercise intensities across their entire performance range they are able to link physiological responses to corresponding and interdependent RPE values. This assumes that maximal RPE (e.g., ten on the OMNI Scale) is linked to attainment of maximal exercise intensity (e.g., POmax, VO_2max, 1RM). However, the achievement of maximal level of exercise-induced muscle pain as required for exercise anchoring procedures is not always possible. In studies by Cook and colleagues (1997, 1998), individuals did not detect muscle pain during load-incremented exercise until they attained 50–60 % of peak exercise capacity, with the pain threshold of some individuals not occurring until 90 %. Peak muscle pain values averaged ~5.5 in females and 8–8.5 in males using the 0–10 Pain Intensity Scale (Cook et al. 1997, 1998). In addition, although an individual may be able to remember a high level of muscle pain sensation experienced during previous exercise, for both clinical and physiological reasons it may not be possible to elicit such a response in certain individuals. Such limitations render the use of exercise anchoring procedures for category pain scales impractical.

14.1.2 Comment on Anchoring for Affective Responses to Exercise

Ratings of affective responses (AR) and PA enjoyment (PAE) during exercise are recognized as psychosocial correlates of perceived exertion. However, category scales to measure these constructs cannot be anchored at very low and very high exercise intensities as is the accepted procedures when anchoring an RPE scale. Individuals cannot be instructed to link AR or PAE values to any specific exercise intensity because these responses have been uniquely shaped over time in each individual. Previous PA experience, subjective behavioral norms, and values pertaining to

PA adherence can vary greatly between individuals. This can result in interindividual differences between specific psychosocial domains that dominate the affective and enjoyment experience to exercise intensity.

Research results are conflicting regarding the intensities of exercise that result in the most positive AR during exercise. Kirkcaldy and Shephard (1990) proposed an inverted-U paradigm, predicting that moderate intensity exercise produces an optimal AR. Similar findings have been reported by Moses et al. (1989). Low exercise intensities may be insufficient to evoke positive changes in AR and high exercise intensities may produce significant negative shifts in AR. More recent evidence refutes this relation such that both high (Tate and Petruzzello 1995) and low intensity (Ekkekakis et al. 2000) exercise programs have led to positive changes in AR.

Ekkekakis' (2003) "dual-mode" model explains the interindividual variability in AR that occurs across exercise intensities, specifically as it relates to the anaerobic threshold (AT). The AR during low to moderate intensity exercise (i.e., below the AT) is primarily shaped by cognitive processes that are unique to the individual. Above the AT, interoceptive cues driven by the increasing demand for energy supplied by anaerobic pathways dominate the AR. Therefore, AR at exercise intensities at or somewhat below the AT are rather heterogeneous, but AR at exercise intensities above the AT become increasingly less positive/more negative and are relatively homogeneous (Ekkekakis 2003; Ekkekakis et al. 2005; Hall et al. 2002).

Research has confirmed the marked interindividual differences in AR during exercise intensities below the AT, especially involving moderate intensity exercise. In a study by Van Lunduyt and colleagues (2000), participants estimated AR during moderate intensity cycle exercise (60 % VO_2peak). Results indicated that 44.4 % of subjects experienced an increase in AR, 41.3 % experienced a decrease in AR, and 14.3 % experienced no change in AR. Other studies have confirmed the shift from heterogeneity in AR at intensities below the AT to homogeneity in AR above the AT. In response to separate 15-min bouts of treadmill exercise, 47 % of subjects exhibited a decline in AR at intensities below the ventilatory threshold (VT) and 80 % of subjects exhibited a decline in AR at intensities above the VT (Ekkekakis et al. 2005). Similar results were found in response to 20 min of treadmill exercise. AR was more positive and stable below the AT with only 25 % of subjects exhibiting a decline in AR during performance at these intensities. Above the AT, 83 % of subjects exhibited a negative shift in AR (Parfitt et al. 2006).

14.1.3 Scaling Procedures: Practice and Feedback for Perceptual and Affective Variables

When exercise-induced muscle pain or affect are part of a perceptual research paradigm, it may be beneficial to ask the individual to practice rating these variables along with perceived exertion during exercise anchoring procedures or a practice exercise test. This will allow the individual to practice rating all three of these independent constructs within a close time-frame during exercise. In addition, such orientation procedures present an opportunity to provide feedback to an individual prior to fitness testing or experimental exercise procedures regarding psychophysical appropriateness of his/her rating responses. It may be especially beneficial for children, enabling them to more accurately link the exercise intensity range to their own pain and affective experience (Robertson et al. 2009).

14.2 Validation of Scales for Measuring Pain and Affect During Exercise

See Chap. 6. Perceived Exertion Scale Validation.

14.2.1 Validity of Exercise-Induced Pain Scales

The neurophysiological mechanisms for naturally occurring, exercise-induced pain in healthy, uninjured individuals involve stimulation of mechanical and biochemical nociceptive systems in skeletal muscle. Pain threshold is defined as the onset of pain sensation and varies between individuals. Once pain threshold is reached, ratings of exercise-induced muscle pain should increase with physical measures of exercise intensity, such as PO and weight lifted. This measure of pain sensation occurs in conjunction with the accumulation of noxious by-products of metabolism such as blood lactate, hydrogen ions, and bradykinin, all of which increase as a function of increasing exercise intensity. Early exercise-induced muscle pain studies used the Borg (0–10) CR10 Scale to measure "aches and pain in the legs" during load-incremented and constant PO cycle exercise (Borg et al. 1985; Ljunggren et al. 1987). The investigations demonstrated evidence of concurrent validity of the CR10 Scale to measure pain sensations. Pain ratings were moderately correlated to blood lactate concentration at high PO's during load-incremented exercise, with $r=0.45$ at 200 W and $r=0.39$ at 240 W (Borg et al. 1985), and at the end of constant PO exercise, with $r=0.54$ (Ljunggren et al. 1987).

Later studies confirmed concurrent validity of the Pain Intensity Scale developed by Cook and colleagues (1997). The Pain Intensity Scale employs construct-specific verbal descriptors that are linked to the same numerical categories as appear on the original Borg CR10 Scale. In Cook's investigation, pain ratings increased as a positively accelerating function of exercise intensity once pain threshold was achieved. It was noted that pain threshold ranged from 9 to 95 % of POpeak, indicating marked interindividual differences during load-incremented cycle ergometry (Cook et al. 1997, 1998). Mean pain threshold was ~50 % POpeak in males (Cook et al. 1997, 1998) and ~60 % POpeak in females (Cook et al. 1998). In males, pain ratings derived from the Pain Intensity Scale increased from a mean of ~2 at 60 % of POpeak to ~8–8.5 at 100 % of POpeak. In females, pain ratings increased from a mean of ~1 at 60 % of POpeak to ~5.5 at 100 % of POpeak

(Cook et al. 1997, 1998). Robertson and colleagues (2009) developed the OMNI-Muscle Hurt Scale to measure exercise-induced muscle pain in children. This investigation found evidence for concurrent scale validity during isotonic resistance exercise performed by young children. High correlations were exhibited between weight lifted and pain ratings for biceps curl resistance exercise and knee extension resistance exercise, with r values across sets ranging from 0.67 to 0.87 (Robertson et al. 2009). In addition, construct validity was evidenced in Cook's original study using the Pain Intensity Scale during load-incremented cycle exercise. High correlations ranging from $r = 0.79–0.94$ were found at intensities from 60 to 100 % POpeak (Cook et al. 1997).

14.2.2 Construct Validity Evidence for the Feeling Scale

Hardy and Rejeski (1989) demonstrated both construct and content validity of the Feeling Scale (FS) in college-aged males and females. The Multiple Affective Adjective Checklist (MAAC) employs a set of 132 adjectives. Subscales of the MAAC were used to compute criterion scores for both positive and negative affect. One group of subjects was instructed to choose adjectives describing a good feeling during exercise, while the other group chose adjectives describing a bad feeling during exercise. The results of the study found that subjects identified different affective states having good and bad feelings during exercise. The AR appropriately represented items at either end of the pleasure–displeasure continuum. The differentiated AR continuum was seen in 97 % of subjects who were asked to identify adjectives matching bad feelings and 94 % of subjects asked to identify adjectives matching good feelings (Hardy and Rejeski 1989). Kenney and colleagues (1987) conducted an investigation that also provided construct validity evidence for the FS in college-aged females. The study involved a cognitive-behavioral distress management training (DMT) program. The DMT program was administered to half of the participants between separate treadmill exercise bouts performed to exhaustion. The subjects who were administered the DMT program rated a more positive AR than subjects who did not receive the DMT when measures were obtained at the end of the treadmill run to exhaustion. However, RPE values were similar between subject groups (Kenney et al. 1987).

14.2.3 Validity of Enjoyment Measures during Exercise

A few investigations have tested the validity of recently developed single-item PA enjoyment (PAE) scales (Haile et al. 2012; Stanley et al. 2009). These investigations correlated PAE ratings with AR measured using the FS. During both a load-incremented cycle ergometer protocol terminating at VO_2peak (Haile et al. 2012) and during a 20-min moderate intensity constant load cycle ergometer protocol

(Stanley et al. 2009), significant positive relations were demonstrated between PAE and FS responses. In the investigation by Haile et al. (2012), PAE was measured using an 11-category scale that used the same format as the FS, with responses ranging from −5 to 5. The observed correlation coefficient between PAE and FS ratings was $r=0.92$. In the investigation by Stanley et al. (2009), PAE was measured using a seven-point scale that employed a format different than the FS. The observed correlation coefficients between PAE and FS ratings ranged from $r=0.48$–0.55.

The comparatively higher correlation coefficients reported by Haile et al. (2012) may be due to their use of a scale with similar format for the measurement of both AR and PAE. This argument has been employed to avoid the measurement of independent perceptual constructs using the same scale (Cook et al. 1997). For example, previous investigations measured both perceived exertion and pain during exercise using the CR10 Scale (Borg et al. 1985; Ljunggren et al. 1987). The resultant high correlation coefficients between RPE and pain intensity ratings may have been a "*demand artifact*" resulting from use of the same perceptual scale format to measure the two independent perceptual constructs (Cook et al. 1997). AR and PAE cannot be labeled as independent constructs similar to perceived exertion and pain. Rather, PAE is a specific domain of overall affect that may dominate the AR to exercise in many individuals. In addition, the PAE rating scale, although having a similar format to the FS, has verbal descriptors specific to enjoyment (Haile et al. 2012). Regardless, since acute exercise enjoyment is a novel construct, further research is necessary to study the measurement of AR and PAE simultaneously during exercise. In some populations in which enjoyment is a primary mediator of the overall affective experience during PA, it may be appropriate to measure PAE only.

14.3 Target Pain and Affect Ratings for Exercise Intensity Prescription

See Chap. 7. Target RPE at the Ventilatory Threshold.

14.3.1 Target Pain Ratings for Exercise Prescription

Symptomatic pain has been used routinely to identify tolerable limits of exercise for clinical populations such as those with peripheral artery disease who experience intermittent claudication in active limbs. However, little research has focused on the use of exercise-induced pain as a target for exercise intensity prescription. O'Connor and Cook (2001) had young female adults perform 20 min of cycle ergometer exercise at a target muscle pain intensity rating of 3 on Cook's (1997) 0–10 Pain Intensity Scale. A rating of 3 corresponds to the verbal descriptor "moderate pain." On average, the target level of muscle pain was associated with a relative aerobic metabolic rate of 73.9 % VO_2peak at 6 min of continuous exercise, decreasing to 68.5 % VO_2peak at 20 min (O'Connor and Cook 2001).

The long-term adherence to exercise prescriptions that are based on muscle pain response are unknown. Pain experience of any kind during exercise may be a major factor contributing to sedentary behavior in many individuals. Therefore, prescribing exercise at intensities below the individual's pain threshold may promote adherence to PA programs. However, previous research has found great interindividual variability in the pain threshold, as evoked during exercise. This makes it difficult to identify a group-normalized pain response that corresponds to a target physiological outcome and applies to a variety of activities for a wide range of individuals in a manner such as been shown for the RPE at the VT (Goss et al. 2003). Studies by Cook and colleagues (1997, 1998) determined that the pain threshold during load-incremented exercise occurred at 50–60 % of peak exercise capacity, but values ranged from 9 to 95 % of POpeak. Therefore, exercise intensity prescription based on pain ratings should take an individual approach, recognizing that the procedure may not be appropriate in those with a low pain threshold. Athletes performing high intensity exercise in which exercise-induced muscle pain is expected are a healthy population for which the prescription of exercise intensity using target muscle pain ratings has the most utility.

14.3.2 Target AR for Exercise Intensity Prescription

It has been shown that the amount of time spent during a given situation can depend on the affect experienced during the activity (Emmons and Diener 1986). Therefore, the acute AR to an initial exercise performance may influence future exercise participation. Exercise perceived as feeling pleasant may promote future participation. On the other hand, exercise perceived as feeling unpleasant could decrease future participation or lead to withdrawal from the activity altogether (Parfitt et al. 2006). The goal, then, is to maximize the positive AR that an individual experiences during exercise. This goal recognizes that a positive affective experience is an important link in the chain between exercise adoption and maintenance (Van Lunduyt et al. 2000).

A study by Da Silva and colleagues (2011) determined the AR corresponding to exercise intensities spanning the VT in sedentary normal weight, overweight and obese women. This application of the AR in exercise prescription was similar to methods used for calculation of RPE-VT. FS ratings were assessed throughout a graded treadmill exercise test to measure VO_2max. The FS ratings corresponding to 90 %, 100 % and 110 % of the VT were identified. Group average FS ratings for the entire sample were ~2.7, ~1.6, and ~0 corresponding to exercise intensities at 90 %, 100 % and 110 % of the VT, respectively. The AR were similar between normal weight and overweight groups at each intensity. The FS ratings were approximately 3, 2, and 1 at 90 %, 100 % and 110 % of the VT, respectively. The obese group had similar FS ratings to the normal weight and overweight groups at 90 % of the VT, but their ratings were significantly less positive at 100 % of the VT (mean FS rating = 0.5) and 110 % of the VT (mean FS rating = −1.95). These data indicate a positive affective experience at intensities spanning the VT in sedentary normal weight and overweight women, but obese women may require exercise intensities below the VT to experience positive AR (Da Silva et al. 2011). It must be noted that there was considerable variability in FS responses at each intensity, so even when the average FS rating was positive some subjects rated a negative affective experience. In addition, these data were collected during graded treadmill exercise, unlike normal continuous intensity or interval exercise bouts prescribed for health-fitness benefits.

Performing at exercise intensities that span the VT has resulted in significant changes in FS ratings during exercise. Ekkekakis and colleagues (2008) studied the AR of young adults during 15 min of continuous treadmill exercise at intensities corresponding to the VT, 20 % below the VT, and 10 % above the VT. In the condition where intensity was below the VT, 50 % of subjects experienced no change in AR throughout exercise while 43 % experienced a decrease in AR. At intensities equal to and above the VT, 77 % and 80 % of subjects experienced a decrease in AR throughout exercise, respectively. In every condition, however, a small number of subjects experienced an increase in AR during exercise (Ekkekakis et al. 2008).

The identification of a group-normalized AR at the VT may prove difficult for exercise intensity prescription due to the marked interindividual variability in AR across exercise intensities. This wide response variability is similar to that evidenced for exercise-induced pain ratings. The variability in AR is especially evident during moderate intensity exercise (Van Lunduyt et al. 2000). It is of note that moderate intensity exercise is recommended by professional organizations as the optimal level for PA programs designed to produce health-fitness benefits (ACSM 2013). According to Ekkekakis' "dual-mode model," exercise intensity above the VT results in the lowest interindividual variability in AR. This is due to the comparative dominance of noxious properties of physiological signals over cognitive processes in shaping the affective experience. Unfortunately, the comparatively more homogenous AR to exercise above the VT is typified by progressively more negative feelings (Ekkekakis et al. 2005). A negative AR during exercise indicating a displeasurable experience most likely contributes to poor program adherence.

Various investigations have shown that optimal AR may occur at low, moderate, or even high exercise intensities (Ekkekakis et al. 2000; Moses et al. 1989; Tate and Petruzzello 1995). As such, the development of an exercise prescription using AR measured separately for each individual may be a necessary approach to maximize PA adherence. Exercise prescriptions should identify the appropriate exercise intensity by choosing a target HR or RPE based on the optimal AR, or even by prescribing exercise intensity using a target FS rating. Rose and Parfitt (2008) asked sedentary women to perform separate 30-min treadmill exercise bouts at target FS ratings of 1 and 3. On average, the women chose an exercise intensity similar to the VT for both target FS ratings, indicating that the women felt the treadmill exercise was pleasurable (Rose and Parfitt 2008). The implications for program adherence using prescribed target FS ratings are unknown, but hold promise from a public health perspective. Monitoring and adjusting PA programs to continually optimize AR may be necessary to promote long-term habitual PA participation.

14.4 Estimation–Production Paradigm and Exercise Intensity Self-Regulation Using Pain and Affect

See Chap. 9. The Estimation–Production Paradigm for Exercise Intensity Self-Regulation.

14.4.1 Use of the Estimation–Production Paradigm for Exercise-Induced Pain

An estimation–production prescription paradigm has been used to assess the validity of exercise intensity self-regulation using ratings of exercise-induced muscle pain intensity (O'Connor and Cook 2001). This prescription procedure recognizes that normally occurring muscle pain during exercise may be an appropriate cue upon which to self-regulate exercise intensity for healthy, injury free individuals. The paradigm employed in the investigation by O'Connor and Cook (2001) differed from investigations of prescription congruence using a target RPE in that physiological responses were not compared between estimation and production protocols. Rather, the estimation trial had two purposes: (1) to allow subjects to experience the range of perceptual responses for both quadriceps muscle pain and RPE from very low to very high cycle ergometer exercise intensity, and (2) to measure the PO that corresponded to the subjects' pain threshold, which was used as the initial intensity during the production protocol (O'Connor and Cook 2001).

It has been proposed that mechanisms underlying exercise-induced muscle pain involve noxious chemical by-products of metabolism, such as bradykinin and hydrogen ions. These by-products will accumulate as exercise duration increases, intensifying muscle pain. As such, sustained exercise at an intensity above the pain threshold may result in an increase in ratings of muscle pain intensity. To maintain a specific pain rating, then, would require a gradual decrease in exercise intensity over time. Over prolonged exercise periods at a moderate pain intensity level, it would be expected that physiological variables such as VO_2 and HR would be lower during the production trial than the estimation trial.

O'Connor and Cook (2001) asked college-aged female subjects to produce a moderate muscle pain intensity level equivalent to a category 3 on the Cook (1997) Pain Intensity Scale during 20 min of cycle ergometry. On average, the women achieved the desired pain intensity level by minute 4, then decreased power output almost 16 % throughout the remaining 16 min of the production trial in order to maintain the target pain level. Moderate pain intensity was associated with an average RPE of approximately 14–15 on the Borg Scale and 70–75 % VO_2peak (O'Connor and Cook 2001).

Due to the interindividual variability in pain intensity responses, as well as differing affective components of pain, exercise intensity prescription based on muscle pain may not be appropriate for some. However, athletes accustomed to experiencing naturally occurring muscle pain during high intensity exercise may find it useful to self-regulate exercise intensity according to a pain intensity scale. Others may prefer to exercise at an intensity below the pain threshold. Identifying the highest exercise intensity an individual can perform without experiencing muscle pain may be a method to improve PA participation. However, some individuals may experience their pain threshold at exercise intensities below the physiological threshold required to produce health-fitness benefits. The measurement of exercise-induced muscle pain during a GXT, along with measurements of RPE and AR, can provide the necessary information to prescribe exercise intensity to optimize PA program adherence.

14.4.2 Exercise Intensity Self-Regulation Using AR

An estimation–production prescription paradigm has been used to assess the validity of exercise intensity self-regulation using FS ratings of AR measured during single exercise bouts (Rejeski et al. 1987; Rose and Parfitt 2008). Affect is a psychosocial construct that mediates the perception of physical exertion. As such, it is an appropriate cue for exercise intensity prescription, ultimately promoting optimal adherence to PA programs. Rose and Parfitt (2008) conducted an investigation in which an estimation protocol was used to familiarize sedentary female subjects with use of the FS and Borg Scale prior to the performance of eight 30-min production protocols over the course of 4 weeks. During four consecutive production protocols, subjects were asked to produce a target FS rating of 1 (fairly good). During the other four consecutive production protocols, subjects were asked to produce a target FS rating of 3 (good). The purpose of four consecutive production protocols for each target FS rating was to test the reproducibility of exercise intensity self-regulation using FS ratings of AR (Rose and Parfitt 2008). However, physiological responses were not compared between estimation and production protocols, negating the opportunity to assess prescription congruence. For each target FS rating, subjects consistently self-regulated exercise intensity across trials. A FS rating of 1 was associated with a group average of 68 % HRmax and Borg Scale RPE of 12. A FS rating of 3 was associated with a group average of 64 % HRmax and RPE of 11.4. Interestingly, the difference between feeling "fairly good" and "good" during exercise was represented by 4 HRmax percentage points and less than one Borg Scale RPE numerical category. This indicates that changes in overall AR can be caused by very small changes in exercise intensity (Rose and Parfitt 2008), with some individuals having a more sensitive AR to exercise than others. This highlights the utility of measuring AR, and even possibly enjoyment during exercise, in order to determine appropriate exercise intensity for PA particiption. For example, using a response-normalized perceptual response to prescribe exercise intensity such as an RPE-VT of 13 on the Borg Scale could result in negative FS ratings in some individuals but positive FS ratings in others. Prescribing intensity based on the specific RPE at which positive AR was experienced during an estimation trial, or by using FS ratings directly such as in the study by Rose and Parfitt (2008), may be a practical method to promote exercise adherence. In addition, using VO_2 estimated from

ACSM metabolic equations, subjects produced intensities averaging greater than the VT for FS ratings of both 1 and 3 (Rose and Parfitt 2008). Therefore, this prescription method may be effective for promoting health-fitness benefits as well.

14.5 Interval Exercise Prescription Using Pain and Affect

See Chap. 10. Exercise Intensity Self-regulation for Interval Exercise.

14.5.1 Regulation of Aerobic Interval Exercise Using Target Exercise-Induced Pain Ratings

O'Connor and Cook (2001) conducted an investigation in which the Pain Intensity Scale was used by females to produce a cycle ergometer exercise intensity corresponding to a target pain rating of 3, indicating moderate intensity pain. The target pain rating was achieved after approximately 4 min of exercise, whereupon the women gradually decreased power output to maintain the target pain rating throughout the remainder of the 20-min exercise bout. The self-regulated intensity corresponded to 68–74 % VO_2peak and Borg Scale RPE's of 14–15 (O'Connor and Cook 2001). These responses indicate that exercise intensity self-regulation using target pain ratings can be a model for an effective exercise program.

Prescribing multiple target pain ratings to regulate an interval exercise format should also be explored. Even though popular exercise programs and video programs promote exercise using phrases like "feel the burn" and "no pain no gain" (Cook et al. 1997), some individuals may not be comfortable exercising at a moderate pain intensity for prolonged periods, such as the 20-min exercise bout used by O'Connor and Cook (2001). Therefore, prescribing exercise intensity using an interval format may be more appropriate for these individuals. For health-fitness programming, exercise bouts could be prescribed using comparatively higher intensity intervals corresponding to moderate pain intensity, such as a 3 on the Pain Intensity Scale or a 4 on the Children's OMNI Muscle Hurt Scale. These exercise intervals are interspersed with active recovery phases performed at intensities below the pain threshold, i.e., pain ratings of 0. The duration of higher intensity intervals could be adjusted based on the individual's tolerance to exercise-induced pain.

14.5.2 Aerobic Interval Exercise: Intensity Discrimination Using AR

Rose and Parfitt (2008) used an estimation–production paradigm to assess the validity of exercise intensity self-regulation using target ratings of AR, specifically the target FS ratings 1 (fairly good) and 3 (good). Sedentary women were asked to produce each target FS rating on four separate occasions while performing 30 min of treadmill exercise each session. On average, self-regulated exercise intensity at a target FS rating of 1 was associated with 68 % HRmax and Borg Scale RPE of 12, while the target FS rating of 3 was associated with 64 % HRmax and RPE of 11.4 (Rose and Parfitt 2008). Statistically, these differences in % HRmax and RPE between target FS ratings 1 and 3 were significant, and therefore can be considered evidence for subjective intensity discrimination. From a practical standpoint, a difference of 4 % of HRmax and less than one numerical RPE category, while statistically significant, may not be functionally important. Nevertheless, these results indicate that small changes in exercise intensity can be quite important in altering the affective exercise response. These relatively small changes in HR and perceived exertion represented the difference between feeling "fairly good" and "good" during four 30-min bouts of exercise for each condition. Using ACSM metabolic equations to estimate VO_2, the investigators found the produced intensities to be slightly higher than the group average VT (average of 6 % for FS rating 3, 8 % for FS rating 1) (Rose and Parfitt 2008). At intensities near the VT, substantial changes in the physiological milieu occur that may result in a heightened sensitivity to changes in AR.

The study by Rose and Parfitt (2008) demonstrated that individuals can self-regulate exercise intensity using target FS ratings. In addition, FS ratings of 1 and 3 were both associated with an exercise intensity slightly above the VT. This result is in line with previous research in which subjects were asked to self-select exercise intensity for use during a cardiorespiratory conditioning program. On average, subjects chose intensities near the VT eliciting positive FS ratings between 2 and 4 (Lind et al. 2005; Parfitt et al. 2006; Rose and Parfitt 2007). This indicates that self-regulating exercise intensity based on target AR can yield an effective exercise program from both a physiological and psychological standpoint. Identifying exercise intensities that "feel good" may be an important characteristic to promote long term adherence to an exercise program. Therefore, an aerobic interval exercise program based on target FS ratings could also be effective in promoting health-fitness outcomes. Choosing FS ratings that correspond to target interval intensities may be difficult for some individuals, especially those who exhibit little change in FS ratings during load-incremented exercise. In addition, due to the marked interindividual differences in AR, especially at exercise intensities at or below the VT, an individual approach to determining target FS ratings may be best. Since many individuals provide increasingly negative FS ratings as exercise intensity exceeds the VT, intervals may have to include both positive target FS ratings (for lower intensities) and negative FS ratings (for higher intensities).

14.5.3 Effect of Aerobic Interval Exercise on AR and PAE

A lack of exercise-related enjoyment has been cited as a major barrier to regular PA participation (Trost et al. 2002). The variations in exercise intensity during aerobic interval exercise may be seen as more enjoyable than traditional, continuous, moderate intensity exercise prescriptions, especially for children. Children are more accustomed to spontaneous PA, such as encountered during sports and unstructured games during recess or after school. These activities involve short bouts of high intensity exercise interspersed with longer periods of light to moderate intensity exercise rather than continuous exercise for a prolonged period (Crisp et al. 2012a, 2012b). Interval exercise results in more enjoyment during exercise in both adults and children participating in a wide range of activities.

Bartlett and colleagues (2011) compared enjoyment responses between high intensity interval running with continuous moderate intensity running in recreationally active men. The moderate intensity running was performed at 70 % VO_2max for 50 min. The high intensity interval running included six 3-min intervals at 90 % VO_2max each followed by a 3-min recovery period at 50 % VO_2max. The interval exercise bout included 7 min of warm-up and cool-down at 70 % VO_2max to match overall time and work performed. Interval exercise resulted in a higher Borg Scale RPE versus moderate exercise (group average of 14 versus 13, respectively). Average VO_2, HR and energy expenditure were similar between trials. PA enjoyment (PAE), measured post-exercise using the PA Enjoyment Scale (PACES), was higher following high intensity interval exercise than moderate intensity continuous exercise (Bartlett et al. 2011).

Crisp and colleagues (2012a, 2012b) conducted two investigations that measured the effects of adding sprint intervals to continuous exercise at the intensity that optimizes fat oxidation in young boys performing cycle ergometry. Each exercise bout was performed for 30 min. In the sprint interval bouts, the boys were asked to perform 4-s, maximal intensity sprints every 2 min (Crisp et al. 2012a, 2012b), every 1 min, or every 30 s (Crisp et al. 2012b). This resulted in 1, 2, or 4 min of sprinting within the entire 30-min bout, respectively (Crisp et al. 2012a, 2012b). In the investigation that only included sprints every 2 min (Crisp et al. 2012a), the sprint intervals increased energy expenditure (via carbohydrate oxidation). However, PAE measured post-exercise using PACES was similar between trials. Investigators also asked the boys to indicate which exercise trial they preferred. Only 2 out of 18 preferred the moderate intensity exercise (Crisp et al. 2012a). In the investigation that included three separate exercise bouts with sprint intervals (Crisp et al. 2012b), energy expenditure was greater using the sprint than continuous format, regardless of the length of the active rest phases between sprint intervals. In addition, adding sprints every 30 or 60 s resulted in greater energy expenditure than adding sprints every 2 min, but the two higher frequencies were similar in energy expenditure. PACES scores were similar between exercise bouts with sprints and without except for the 30-s sprint frequency trial, which resulted in lower PAE (Crisp et al. 2012b). Overall, the results of the two studies indicate that adding sprints to a standard exercise intensity protocol at which maximal fat oxidation occurs could improve weight loss or weight maintenance. The sprint intervals added to moderate intensity exercise increased the overall caloric expenditure of acute exercise. PACES scores of PAE measured post-exercise were largely similar between bouts, but sprinting every 30 s was reported as unenjoyable (Crisp et al. 2012b). Since the boys indicated that they preferred the sprint interval exercise over the moderate intensity exercise bout

(Crisp et al. 2012a), perhaps a more simplified rating scale such as the FS or a single-item PAE scale could be used to measure basic AR or PAE during and after aerobic or sprint interval exercise to explore the acute exercise responses of these constructs. Due to the time needed for administration, a questionnaire such as PACES could not be used to measure overall PAE *during* exercise.

14.6 JND Methods for Exercise-Induced Muscle Pain and AR

See Chap. 11. Exercise Intensity Self-regulation using the Perceived Exertion JND.

Methods developed to determine the perceived exertion JND are appropriate for use with exercise-induced muscle pain ratings or FS ratings of AR in most individuals. As noted previously, marked interindividual differences have been found for the relation of these variables with exercise intensity. As such, a pre-participation GXT (i.e., estimation protocol) that includes measurement of pain and affect could help identify whether or not the use of these variables to prescribe exercise intensity is appropriate. For example, an individual may experience his/her pain threshold at or below an exercise intensity that elicits an optimal overload training stimulus. If during exercise, pain intensity ratings gradually increase over time, then a target pain rating could be used for exercise intensity prescription. In this instance, the JND for muscle pain intensity could be measured and used to assess the accuracy of exercise intensity self-regulation error in a subsequent production trial. Likewise, an individual's FS ratings of affect may gradually change (i.e., decrease) across the range of exercise intensities that are used as targets for exercise intensity prescription. When this occurs, a target FS rating could be used for exercise prescription and the JND for the AR during exercise could be measured. It is important that there be a gradual change in pain or affect across a range of exercise intensities so the target rating corresponds to a specific exercise intensity that is to be self-regulated.

However, standard methods to determine the JND may not be appropriate for use with exercise-induced muscle pain ratings or FS ratings of AR for some individuals. In certain cases, the inappropriateness of these variables for use in exercise prescription can be identified during the estimation trial. For example, an individual may not reach the pain threshold until exercise intensity is higher than the VT. A prescribed target intensity equivalent to the pain threshold likely could not be sustained for a sufficient period to achieve health-fitness goals. As such, exercise intensity prescription based on a target pain intensity rating would not be appropriate and JND methods could not be applied to exercise-induced muscle pain. In addition, some individuals may not experience gradual changes in FS ratings across the range of exercise intensities that are part of the prescribed exercise program (i.e., the subject reports each intensity as feeling "very good" or "very bad"). In these instances, a comparatively large range of exercise intensities is linked by a single rating of exercise-induced pain intensity or AR. For such individuals, it would be best to prescribe exercise intensity using RPE, which should gradually increase with exercise intensity regardless of interindividual differences in pain and AR.

14.7 Effect of Self-Selected Versus Imposed Exercise Intensity on Affect

See Chap. 12. Self-Selected versus *Imposed Exercise Intensities.*

Research that has compared FS ratings of AR between self-selected and imposed exercise intensities provides promising results, validating the procedures for use in prescribing exercise programs that are both physiologically effective and also promote long term adherence. It has been shown that many individuals experience a similar AR when performing exercise trials involving self-selected exercise intensity and imposed intensity, even though the intensity is actually higher for the self-selected condition (Ekkekakis and Lind 2006; Lind et al. 2008; Parfitt et al. 2006; Rose and Parfitt 2007). This indicates that subjects may be willing to perform a higher exercise intensity when it is self-selected to achieve a preferred level as compared to a prescription where exercise intensity is imposed. In addition, many individuals will self-select exercise intensities within ACSM guidelines for improvements in cardiorespiratory fitness (Dishman et al. 1994; Lind et al. 2005; Lind et al. 2008; Parfitt et al. 2006; Rose and Parfitt 2007). Therefore, the prescription of self-selected exercise may not only optimize AR and result in improved adherence to exercise programs, but may produce physiological benefit as well.

A number of recent studies have compared FS ratings of AR that were measured during self-selected and imposed exercise intensities. In these investigations, imposed exercise intensities have included those corresponding to levels below AT, above AT (Parfitt et al. 2006; Rose and Parfitt 2007; Sheppard and Parfitt 2008), equal to the AT (Rose and Parfitt 2007), as well as 10 % higher than the self-selected intensity (Ekkekakis and Lind 2006; Lind et al. 2008). Parfitt and colleagues (2006) compared FS ratings between 20 min of self-selected treadmill exercise and imposed exercise at intensities below and above the lactate threshold (LT) in sedentary males. Self-selected intensity was similar to that corresponding to the LT. FS ratings were similar (~3) between self-selected exercise and the imposed intensity below the LT. Self-selected exercise intensity and the imposed exercise intensity below the LT were performed at intensities equivalent to estimated VO_2 levels of 54.1 % and 39.8 % VO_2max, respectively. FS ratings during the imposed exercise condition at an intensity above the LT declined significantly over time, with the mean value eventually becoming negative by the 20-min time point (Parfitt et al. 2006).

Rose and Parfitt (2007) compared FS ratings during 20 min of self-selected treadmill exercise to imposed exercise at intensities below, above, and equal to the LT in sedentary women. Mean blood lactate concentration was similar between the self-selected exercise and the imposed intensities equal to and below the LT. The self-selected intensity resulted in FS ratings (ranging from 2.4 to 2.8) that were similar to those during an imposed exercise intensity which was below the LT. However, significantly more positive FS ratings were observed for self-selected exercise compared to an imposed exercise intensity equal to the LT (FS ratings ranged from 1.0 to 1.3). Imposed exercise at an intensity above the LT resulted mean FS ratings that declined significantly and remained negative throughout exercise with values ranging from −0.3 to −1.9 (Rose and Parfitt 2007).

Sheppard and Parfitt (2008) compared FS ratings between 15 min of self-selected and imposed cycle ergometer exercise intensities in young adolescent boys and girls who self-reported that they were physically active and moderately fit. The imposed intensities were below the VT (80 % of PO corresponding to VT) and above the VT (130 % of PO corresponding to VT). The imposed exercise intensity above the VT resulted in FS ratings that declined significantly over time and were significantly lower compared to those measured for both the imposed intensity that was performed below the VT and to those reported for the self-selected intensity condition. FS ratings were similar and stable over time during both the imposed intensity below the VT and the self-selected intensity condition. The mean FS ratings for the imposed intensities above and below the VT were ~0.4 and 2.5, respectively. Self-selected exercise intensity elicited an average FS rating of ~2.8 (Sheppard and Parfitt 2008).

Lind et al. (2008) compared FS ratings between 20 min of self-selected treadmill exercise to those reported during an imposed intensity that was 10 % higher than the self-selected intensity in sedentary women. At the 20-min time-point, the average exercise intensity was 98 % of VT for self-selected exercise and 115 % of VT for imposed exercise intensity. Subjects maintained a stable, positive AR during the self-selected condition. However, FS ratings of AR declined significantly during the imposed intensity that was only 10 % higher than the self-selected condition. The 10 % increase in intensity for the imposed condition, though comparatively small, was sufficient to prevent attainment of both a physiological and affective steady state (Lind et al. 2008). Ekkekakis and Lind (2006) compared FS ratings measured during 20 min of self-selected treadmill exercise and those measured during imposed exercise at an intensity 10 % higher than self-selected intensity in normal-weight and overweight sedentary women. Average self-selected intensity was below the VT while average imposed intensity was above the VT. However, FS ratings were similar between conditions. In both groups, average FS ratings were between 2 and 3 (Ekkekakis and Lind 2006).

14.8 Predicted and Session Measures of Pain and Affect

See Chap. 13. Predicted, Momentary and Session RPE.

14.8.1 Predicted and Session Exercise-Induced Pain

Hunt et al. (2007) and Haile at al. (2008) compared the predicted and momentary exercise-induced pain responses to load-incremented cycle ergometer exercise in young female and male adults, respectively. In both investigations, pain ratings were measured with Cook's (1997) Pain Intensity Scale. Both female and male subjects overpredicted their overall muscle pain response when compared to the momentary response actually experienced during exercise. The overprediction of pain found in these investigations may be due to the physiological demands of a load-incremented graded exercise test since few individuals perform maximal exercise on a regular basis. In addition, Kane and colleagues (2010) measured predicted exercise-induced muscle pain in middle school children prior to the performance of the PACER shuttle run test. Measures of muscle pain were determined using the Children's OMNI Muscle Hurt Scale. The children significantly over-predicted muscle pain by a value greater than 1 OMNI Scale rating category (Kane et al. 2010). It is common for individuals to overpredict an expected pain experience (Rachman and Arntz 1991). This has been suggested as a protective mechanism to avoid activities having the potential to cause tissue damage (Rachman and Lopatka 1988).

The investigations by Hunt et al. (2007) and Haile et al. (2008) also compared momentary and session pain responses. Subjects' session pain response was greater than the momentary response but was similar to predicted pain intensity. The rebound effect was most likely due to the influence of the most recently performed exercise intensity on the pain response, i.e., the intensity at which VO_2peak was achieved (Haile et al. 2008; Hunt et al. 2007).

14.8.2 Predicted and Session AR

Using the FS, Hardy and Rejeski (1989) asked subjects to predict the AR that would be experienced during running at specific Borg Scale RPEs. However, the question was asked hypothetically, that is, no exercise was performed following the prediction of AR. Average predicted AR was 2.6, 0.6, and −1.0 for Borg RPEs 11, 15, and 19, respectively. As RPE increased, FS ratings decreased. This inverse relation was consistent with predictions of the dual-mode model of Ekkekakis (2003) since the RPE zone encompassing the VT includes a Borg RPE of 11. The predicted FS values were significantly correlated to past and present levels of PA (i.e., past grade school, high school, and college PA, current PA frequency, current miles jogged per week) (Hardy and Rejeski 1989). The results indicate that predicted AR can provide valuable information that may help identify individuals who struggle with the adoption and maintenance of regular PA. Further research using a match–mismatch paradigm to compare the predicted and momentary AR associated with exercise may provide further information that can be useful for PA behavior change interventions.

Haile and colleagues (2013b) conducted an investigation that compared momentary and session FS ratings measured during 20 min of self-selected and imposed cycle ergometer exercise in young adult males. In this study, the self-selected exercise session was undertaken first so that subjects could perform the same intensity in the imposed condition, although they were not aware that the intensity was the same. Session AR was significantly greater than momentary AR for the self-selected exercise, but not the imposed exercise. In either case, however, the difference between momentary and session AR was less than 1 FS unit (Haile et al. 2013b). In another investigation, Haile and colleagues (2013a) compared momentary AR with both session and segmented session AR values for 20 min of self-selected treadmill exercise. In this investigation, both session RPE and segmented session RPE (expressed as the mean of the two segmented session RPE values, one for each half of exercise) were similar to the mean of the momentary RPE's measured during exercise. In addition, each separate segmented session RPE value was similar to the mean of momentary RPE's measured during that respective half of the exercise session (Haile et al. 2013a). Asking individuals to reflect upon specific segments of a previous exercise bout may improve their ability to accurately rate the perceived exertion experienced during previous exercise.

14.8.3 The Exercise Discomfort Index

In a study involving children, Kane et al. (2010) calculated an Exercise Discomfort Index (EDI) as the product of OMNI RPE-O and OMNI Muscle Hurt ratings (EDI = RPE-O × muscle hurt). Comparisons were made between predicted EDI and momentary EDI. The children significantly overpredicted EDI, but this response was primarily driven by the overprediction of muscle hurt ratings (Kane et al. 2010). Session EDI has not been investigated. An index such as EDI may provide a more in-depth explanation of an individual's perceptual expectations of exercise than either perceived exertion or exercise-induced muscle pain alone. A match–mismatch paradigm can be used to compare predicted EDI and momentary EDI. The purpose of this paradigm would be to identify those individuals with a response mismatch who may require cognitive or behavioral intervention to learn appropriate expectations of exertional perceptions during exercise (Kane et al. 2010). Such information may be crucial in helping such individuals to adopt and maintain regular PA. In addition, recall or session EDI can be used in PA questionnaires to describe the perceived exertion and muscle pain response to previous exercise. EDI may provide a more accurate description of an individual's recalled perceptual experience than either perceived exertion or exercise-induced muscle pain alone, whether the exercise was performed minutes ago (session EDI) or over the past few months (recall EDI).

The application of EDI could be expanded to include AR measured using the FS. Positive FS ratings indicate an individual is feeling good during exercise. Such feelings help to minimize exercise discomfort, promote the continuation of an exercise bout, and make it more likely that the individual would perform that exercise bout again. Negative FS ratings indicate an individual is feeling bad during exercise. Such feelings may exacerbate exercise discomfort, lead to premature termination of an exercise bout, and make it less likely that the individual would choose to perform that exercise bout again. Therefore, subtracting the FS rating from the EDI would be appropriate, allowing the formation of a revised EDI (EDI = OMNI RPE × muscle pain/hurt − FS rating). This newly proposed EDI can employ measures of either the undifferentiated or differentiated OMNI RPE, depending on the specific type of exercise evaluated. The modified EDI may also include ratings from the Cook Pain Intensity Scale which can be substituted for ratings obtained from the Children's OMNI Muscle Hurt Scale when adults are evaluated.

References

American College of Sports Medicine. ACSM's guidelines for exercise testing and prescription. 9th ed. Philadelphia, PA: Lippincott Williams & Wilkins; 2013.

Bartlett JD, Close GL, MacLaren DPM, Gregson W, Drust B, Morton JP. High-intensity interval running is perceived to be more enjoyable than moderate-intensity continuous exercise: implications for exercise adherence. J Sports Sci. 2011;29:547–53.

Borg G, Ljunggren G, Ceci R. The increase of perceived exertion, aches and pain in the legs, heart rate and blood lactate during exercise on a bicycle ergometer. Eur J Appl Physiol. 1985;54: 343–9.

Cook DB, O'Connor PJ, Eubanks SA, Smith JC, Lee M. Naturally occurring muscle pain during exercise: assessment and experimental evidence. Med Sci Sports Exerc. 1997;29:999–1012.

Cook DB, O'Connor PJ, Oliver SE, Lee Y. Sex differences in naturally occurring muscle pain and exertion during maximal cycle ergometry. Int J Neurosci. 1998;95:183–202.

Crisp NA, Fournier PA, Licari MK, Braham R, Guelfi KJ. Adding sprints to continuous exercise at the intensity that maximises fat oxidation: implications for acute energy balance and enjoyment. Metabolism. 2012a;61:1280–8.

Crisp NA, Fournier PA, Licari MK, Braham R, Guelfi KJ. Optimising sprint interval exercise to maximise energy expenditure and enjoyment in overweight boys. Appl Physiol Nutr Metab. 2012b;37:1222–31.

Da Silva SG, Elsangedy HM, Krinski K, De Campos W. Effect of body mass index on affect at intensities spanning the ventilatory threshold. Percept Mot Skills. 2011;113:575–88.

Dishman RK, Farquhar RP, Curetone KG. Responses to preferred intensities of exertion in men differing in activity levels. Med Sci Sports Exerc. 1994;26:783–90.

Ekkekakis P. Pleasure and displeasure from the body: perspectives from exercise. Cogn Emot. 2003;17:213–39.

Ekkekakis P, Hall EE, Van Lunduyt LM, Petruzzello SJ. Walking in (affective) circles. Can short walks enhance affect? J Behav Med. 2000;23:245–75.

Ekkekakis P, Hall EE, Petruzzello SJ. Variations and homogeneity in affective responses to physical activity of varying intensities: an alternative perspective on dose–response based on evolutionary considerations. J Sport Sci. 2005;23:477–500.

Ekkekakis P, Lind E. Exercise does not feel the same when you are overweight: the impact of self-selected and imposed intensity on affect and exertion. Int J Obes. 2006;30:652–60.

Ekkekakis P, Hall EE, Petruzzello SJ. The relationship between exercise intensity and affective responses demystified: to crack the 40-year-old nut, replace the 40-year-old nutcracker! Ann Behav Med. 2008;35:136–49.

Emmons RA, Diener E. A goal-affect analysis of everyday situational choices. J Res Pers. 1986;20:309–26.

Goss F, Robertson R, DaSilva S, Suminski R, Kang J, Metz K. Ratings of perceived exertion and energy expenditure during light to moderate activity. Percept Mot Skills. 2003;96:739–47.

Haile L, Ledezma CM, Koch KA, Shouey LB, Aaron DJ, Goss FL, Robertson RJ. Predicted, actual and session muscle pain and perceived exertion during cycle exercise in young men. Med Sci Sports Exerc. 2008;40:S301.

Haile L, Gallagher M, Haile AM, Dixon CB, Goss FL, Robertson RJ. Session, segmented session, and acute RPE and affective responses to self-selected treadmill exercise. Med Sci Sports Exerc. 2013a;45:S167.

Haile L, Goss FL, Robertson RJ, Andreacci JL, Gallagher Jr M, Nagle EF. Session perceived exertion and affective responses to self-selected and imposed cycle exercise of the same intensity ion young men. Eur J Appl Physiol. 2013b;116:1755–65.

Haile AM, Haile L, Taylor M, Shafer A, Wisniewski K, Deldin A, Panzak G, Goss FL, Nagle E, Robertson RJ. Concurrent validity of an exercise enjoyment scale using physiological and psychological criteria. Med Sci Sports Exerc. 2012;44:S645.

Hall EE, Ekkekakis P, Petruzzello SP. The affective beneficence of vigorous exercise revisited. Br J Health Psychol. 2002;7:47–66.
Hardy CJ, Rejeski WJ. Not what but how one feels: the measurement of affect during exercise. J Sport Exerc Psychol. 1989;11:304–17.
Hunt SE, DiAlesandro A, Lambright G, Williams D, Aaron D, Goss F, Robertson R. Predicted and actual leg pain and perceived exertion during cycle exercise in young women. Med Sci Sports Exerc. 2007;39:S485.
Kane I, Robertson RJ, Fertman CI, McConnaha WR, Nagle EF, Rabin BS, Rubinstein EN. Predicted and actual exercise discomfort in middle school children. Med Sci Sports Exerc. 2010;42: 1013–21.
Kenney E, Rejeski WJ, Messier SP. Managing exercise distress: the effect of broad spectrum intervention on affect, RPE, and running efficiency. Can J Sport Sci. 1987;12:97–105.
Kirkcaldy BC, Shephard RJ. Therapeutic implications of exercise. J Sport Psychol. 1990;21: 165–84.
Lind E, Joens-Matre RR, Ekkekakis P. What intensity of physical activity do previously sedentary middle-aged women select? evidence of a coherent pattern from physiological, perceptual, and affective markers. Prev Med. 2005;40:407–19.
Lind E, Ekkekakis P, Vazou S. The affective impact of exercise intensity that slightly exceeds the preferred level: 'pain' for no additional 'gain'. J Health Psychol. 2008;13:464–8.
Ljunggren G, Ceci R, Karlsson J. Prolonged exercise at a constant load on a bicycle ergometer: ratings of perceived exertion and leg aches and pain as well as measurements of blood lactate accumulation and heart rate. Int J Sports Med. 1987;8:109–16.
Moses J, Steptoe A, Mathews A, Edwards S. The effects of exercise training on mental well-being in the normal population: a controlled trial. J Psychosom Res. 1989;33:47–61.
O'Connor PJ, Cook DB. Moderate-intensity muscle pain can be produced and sustained during cycle ergometry. Med Sci Sports Exerc. 2001;33:1046–51.
Parfitt G, Rose EA, Burgess WM. The psychological and physiological responses of sedentary individuals to prescribed and preferred intensity exercise. Br J Health Psychol. 2006;11:39–53.
Rachman S, Lopatka C. Accurate and inaccurate predictions of pain. Behav Res Ther. 1988; 26:291–96.
Rachman S, Arntz A. The overprediction and underprediction of pain. Clin Psychol Rev. 1991; 11:339–55.
Rejeski WJ, Best D, Griffith P, Kenney E. Sex-role orientation and the responses of men to exercise stress. Res Q. 1987;58:260–4.
Robertson RJ, Goss FL, Aaron DJ, Nagle EF, Gallagher Jr M, Kane IR, Tessmer KA, Schafer MA, Hunt SE. Concurrent muscle hurt and perceived exertion of children during resistance exercise. Med Sci Sports Exerc. 2009;41:1146–54.
Rose EA, Parfitt G. A quantitative analysis and qualitative explanation of the individual differences in affective responses to prescribed and self-selected exercise intensities. J Sport Exerc Psychol. 2007;29:281–309.
Rose EA, Parfitt G. Can the feeling scale be used to regulate exercise intensity? Med Sci Sports Exerc. 2008;40:1852–60.
Sheppard KE, Parfitt G. Acute affective responses to prescribed and self-selected exercise intensities in young adolescent boys and girls. Pediatr Exerc Sci. 2008;20:129–41.
Stanley DM, Williams SE, Cumming J. Preliminary validation of a single-item measure of exercise enjoyment: the exercise enjoyment scale. J Sport Exerc Psychol. 2009;31:S138–9.
Tate AK, Petruzzello SJ. Varying the intensity of acute exercise: implications for changes in affect. J Sports Med Phys Fitness. 1995;35:295–302.
Trost SG, Owen N, Bauman AE, Sallis JF, Brown W. Correlates of adults' participation in physical activity: review and update. Med Sci Sports Exerc. 2002;34:1996–2001.
Van Lunduyt LM, Ekkekakis P, Hall EE, Petruzzello SJ. Throwing the mountains into the lakes: on the perils of nomothetic conceptions of the exercise-affect relationship. J Sport Exerc Psychol. 2000;24:151–69.

Chapter 15
Effects of Caffeine on Perceptual and Affective Responses to Exercise

An ergogenic aid is defined as a substance, technique or device that *directly* improves exercise performance or *indirectly* removes constraints to exercise performance. Caffeine is one such substance and is a proven ergogenic aid for performance during endurance, load-incremented, short-term high-intensity and resistance exercise modalities. The primary mechanism of action seemingly responsible for caffeine's ergogenic effect may be due to its antagonism of receptors for the neurotransmitter adenosine. Biochemically blocking adenosine receptors may have widespread effects on the body as a result of an increase in dopamine secretion. Dopamine increases attention, memory, motivation, and feelings of reward. Of importance to the present chapter, appropriate doses of caffeine can affect exercise performance while decreasing perceived exertion. All of these responses to caffeine ingestion can potentially improve exercise performance and PA adherence. In addition, blocking adenosine can decrease the activation of nociceptors, resulting in a blunted pain response to exercise. This chapter presents recent evidence for the ergogenic effects of caffeine during exercise as they pertain to perceived exertion, pain and affective responses. In addition, selected laboratory applications of the perceptual methodologies presented in previous chapters are proposed for use in studying the ergogenic effects of caffeine ingestion during exercise.

15.1 Ergogenic Effect of Caffeine

Studies have shown that caffeine ingestion in amounts ranging from 2 to 10 mg/kg body mass has resulted in significant improvements in endurance, load-incremented, short-term high-intensity and resistance exercise performance (Astorino and Roberson 2010; Doherty and Smith 2004). A meta-analysis by Doherty and Smith (2004) examined the results of 40 different studies that compared the ergogenic properties of caffeine ingestion to a placebo condition. A placebo control design is important because consumption of a caffeine free placebo (i.e., solid or liquid)

L. Haile et al., *Perceived Exertion Laboratory Manual*,
DOI 10.1007/978-1-4939-1917-8_15

alone may improve exercise performance (Beedie et al. 2006). Overall, caffeine improved test outcome by 12.3 % when data from all exercise types were pooled, including endurance exercise, graded exercise tests and short-term high intensity exercise. A significantly greater ergogenic effect was found for endurance exercise compared with graded exercise tests and short-term high intensity exercise when reasonably similar caffeine doses were employed. In addition, the greatest improvement in endurance exercise was found for time-to-exhaustion protocols, usually performed at intensities ranging from 75 to 85 % of VO_2max/peak as compared with performance for a given time or distance (Doherty and Smith 2004). However, time-to-exhaustion protocols are not similar to the common demands found in sport and therefore, have a low ecological validity compared to time- or distance-trials. In addition, the reliability of time and distance protocols has been found to be greater than time-to-exhaustion tests (Doyle and Martinez 1998; Jeukendrup et al. 1996; Laursen et al. 2003; Marino et al. 2002; Schabort et al. 1998a, 1998b).

A systematic review by Astorino and Roberson (2010) examined the results of 28 investigations that studied the effects of caffeine ingestion on short-term high intensity exercise and resistance exercise performance. Of 17 studies involving short-term high intensity exercise, which included mostly sprinting and power-based performances, 11 found significant improvements with caffeine ingestion. Of 11 studies involving resistance exercise, 6 found significant performance improvements (Astorino and Roberson 2010). Overall, the literature reveals strong agreement that caffeine has an ergogenic effect on endurance exercise. Whereas studies of short-term high intensity exercise and resistance exercise have had mixed results regarding the ergogenic properties of caffeine ingestion.

15.2 Mechanisms for the Ergogenic Effects of Caffeine

15.2.1 Glycogen Sparing

An underlying mechanism for the ergogenic effect of caffeine ingestion on endurance exercise performance involves glycogen sparing. Caffeine increases plasma epinephrine concentration (Arciero et al. 1995; Graham and Spriet 1995; Robertson et al. 1981; Van Soeren et al. 1993) which can result in an increased release of free fatty acids (FFA) from adipose tissue triglycerides. An increase in FFA availability for skeletal muscle metabolism, then, should increase fat use as an energy substrate, sparing muscle glycogen (Costill et al. 1978; Essig et al. 1980; Ivy et al. 1979).

In contradiction to this line of thinking, studies have shown that both caffeine ingestion and increased plasma epinephrine do not always result in increased plasma FFA (Arogyasami et al. 1989a; Graham 2001; Graham and Spriet 1995; Winder 1986). In addition, while muscle glycogen sparing following caffeine ingestion has been shown in some studies (Erickson et al. 1987; Essig et al. 1980; Spriet et al. 1992), the findings of most investigations contradict this theory (Arogyasami et al. 1989a, 1989b; Chesley et al. 1995, 1998; Graham et al. 2000; Graham 2001; Greer et al. 2000;

Jackman et al. 1996; Laurent et al. 2000; Raguso et al. 1996; Roy et al. 2001). Graham and colleagues (2008) pooled data from multiple studies involving human subjects (Chesley et al. 1998; Erickson et al. 1987; Essig et al. 1980; Graham et al. 2000; Greer et al. 2000; Jackman et al. 1996; Laurent et al. 2000; Spriet et al. 1992) to achieve an overall sample size of 37 subjects from whom muscle glycogen content during exercise was measured. The compiled data found no significant glycogen sparing due to caffeine ingestion. Other studies have found no overall effect of an increase in plasma epinephrine on endurance exercise performance (Graham and Spriet 1995; Kovacs et al. 1998).

Potential reasons for the foregoing conflicting results are that an increase in plasma epinephrine has been shown to increase muscle glycogen breakdown. This action counteracts the effect of increased FFA availability, and increases the production of lactate, a progenitor of muscle fatigue (Arogyasami et al. 1989a; Kovacs et al. 1998; Jackman et al. 1996; Laurent et al. 2000). In addition, although most evidence shows no significant effect of caffeine on carbohydrate or fat metabolism within skeletal muscle, marked interindividual differences have been found. These data indicate that some individuals seem to be *responders* to caffeine and/or epinephrine resulting in glycogen sparing while most individuals are *non-responders* (Battram et al. 2007; Chesley et al. 1998; Graham et al. 2008; Martin et al. 2006).

15.2.2 Adenosine Antagonism

The primary mechanism of action of caffeine that results in its ergogenic effects is the blocking of receptors for the neurotransmitter adenosine (Davis et al. 2003; Fredholm et al. 1999). Structurally, adenosine is classified as a purine along with a larger molecule of which it is a component, adenosine triphosphate (ATP) (Marieb and Hoehn 2013). Adenosine alone is a potent inhibitor of neurotransmission in the brain and can have widespread effects throughout both the central and peripheral nervous systems (Marieb and Hoehn 2013). These include decreasing the release of neurotransmitters such as dopamine and decreasing overall brain arousal (Davis et al. 2003; Fredholm et al. 1999). While adenosine increases in skeletal muscle and the blood with muscular contraction (Davis et al. 2003), the ingestion of caffeine can counteract its effects, thereby allowing continued release of dopamine and heightened arousal.

Dopamine is known for its involvement in attention, memory, motivation and reward (Meeusen et al. 2006a, 2006b). By extension, it is likely that dopamine plays an important role in exercise performance. In fact, three studies have shown that pharmacologically inhibiting the reuptake of dopamine results in an increased dopamine concentration and improved endurance exercise performance (Bridge et al. 2003; Roelands et al. 2012; Watson et al. 2005). Therefore, the effect of caffeine on dopamine may allow increased attention and memory, leading to more accurate pacing strategy, as well as increased motivation and feelings of reward, all of which can contribute singularly or collectively to enhance exercise performance capacity (Roelands et al. 2013).

The ergogenic effect of caffeine has been attributed to its ability to blunt perceptual responses to exercise, as measured by RPE and naturally occurring muscle pain. The attenuating effect of caffeine on both RPE and muscle pain may be due to its antagonism of adenosine receptors. First, adenosine antagonism may allow an increased secretion of dopamine (Davis et al. 2003; Fredholm et al. 1999). Dopamine has been found to be inversely related to central fatigue during exercise, possibly because of its ability to counteract the lethargy and tiredness caused by elevated levels of serotonin (Davis and Bailey 1997). Studies have shown that pharmacologically blocking dopamine reuptake improved the intensity of exercise as performance without concomitant increases in the RPE response (Watson et al. 2005; Roelands et al. 2012). In addition, studies of repeated submaximal isometric contractions postulated that caffeine ingestion improved time to fatigue through an attenuation of force sensation, not through alterations in neuromuscular contractile properties (Meyers and Cafarelli 2005; Plaskett and Cafarelli 2001). Therefore, caffeine may delay the onset of fatigue and attenuate the perception of force, blunting the RPE response to exercise.

Second, adenosine antagonism may decrease the activation of nociceptors, i.e., pain receptors. Adenosine is one of many chemicals that can activate nociceptors in both the central and peripheral nervous systems (Sawynok and Liu 2003). Its increase during exercise naturally produces muscle pain, similar to hydrogen ions, bradykinin, and substance P, among others (O'Connor and Cook 1999). However, caffeine does not have a known effect on these chemicals. Therefore, caffeine may delay the onset of naturally occurring muscle pain and blunt the pain response to exercise as the intensity and/or duration of the performance increases. Improvements in exercise test performance and concomitant attenuation in the RPE and/or pain response, both owing to caffeine ingestion, have been shown during endurance (time-to-exhaustion and time-trials) (Backhouse et al. 2011; Cole et al. 1996; Demura et al. 2007; Doherty and Smith 2005; Gliottoni and Motl 2008; Hadjicharalambous et al. 2006; Ivy et al. 1979; Jenkins et al. 2008; Laurence et al. 2012; Motl et al. 2006; O'Connor et al. 2004) and resistance exercise (Bellar et al. 2011; Green et al. 2007; Hudson et al. 2008).

15.3 Effect of Caffeine on RPE During Exercise

A meta-analysis by Doherty and Smith (2005) tested the results of 21 studies that compared RPE between caffeine ingestion and placebo conditions during constant load exercise or following exhaustive exercise. The constant-load exercise protocols in these investigations generally required performance at intensities between 50 and 80 % VO_2max/peak. For constant load exercise, RPE was significantly lower in the caffeine than placebo condition by an average of 5.6 %. This corresponded to an average increase of 11.2 % in exercise test performance. Regression analysis revealed that the mean exercise RPE explained 29 % of the variance in the difference in performance between caffeine and placebo conditions. Therefore, it seems that caffeine blunts the RPE response during constant load exercise, allowing individuals to exercise for a longer period of time before subjective fatigue becomes intolerable (Doherty and Smith 2005). This is in agreement with investigations that have studied the effect of caffeine during exercise performed at a constant RPE, i.e., using a perceptual production protocol, rather than a constant workload. Investigations by Ivy et al. (1979) and Cole et al. (1996) found that subjects chose to exercise at higher intensities after ingesting caffeine compared to a placebo condition, yet subjects were instructed to self-regulate exercise intensity at the same target RPE for both exercise conditions. Another interesting finding of the meta-analysis conducted by Doherty and Smith (2005) was that RPE did not differ between caffeine and placebo conditions following exhaustive exercise. This perceptual response at the end of exhaustive exercise is intuitive, given that the individuals were tasked to perform the exercise to the point of complete exhaustion necessitating exercise termination. By definition, such exhaustive exercise should result in a maximal or at least near maximal RPE.

More recent investigations have continued to support the relation between caffeine and reduced perceptions of fatigue during both endurance and resistance exercise (Backhouse et al. 2011; Demura et al. 2007; Green et al. 2007; Hadjicharalambous et al. 2006; Hudson et al. 2008; Laurence et al. 2012). Demura and colleagues (2007) studied the effect of 6 mg/kg caffeine ingestion on physiological variables and RPE during 60 min of submaximal endurance cycling at 60 % VO_2peak. The only difference between the caffeine and placebo conditions was a significantly lower RPE at a given submaximal cycle PO as a result of caffeine ingestion (Demura et al. 2007).

Hadjicharalambous and colleagues (2006) studied the effect of 7–7.5 mg/kg caffeine ingestion on differentiated RPE (legs and chest/breathing) and performance during both constant load (73 % VO_2max) and incremental exercise after consumption of a high fat meal in endurance trained men. The purpose of the high fat meal was to remove the potential ergogenic effect of an increase in FFA, cited earlier as a possible result of caffeine supplementation. Elevated FFA concentration produces glycogen sparing and subsequently increases exercise performance. However, it is important to note that such a response may only be seen in certain individuals (Battram et al. 2007; Chesley et al. 1998; Graham et al. 2008; Martin et al. 2006). In the investigation by Hadjicharalambous et al. (2006), results demonstrated a significantly lower RPE for the legs during both exercise tests and a significantly lower RPE for the chest/breathing during incremental exercise in the caffeine condition only. However, performance was not improved by caffeine supplementation (Hadjicharalambous et al. 2006).

Laurence and colleagues (2012) studied the effects of 6 mg/kg caffeine ingestion on maximal 30-min cycling performance, RPE and RER in sedentary men. Performance (i.e., total work) was significantly greater after caffeine ingestion compared to the placebo condition. RPE and RER were similar between trials across time-points. The improved performance and similar RPE indicate that consequent to caffeine ingestion, an increase in work rate was not accompanied by a corresponding increase in RPE. The men were able to accomplish a higher intensity while reporting the same level of perceived exertion. In addition, by measuring RER, the investigators were able to examine potential changes in energy substrate utilization subsequent to caffeine ingestion. The similar RER between the two work rates may indicate that the relative contribution of carbohydrate and fat as fuel for exercising muscle was similar between experimental conditions even though the work rates were different (Laurence et al. 2012).

Green and colleagues (2007) studied the effects of 6 mg/kg caffeine ingestion on resistance exercise performance (bench press and leg press) and differentiated RPE (specific to active muscles) in men and women. Caffeine ingestion resulted in significantly greater performance for leg press exercise as indicated by an increased number of repetitions to failure at a 10RM resistance. The size of this ergogenic effect was similar to that reported by Laurence et al. (2012) for a 30-min cycling performance. The increased performance for leg press exercise was not accompanied by an increase in RPE, indicating a delay in fatigue induced by caffeine. However, these results were not found for bench press exercise (Green et al. 2007).

Two investigations studied the effects 5 and 10 mg/kg caffeine ingestion on moderate intensity cycling performance (60 % VO_2peak) and leg muscle pain in subjects who reported low habitual levels of caffeine consumption. One investigation employed males as subjects (O'Connor et al. 2004), and the other employed females (Motl et al. 2006). Both investigations revealed significant decreases in leg muscle pain ratings during exercise after caffeine ingestion (5 and 10 mg/kg) compared to placebo measurements, without any changes in such physiological variables as BP, HR, and VO_2. In addition there was no statistically significant difference in leg muscle pain between the 5 and 10 mg/kg caffeine conditions. In the study by O'Connor et al. (2004), each male subject rated perceived pain lower during the 10 mg/kg condition compared to the 5 mg/kg condition, yet the overall group mean difference was not statistically significant. These results indicate there may be a dose–response relation between caffeine ingestion and the attenuation of exercise-induced muscle pain in males. Another investigation compared ingestion of 5 mg/kg of caffeine to a placebo condition during high intensity cycling (80 % VO_2peak) in women. Leg muscle pain was significantly lower during the caffeine than placebo conditions (Gliottoni and Motl 2008). The size of this caffeine-induced ergogenic effect was similar to that reported for cycle performance at 60 % VO_2peak intensity (Motl et al. 2006; O'Connor et al. 2004).

15.4 Effect of Caffeine on RPE and Pain During Exercise

A number of investigations studied both RPE and pain responses to exercise following caffeine ingestion (Astorino et al. 2011; Hudson et al. 2008; Jenkins et al. 2008). Jenkins and colleagues (2008) examined the effects of 1, 2 and 3 mg/kg caffeine on cycling performance, RPE (measured for the overall body, legs and chest/breathing) and muscle pain in trained men. The exercise trial involved 15 min of cycling at 80 % VO_2peak followed by 4 min of active recovery. The initial phase of the protocol was followed by a second phase whereby exercise progressed sequentially to maximal performance in a 15-min time trial that simulated the end of a race (Jenkins et al. 2008). Cycling performance was improved following ingestion of 2 and 3 mg/kg caffeine compared to placebo with no effects on RPE (differentiated and undifferentiated) or muscle pain (Jenkins et al. 2008). Hudson and colleagues (2008) studied the effects of 6 mg/kg caffeine on light resistance training performance (leg extension and arm curl), RPE and active muscle pain perception. Caffeine ingestion resulted in significantly greater performance compared to a placebo condition without increases in RPE or pain (Hudson et al. 2008). Astorino and colleagues (2011) studied the effects of 2 and 5 mg/kg caffeine ingestion during high intensity isokinetic knee extension and flexion exercise. Although 5 mg/kg caffeine resulted in significant improvements of muscle function (i.e., peak and average torque, total work, PO), neither RPE nor pain responses were affected by supplementation (Astorino et al. 2011). The results of all three studies indicate blunted perceptual responses (both RPE and pain) with increased work rates and performance as a result of caffeine ingestion (Astorino et al. 2011; Hudson et al. 2008; Jenkins et al. 2008).

15.5 Effect of Caffeine on Mood During Exercise

Another dimension of the ergogenic effect of caffeine that deserves attention is its ability to produce positive mood shifts during exercise (Smith 2002). The mechanism underlying this effect is most likely similar to the effect of caffeine ingestion on RPE and naturally occurring muscle pain. The mechanism involves the antagonism of adenosine by caffeine, leading to the maintenance or enhancement of dopamine. Specifically, the roles of dopamine in motivation and reward have been systematically documented, but changes in exertion and pain are also important moderators of mood. Studies involving the AR to exercise have shown marked inter-individual differences during exercise (Ekkekakis et al. 2000; Tate and Petruzzello 1995). Such variability notwithstanding, once exercise intensity exceeds a preferred level, generally corresponding to the VT, AR tends to deteriorate as both RPE and muscle pain ratings continue to increase (Ekkekakis 2003; Parfitt et al. 2006). It has been shown that caffeine induces increased feelings of well-being and happiness when examined under non-exercise conditions (Zwyghuizen-Doorenbos et al. 1990), but little attention has been paid to the effects of caffeine on mood during exercise. In a study by Backhouse and colleagues (2011), both FS ratings of AR and RPE were studied after ingestion of 6 mg/kg caffeine compared to placebo during cycling at 70 % VO_2max in trained cyclists. In the caffeine condition, affect was significantly higher and RPE was significantly lower compared to the placebo condition (Backhouse et al. 2011).

15.6 Selected Applications of Perceptual Methodology

The ergogenic effects of caffeine ingestion during exercise may have public health implications. It is proposed that optimal doses of caffeine may increase motivation and feelings of reward, decrease perceptions of effort and pain, and improve mood during acute exercise. Therefore, ergogenic application of low to moderate doses of caffeine may have the ability to extend those effects across multiple exercise sessions typically employed in an exercise or PA behavioral intervention. Could caffeine help to promote exercise adherence? The results from available literature are promising, but more research is needed to investigate the mechanisms underlying caffeine's ergogenic properties, especially its long-term effects as a function of habitual use. This includes research investigating variations of habitual caffeine consumption and its effects on performance, physiological, perceptual and psychosocial variables over time.

The following are selected applications of the perceptual methodologies presented in previous chapters of this book that can be applied to study the ergogenic effects of caffeine ingestion during exercise. In addition, methodological notes are presented that should be taken into consideration prior to developing a laboratory design that explores the ergogenic properties of caffeine consumption.

15.6.1 Perceived Exertion Scale Validation

Rationale: Load-incremented endurance and resistance exercise protocols have been used to validate perceptual scales via concurrent measurement of physiological and physical variables which are expected to increase linearly with exercise intensity. Caffeine ingestion prior to exercise may blunt submaximal perceptual responses and increase maximal exercise performance. *Research question*: Does caffeine ingestion alter the validity of category scales intended to measure the perceptions of physical exertion, pain, and affective responses to exercise participation?

15.6.2 Target RPE at the VT

Rationale: A target RPE corresponding to a specific physiological intensity, such as the VT, can be calculated following performance of a perceptual estimation protocol that includes aerobic metabolic measurements. The target RPE can then be used as an effective method of prescribing and self-regulating intensity for an exercise program without the need to calculate a target HR. However, caffeine may blunt the perceptual response to submaximal workloads and may increase load-incremented exercise test performance. *Research question*: Does caffeine ingestion have an effect on the RPE at the VT?

15.6.3 Prediction of VO_2max and 1RM Using RPE

Rationale: VO_2max and 1RM can be predicted from RPE assessed during submaximal exercise intensities. Caffeine ingestion prior to exercise has been shown to blunt the RPE response to submaximal intensities during aerobic and resistance exercise. *Research question*: Does caffeine ingestion affect the prediction of VO_2max and 1RM using RPE responses to submaximal exercise test protocols?

15.6.4 Estimation–Production Paradigms for Exercise Intensity Self-Regulation

Rationale: Estimation–production paradigms are used to test individuals' ability to perceptually self-regulate exercise intensity. This is done by comparing physiological variables corresponding to specific target RPE's as measured during the estimation and production protocols. The production protocol can use a single target RPE when self-regulating continuous exercise and two target RPE's when self-regulating exercise for an interval exercise format. The effect of caffeine on dopamine may allow increased attention and memory, leading to more accurate exercise intensity self-regulation. *Research question*: Does caffeine ingestion reduce exercise intensity self-regulation error when using a continuous RPE production protocol or an interval RPE production protocol?

15.6.5 Teleoanticipation and the Perceived Exertion JND

Rationale: The effect of caffeine on attention and memory may influence teleoanticipation and the perceived exertion JND. An increase in attention and memory may allow an improved response to teleoanticipation and a heightened perceptual acuity. This potential effect may be different across the exercise intensity range. *Research questions*: Does caffeine ingestion improve the effect of teleoanticipation? Does caffeine ingestion alter the perceived exertion JND for exercise intensities that are below, equal to, or above the individual's VT?

15.6.6 Self-Selected Versus Imposed Exercise Intensities

Rationale: Self-selected exercise intensity produces a comparatively more positive AR than imposed exercise intensity. In addition, many individuals self-select an exercise intensity for health-fitness conditioning that is near their VT. Therefore, self-selected exercise intensity has been posited as a method to optimize AR to exercise and promote program adherence. In addition, caffeine has been shown to improve exercise performance and blunt perceptual responses to exercise. *Research questions*: Does caffeine ingestion alter perceptual and psychosocial responses to self-selected as compared to imposed exercise intensities? Does caffeine ingestion change the preferred intensity of exercise conditioning?

15.6.7 Predicted, Momentary and Session RPE

Rationale: Caffeine ingestion may improve AR during exercise and blunt perceptual responses to exercise. *Research question*: Does caffeine ingestion alter predicted, momentary, or session RPE, pain, and AR during exercise.

15.6.8 Research Methodology Notes

1. In addition to standard health-risk screening questionnaires, such as the Physical Activity Readiness Questionnaire (PAR-Q), studies involving caffeine administration should include a caffeine-sensitivity questionnaire to screen participants for risk of negative side effects associated with caffeine ingestion (Motl et al. 2003).
2. Will you choose subjects who are or are not habitual caffeine users? Research findings are not certain that differences exist in athletic performance after caffeine administration between those with varying levels of caffeine habituation (Graham 2001). In addition, even among those who are habitual caffeine users, there can be varying effects of caffeine on mood (Attwood et al. 2007) which, in turn, can have differential effects on exercise performance.
3. How will you define the level of habitual caffeine use? Previous studies of the ergogenic properties of caffeine during exercise have used dietary-caffeine recall questionnaires (Jenkins et al. 2008) and lists of common beverages, foods and medications that contain caffeine to estimate subjects' daily consumption (Bunker and McWilliams 1989).
4. Will you require caffeine abstinence prior to exercise performance? If so, how long? Requesting pretest abstinence from caffeine is commonplace in the literature. In the studies employing a meta-analysis and a systematic review by Doherty and Smith (2004, 2005), caffeine abstinence ranged from 0 to 168 h with medians of 24–48 h. However, research has also shown that a pre-experimental abstinence period ranging from 0 to 96 h, may or may not alter the ergogenic effect of caffeine ingestion (Graham 2001) depending on the subject's level of habitual caffeine use.
5. When will the caffeine be administered prior to exercise testing? In the studies by Doherty and Smith (2004, 2005) the median time was 60 min prior to exercise testing and ranged from 30 to 360 min.

6. How much caffeine will you administer to the subjects? Amounts ranging from 2 to 13 mg/kg body mass have resulted in significant improvements in endurance, load-incremented, short-term high-intensity and resistance exercise performance (Astorino and Roberson 2010; Doherty and Smith 2004; Jenkins et al. 2008). However, doses greater than 9 or 10 mg/kg may not be necessary for significant ergogenic effects and often result in side effects such as anxiety, restlessness, and headaches that could hinder these improvements (Astorino and Roberson 2010; Graham and Spriet 1995; Lindinger et al. 1993).
7. Research indicates there may be interactive effects of caffeine dose and habitual consumption on perceptual and psychosocial responses to exercise. Therefore, investigations involving these responses but that are not studying the direct ergogenic effect of caffeine ingestion should request that subjects engage in normal caffeine consumption prior to experimentation. This requirement should also be extended to the experimental period involving scale anchoring instructions and orientation to exercise procedures.

References

Arciero PJ, Gardner AW, Calles-Escandon J, Benowits NL, Poehlman ET. Effects of caffeine ingestion on NE kinetics, fat oxidation, and energy expenditure in younger and older men. Am J Physiol. 1995;84:908–13.

Arogyasami J, Yang HT, Winder WW. Effect of caffeine on glycogenolysis during exercise in endurance trained rats. Med Sci Sports Exerc. 1989a;21:173–7.

Arogyasami J, Yang HT, Winder WW. Effect of intravenous caffeine on muscle glycogenolysis in fasted exercising rats. Med Sci Sports Exerc. 1989b;21:167–72.

Astorino TA, Roberson DW. Efficacy of acute caffeine ingestion for short-term high-intensity exercise performance: a systematic review. J Strength Cond Res. 2010;24:257–65.

Astorino TA, Terzi MN, Roberson DW, Burnett TR. Effect of caffeine intake on pain perception during high-intensity exercise. Int J Sport Nutr Exerc Metab. 2011;21:27–32.

Attwood AS, Higgs S, Terry P. Differential responsiveness to caffeine and perceived effects of caffeine in moderate and high regular caffeine consumers. Psychopharmacology (Berl). 2007; 190:469–77.

Backhouse SH, Biddle SJH, Bishop NC, Williams C. Caffeine ingestion, affect and perceived exertion during prolonged cycling. Appetite. 2011;57:247–52.

Battram DS, Graham TE, Dela F. Caffeine's impairment of insulin-mediated glucose tolerance in persons with tetraplegia. J Appl Physiol. 2007;569:347–55.

Beedie CJ, Stuart EM, Coleman DA, Foad AJ. Placebo effects of caffeine on cycling performance. Med Sci Sports Exerc. 2006;38:2159–64.

Bellar D, Kamimori G, Glickman EL. The effects of low-dose caffeine on perceived pain during a grip to exhaustion task. J Strength Cond Res. 2011;25:1225–8.

Bridge MW, Weller AS, Rayson M, Jones DA. Responses to exercise in the heat related to measures of hypothalamic serotonergic and dopaminergic function. Eur J Appl Physiol. 2003;89: 451–9.

Bunker ML, McWilliams M. Caffeine content of common beverages. J Am Diet Assoc. 1989;74: 28–32.

Chesley A, Hultman E, Spriet LL. Effects of epinephrine infusion on muscle glycogenolysis during intense aerobic exercise. Am J Physiol. 1995;268:E127–34.

Chesley A, Howlett RA, Heigenhauser GJ, Hultman E, Spriet LL. Regulation of muscle glycogenolytic flux during intense aerobic exercise after caffeine ingestion. Am J Physiol Regul Integr Comp Physiol. 1998;275:R596–603.

Cole KJ, Costill DL, Starling RD, Goodpaster BH, Trappe SW, Fink WJ. Effect of caffeine ingestion on perception of effort and subsequent work production. Int J Sport Nutr. 1996;6:14–23.

Costill DL, Dalsky GP, Fink WJ. Effects of caffeine ingestion on metabolism and exercise performance. Med Sci Sports. 1978;10:155–8.

Davis JM, Bailey SP. Possible mechanisms of central nervous system fatigue during exercise. Med Sci Sports Exerc. 1997;29:45–57.

Davis JM, Zhao Z, Stock HS, Mehl KA, Buggy J, Hand GA. Central nervous system effects of caffeine and adenosine on fatigue. Am J Physiol Regul Integr Comp Physiol. 2003;284: R399–404.

Demura S, Yamada T, Terasawa N. Effect of coffee ingestion on physiological responses and ratings of perceived exertion during submaximal endurance exercise. Percept Mot Skills. 2007;105:1109–16.

Doherty M, Smith PM. Effects of caffeine ingestion on exercise testing: a meta-analysis. Int J Sport Nutr Exerc Metab. 2004;14:626–46.

Doherty M, Smith PM. Effects of caffeine ingestion on rating of perceived exertion during and after exercise: a meta-analysis. Scand J Med Sci Sports. 2005;15:69–78.

Doyle JA, Martinez AL. Reliability of a protocol for testing endurance performance in runners and cyclists. Res Q Exerc Sport. 1998;69:304–7.

Ekkekakis P. Pleasure and displeasure from the body: perspectives from exercise. Cogn Emot. 2003;17:213–39.

Ekkekakis P, Hall EE, Van Lunduyt LM, Petruzzello SJ. Walking in (affective) circles: Can short walks enhance affect? J Behav Med. 2000;23:245–75.

Erickson MA, Schwarzkopf RJ, McKenzie RD. Effects of caffeine, fructose, and glucose ingestion on muscle glycogen utilization during exercise. Med Sci Sports Exerc. 1987;19:579–83.

Essig D, Costill DL, Van Handel PJ. Effects of caffeine ingestion on utilization of muscle glycogen and lipid during leg ergometer cycling. Int J Sports Med. 1980;1:86–90.

Fredholm BB, Battig K, Holmen J, Nehlig A, Zvartau EE. Actions of caffeine in the brain with special reference to factors that contribute to its widespread use. Pharmacol Rev. 1999;51: 83–133.

Gliottoni RC, Motl RW. Effect of caffeine on leg-muscle pain during intense cycling exercise possible role of anxiety sensitivity. Int J Sport Nutr Exerc Metab. 2008;18:103–15.

Graham TE. Caffeine and exercise: metabolism, endurance, and performance. Sports Med. 2001;31:785–807.

Graham TE, Spriet LL. Metabolic, catecholamine, and exercise performance responses to various doses of caffeine. J Appl Physiol. 1995;78:867–74.

Graham TE, Helge JW, MacLean DA, Kiens B, Richter EA. Caffeine ingestion does not alter carbohydrate or fat metabolism in human skeletal muscle during exercise. J Physiol. 2000;529: 837–47.

Graham TE, Battram DS, Dela F, El-Sohemy A, Thong FSL. Does caffeine alter muscle carbohydrate and fat metabolism during exercise? Appl Physiol Nutr Metab. 2008;33:1311–8.

Green JM, Wickwire PJ, McLester JR, Gendle S, Hudson G, Pritchett RC, Laurent CM. Effects of caffeine in repetitions to failure and ratings of perceived exertion during resistance training. Int J Sports Physiol Perform. 2007;2:250–9.

Greer F, Friars D, Graham TE. Comparison of caffeine and theophylline ingestion: exercise metabolism and endurance. J Appl Physiol. 2000;89:1837–44.

Hadjicharalambous M, Georgiades E, Kilduff LP, Turner AP, Tsofliou F, Pitsiladis YP. Influence of caffeine on perception of effort, metabolism and exercise performance following a high-fat meal. J Sports Sci. 2006;24:875–87.

Hudson GM, Green JM, Bishop PA, Richardson MT. Effects of caffeine and asprin on light resistance training performance, perceived exertion, and pain perception. J Strength Cond Res. 2008;22:1950–7.

Ivy JL, Costill DL, Fink WJ, Lower RW. Influence of caffeine and carbohydrate feedings on endurance performance. Med Sci Sports. 1979;11:6–11.

Jackman MR, Wendling P, Friars D, Graham TE. Metabolic, catecholamine, and endurance responses to caffeine during intense exercise. J Appl Physiol. 1996;81:1658–63.

Jenkins NT, Trilk JL, Singhal A, O'Connor PJ, Cureton KJ. Ergogenic effects of low doses of caffeine on cycling performance. Int J Sport Nutr Exerc Metab. 2008;18:328–42.

Jeukendrup A, Saris WHM, Brouns F, Kester ADM. A new validated endurance performance test. Med Sci Sports Exerc. 1996;28:266–70.

Kovacs EMR, Stengen JHC, Brouns F. Effect of caffeinated drinks on substrate metabolism, caffeine excretion, and performance. J Appl Physiol. 1998;85:709–11.

Laurence G, Wallman K, Guelfi K. Effects of caffeine on time trial performance in sedentary men. J Sports Sci. 2012;30:1235–40.

Laurent D, Schneider KE, Prusaczyk WK, Franklin C, Vogel SM, Krssak M, Petersen KF, Shulman GI. Effects of caffeine on muscle glycogen utilization and the neuroendocrine axis during exercise. J Clin Endocrinol Metab. 2000;85:2170–5.

Laursen PB, Shing CM, Jenkins DG. Reproducibility of the cycling time to exhaustion at VO2peak in highly trained cyclists. Can J Appl Physiol. 2003;28:605–15.

Lindinger MI, Graham TE, Spriet LL. Caffeine attenuates the exercise-induced increase in plasma $[K^+]$ in humans. J Appl Physiol. 1993;74:1149–55.

Marieb EN, Hoehn K. Human anatomy & physiology. 9th ed. Boston, MA: Pearson Education; 2013.

Marino FE, Kay D, Cannon J, Serwach N, Hilder M. A reproducible and variable intensity cycling performance protocol for warm conditions. J Sci Med Sport. 2002;5:95–107.

Martin EA, Nicholson WT, Eisenach JH, Charkoudian N, Joyner MJ. Bimodal distribution of vasodilator responsiveness to adenosine due to difference in nitric oxide contribution: implications for exercise hyperemia. Appl Physiol. 2006;101:492–9.

Meeusen R, Watson P, Dvorak J. The brain and fatigue: new opportunities for nutritional interventions? J Sports Sci. 2006a;24:1–10.

Meeusen R, Watson P, Hasegawa H, Roelands B, Piacentini MF. Central fatigue: the serotonin hypothesis and beyond. Sports Med. 2006b;36:881–909.

Meyers BM, Cafarelli E. Caffeine time to fatigue by maintaining force and not by altering firing rates during submaximal isometric contractions. J Appl Physiol. 2005;99:1056–63.

Motl RW, O'Connor PJ, Dishman RK. Effect of caffeine on perceptions of leg muscle pain during moderate intensity cycling exercise. J Pain. 2003;4:316–21.

Motl RW, O'Connor PJ, Turbandt L, Puetz T, Ely MR. Effect of caffeine on leg muscle pain during cycling exercise among females. Med Sci Sports Exerc. 2006;38:598–604.

O'Connor PJ, Cook DB. Exercise and pain: the neurobiology, measurement, and laboratory study of pain in relation to exercise in humans. Exerc Sports Sci Rev. 1999;29:119–66.

O'Connor PJ, Motl RW, Broglio SP, Ely MR. Dose-dependent effect of caffeine on reducing leg muscle pain during cycling exercise is unrelated to systolic blood pressure. Pain. 2004;109: 291–8.

Parfitt G, Rose EA, Burgess WM. The psychological and physiological responses of sedentary individuals to prescribed and preferred intensity exercise. Br J Health Psychol. 2006;11:39–53.

Plaskett CJ, Cafarelli E. Caffeine increases endurance and attenuates force sensation during submaximal isometric contractions. J Appl Physiol. 2001;91:1535–44.

Raguso CA, Coggan AR, Sidossis LS, Gastaldelli A, Wolfe RR. Effect of theophylline on substrate metabolism during exercise. Metabolism. 1996;45:1153–60.

Robertson D, Wade D, Workman R, Woosley RL, Oates JA. Tolerance to the humoral and hemodynamic effects of caffeine in man. J Clin Invest. 1981;67:1111–7.

Roelands B, Watson P, Cordery P, Decoster S, Debaste E, Maughan R, Meeusen R. A dopamine/noradrenaline reuptake inhibitor improves performance in the heat, but only at the maximum therapeutic dose. Scand J Med Sci Sports. 2012;22:e93–8.

Roelands B, de Koning J, Foster C, Hettinga F, Meeusen R. Neurophysiological determinant of theoretical concepts and mechanisms involved in pacing. Sports Med. 2013;43:301–11.

Roy B, Bosman M, Tarnopolsky MA. An acute dose of caffeine does not alter glucose kinetics during prolonged dynamic exercise in trained endurance athletes. Eur J Appl Physiol. 2001;85:280–6.

Sawynok J, Liu XJ. Adenosine in the spinal cord and periphery: release and regulation of pain. Prog Neurobiol. 2003;69:313–40.

Schabort EJ, Hawley JA, Hopkins WG, Mujika I, Noakes TD. A new reliable laboratory test of endurance performance for road cyclists. Med Sci Sports Exerc. 1998a;30:1744–50.

Schabort EJ, Hopkins WG, Hawley JA. Reproducibility of self-paced treadmill performance of trained endurance runners. Int J Sports Med. 1998b;19:48–51.

Smith A. Effects of caffeine on human behaviour. Food Chem Toxicol. 2002;40:1243–55.

Spriet LL, MacLean DA, Dyck DJ, Hultman E, Cederbald G, Graham TE. Caffeine ingestion and muscle metabolism during prolonged exercise in humans. Am J Physiol. 1992;262:E891–8.

Tate AK, Petruzzello SJ. Varying the intensity of acute exercise: implications for changes in affect. J Sports Med Phys Fitness. 1995;35:295–302.

Van Soeren MH, Sathasivam P, Spriet LL, Graham TE. Caffeine metabolism and epinephrine responses during exercise in users and nonusers. J Appl Physiol. 1993;75:805–12.

Watson P, Hasegawa H, Roelands B, Piacentini MF, Looverie R, Meeusen R. Acute dopamine/noradrenaline reuptake inhibition enhances human exercise performance in warm, but not temperate conditions. J Physiol. 2005;565:873–83.

Winder WW. Effect of intravenous caffeine on liver glycogenolysis during prolonged exercise. Med Sci Sports Exerc. 1986;18:192–6.

Zwyghuizen-Doorenbos A, Roehrs TA, Lipschutz L, Timms V, Roth T. Effects of caffeine alertness. Psychopharmacology (Berl). 1990;100:36–9.

Chapter 16
Effects of Carbohydrate on Perceptual and Affective Responses to Exercise

Carbohydrate is arguably the most important dietary component that has the potential to improve endurance exercise performance. A diet high in carbohydrate that is consumed prior to endurance exercise increases glycogen storage in skeletal muscle and the liver, subsequently producing an ergogenic effect. In addition, carbohydrate ingestion during prolonged exercise maintains blood glucose and carbohydrate oxidation rates such that muscle and liver glycogen are spared. The improved carbohydrate availability during exercise has been shown to attenuate perceived exertion responses, especially near the end of high intensity exercise performance. Carbohydrate ingestion during exercise has also been shown to improve short-term exercise performance of 1 h or less. Carbohydrate ingested during such exercise could not reach the bloodstream in amounts needed to significantly improve carbohydrate availability. The potential mechanism of such a seemingly contradictory effect may be that the stimulation of glucose receptors in the mouth has central nervous system responses such as the activation of reward centers and an increase in central drive/motivation. Studies employing carbohydrate ingestion or mouth rinses during short-term exercise performance have reported mixed results. Nevertheless, blunted RPE and improved AR have accompanied an increased performance in some of these investigations. This chapter presents recent evidence for the ergogenic effects of carbohydrate ingestion *during* exercise as they pertain to perceptual and affective responses. In addition, selected laboratory applications of the perceptual methodologies presented in previous chapters are again used to study the ergogenic effects of carbohydrate ingestion during exercise.

L. Haile et al., *Perceived Exertion Laboratory Manual*,
DOI 10.1007/978-1-4939-1917-8_16

16.1 Ergogenic Effect of Carbohydrate Ingestion Prior to Exercise

Carbohydrate supercompensation protocols, or carbohydrate loading, involve diet and exercise routines over the course of the week preceding an endurance exercise performance. The function of these dietary manipulations is to maximize muscle glycogen stores and improve performance in both time-to-fatigue and time trial protocols (Hawley et al. 1997). A high-carbohydrate meal consumed before an exercise performance, generally 3–5 h prior, has been shown to contribute to muscle glycogen stores and improve endurance performance (Hargreaves 2004). However, carbohydrate ingestion at this point may be more important for maximizing liver glycogen stores, especially if the enhanced diet is consumed at breakfast after an overnight fast. Ample liver glycogen is crucial to the maintenance of blood glucose levels necessary for prolonged exercise. The ergogenic effect of 3- to 5-day dietary supercompensation can then be enhanced with further carbohydrate ingestion in the hour before and throughout a prolonged bout of exercise (Jeukendrup and Gleeson 2010).

16.2 Mechanisms of the Ergogenic Effect of Carbohydrate Ingestion During Exercise

Carbohydrate ingestion during exercise is proposed to improve endurance exercise performance via a number of metabolic mechanisms (Jeukendrup and Gleeson 2010). Compared to a placebo, blood glucose levels and carbohydrate oxidation rates are maintained with carbohydrate ingestion, prolonging endurance performance (Coyle et al. 1986). Supplementing blood glucose with ingested carbohydrate decreases the rate of liver glycogen breakdown during exercise, sparing glycogen for later use as an energy source (Jeunkendrup et al. 1999). This metabolic pathway secondary to carbohydrate ingestion occurs for muscle glycogen use during running (Tsintzas et al. 1995), but possibly not cycling (Jeunkendrup et al. 1999).

Part of the ergogenic effect of carbohydrate ingestion in prolonging endurance exercise may be related to the role of carbohydrate substrate availability as a physiological exertional mediator. An increased carbohydrate availability, indicated by increased blood glucose levels and subsequent carbohydrate oxidation rates, may be an important peripheral mediator of perceived exertion (Pandolf 1982). The maintenance of neurological function and skeletal muscle contraction through enhanced glucose availability could help sustain exercise performance (Utter et al. 1997). The attenuation of RPE with carbohydrate ingestion as compared to a placebo has been shown to coincide with improved endurance performance in studies involving time trial (Burgess et al. 1991; Kang et al. 1996; Utter et al. 1997, 1999) and time-to-exhaustion protocols (Wilber and Moffatt 1992). The ergogenic effect has been most notable near the end of prolonged exercise involving 2–2.5 h of moderately high intensity (70–80 % VO_2max/peak) running or cycling (Burgess et al. 1991; Kang et al. 1996; Utter et al. 1997, 1999).

16.3 Mechanisms of the Ergogenic Effect of Carbohydrate Ingestion and Mouth Rinses During Short-Term Exercise

Historically, carbohydrate was not thought to induce an ergogenic effect during shorter-term exercise lasting 1 h or less. This occurred because little of the specific carbohydrate ingested during exercise was able to enter the bloodstream in time to prevent time-dependent decreases in blood glucose. This may explain why at least one recent study has confirmed the lack of ergogenic properties of acute carbohydrate ingestion for comparatively short-term, high intensity exercise (Timmons and Bar-Or 2003). However, carbohydrate ingestion during exercise has resulted in significant improvements in 1-h endurance and intermittent high-intensity exercise performance (Jeukendrup et al. 1997; Winnick et al. 2005). Ingested carbohydrate may interact with receptors in the mouth or stomach, inducing an effect on the central nervous system long before the carbohydrate reaches the blood. This effect, which appears to reflect on attenuation of fatigue perception, has been shown in studies of hypoglycemia in which significant relief is experienced almost immediately after carbohydrate ingestion (Jeukendrup and Gleeson 2010). During exercise, the stimulation of glucose receptors in the mouth may cause central nervous system effects such as the activation of CNS reward centers and an increase in central drive/motivation. This neurosensory response results in attenuated exertional perceptions and a less negative mood shifts that may have salutary effects on exercise performance independent of carbohydrate substrate availability (Carter et al. 2004; Jeukendrup and Gleeson 2010). Experimental evidence supporting such a central nervous system mechanism is provided via studies that employed a carbohydrate mouth rinse during exercise rather than actual ingestion (i.e., the carbohydrate drink is spat out rather than swallowed) (e Silva et al. 2014). The mouth rinse has resulted in similar improvements in 1-h time-trial performance as observed for traditional dietary carbohydrate ingestion (Carter et al. 2004; Pottier et al. 2010). However, not all studies support this neurological mechanism for the observed ergogenic effect (Beleen et al. 2009; Whitham and McKinney 2007). Acute carbohydrate ingestion or a carbohydrate mouth rinse may be more likely to affect the central nervous system when subjects are performing high-intensity exercise in a fasted state (Carter et al. 2004; O'Neal et al. 2013), a less practical situation with lower ecological validity than a fed, or postprandial, state (Beleen et al. 2009).

16.4 Carbohydrate Ingestion or Mouth Rinses During Exercise on RPE and Mood

A number of recent investigations involving various types and intensities of exercise lasting 1 h or less have compared the perceptual and psychosocial effects of carbohydrate ingestion or mouth rinses to placebo conditions. For example, Backhouse and colleagues (2005) compared the effects of in-task consumption of a 6.4 % carbohydrate–electrolyte beverage to placebo during 2 h of cycling at 70 % VO_2max in endurance trained males after an overnight fast. Throughout exercise, RPE and FS ratings of AR were regularly measured. RPE was significantly lower in the carbohydrate than the placebo condition, but not until the 75th minute of exercise. There was an overall main effect reported for affect such that the carbohydrate condition yielded a more positive response compared to the placebo condition throughout exercise. The more positive mood with carbohydrate ingestion was evident beginning at the 30-min time point (Backhouse et al. 2005).

In contrast, O'Neal and colleagues (2013) compared the ergogenic effect of in-task consumption of a 6 % carbohydrate–electrolyte beverage to a non-caloric electrolyte beverage during 50 min of cycling at 60–65 % of heart rate reserve followed by three Wingate anaerobic tests. A Wingate test involves 30 s of cycling at maximum speed against a set resistance. The subjects were active young adults who reported to the laboratory at least 2 h after a meal (i.e., in a postprandial state). The study found no differences between beverage conditions for any performance outcome, momentary RPE during submaximal exercise or session RPE following the final Wingate test (O'Neal et al. 2013).

Winnick and colleagues (2005) compared the ergogenic effects of in-task consumption of a 6 % carbohydrate beverage to a placebo condition during 1 h of intermittent high-intensity exercise intended to mimic a basketball game. Subjects began the exercise protocol after a 12-h fast. Healthy college-aged men and women performed the exercise in 15-min quarters with 5-min breaks separating first-second and third-fourth quarters and a 20-min halftime period. Improved performances were noted in carbohydrate versus placebo condition during the last 15 min of exercise, including faster sprint times and higher jump heights. The Profile of Mood States questionnaire was used to assess overall mood changes as well as alterations in specific feelings. The questionnaire was administered before the simulated basketball game, at halftime, and after exercise. The questionnaire results indicated that overall mood declined significantly throughout exercise in the placebo condition in comparison to the carbohydrate condition. The between group differences were particularly notable for feelings of fatigue and vigor. Whole response levels for these mood constructs were maintained throughout exercise with carbohydrate supplementation (Winnick et al. 2005).

Rollo and colleagues (2008) compared the performance and affective effects of a carbohydrate mouth rinse to a placebo during 30 min of treadmill exercise at a self-regulated intensity corresponding to a 15 on the Borg (6–20) RPE Scale. The subjects were endurance-trained runners who arrived at the laboratory after an overnight fast. The 6 % carbohydrate solution or placebo mouth rinses were administered every 5 min during exercise. The carbohydrate condition resulted in significantly greater total distance covered and higher FS ratings of AR at the beginning of exercise (Rollo et al. 2008).

16.5 Select Applications of Perceptual Methodology

The potential ergogenic effect of carbohydrate ingestion on the central nervous system has been demonstrated for shorter-term exercise conditions that do not rely on comparatively stable carbohydrate substrate availability. These findings may have important implications for individuals other than endurance-trained athletes such as marathoners, cyclists or triathletes who often comprised the experimental samples that were studied in the aforementioned investigations (Beleen et al. 2009; Utter et al. 1997, 1999). An improvement (i.e., positive shift) in mood, indicated by the AR to exercise, or an improvement in exercise enjoyment may have the potential to increase the duration of an exercise bout or even increase overall exercise adherence. Further research is necessary to study the effects of carbohydrate ingestion on perceptual and affective responses to exercise in individuals who are unable to adopt and maintain regular PA. The proper application of high-carbohydrate feeding prior to and during regular exercise participation may be an important link in the psychobehavioral chain between exercise adoption and maintenance. Improved performance, blunted perceived exertion responses, and improved affective responses could collectively improve the motivation to increase PA behavior in order to achieve health-fitness benefits.

The following are selected applications of the perceptual methodologies presented in previous chapters of this book that can be applied to study the ergogenic effects of carbohydrate ingestion during exercise. In addition, methodological notes are presented that should be taken into consideration prior to design of a laboratory investigation of the ergogenic properties of carbohydrate ingestion.

16.5.1 Perceived Exertion Scale Validation

Rationale: Load-incremented endurance and resistance exercise protocols have been used to validate perceptual scales via concurrent measurement of physiological and physical variables expected to increase linearly with exercise intensity. Carbohydrate ingestion during exercise may blunt perceptual responses and increase both submaximal endurance and maximal exercise performance. *Research question*: Does carbohydrate ingestion prior to and during exercise performance affect the validity of category scales to measure exertional perceptions and AR?

16.5.2 Target RPE at the VT

Rationale: A target RPE corresponding to a specific physiological intensity, such as the VT, can be calculated following performance of a perceptual estimation protocol that includes both metabolic and perceptual measurements. The target RPE equivalent to the individually determined VT can then be used as an effective method of prescribing intensity for an exercise program without the need to calculate a target HR. However, carbohydrate ingestion during exercise may blunt the perceptual response to submaximal workloads and may increase load-incremented exercise test performance. *Research question*: Does carbohydrate ingestion have an effect on the RPE at the VT?

16.5.3 Prediction of VO_2max Using RPE

Rationale: VO_2max/peak can be predicted from RPE assessed during submaximal exercise workloads. Carbohydrate ingestion during exercise has been shown to blunt the RPE response to submaximal workloads during aerobic exercise. *Research question*: Does carbohydrate ingestion affect the prediction of VO_2max/peak using RPE responses to submaximal exercise?

16.5.4 Estimation–Production Paradigms for Exercise Intensity Self-Regulation

Rationale: Estimation–production paradigms are used to test individuals' ability to perceptually self-regulate exercise intensity by comparing physiological variables corresponding to a specified target RPE where measures are obtained separately for the estimation and production exercise tests. Carbohydrate ingestion during exercise may influence perceptual and affective responses through central nervous system effects on motivation. *Research question*: Does carbohydrate ingestion have an effect on exercise intensity self-regulation error using a continuous RPE production protocol and an interval RPE production protocol?

16.5.5 Teleoanticipation and the Perceived Exertion JND

Rationale: The potential effect of carbohydrate ingestion on central motivation and AR may influence teleoanticipation and the perceived exertion JND. An increase in motivation may allow a greater response to teleoanticipation and a heightened perceptual acuity. This potential effect may be different across the exercise intensity range.

Research questions: Does carbohydrate ingestion during exercise improve teleoanticipation and the accuracy of exercise intensity self-regulation using the perceived exertion JND?

16.5.6 Self-Selected Versus Imposed Exercise Intensities

Rationale: Self-selected exercise intensity produces a comparatively more positive AR than imposed exercise intensity. In addition, many individuals self-select exercise intensity near their VT, often an appropriate intensity for exercise programming to promote health-fitness. Therefore, self-selected exercise intensity has been posited as a method to optimize AR to exercise and promote program adherence. In addition, carbohydrate ingestion during exercise has been shown to improve exercise performance, blunt perceptual responses and improve AR to exercise. *Research questions*: Does carbohydrate ingestion alter perceptual and psychosocial responses to self-selected and imposed exercise intensities? Does carbohydrate ingestion change the preferred intensity of exercise?

16.5.7 Predicted, Momentary and Session RPE

Rationale: Carbohydrate ingestion may improve mood immediately prior to, during, and following exercise. In addition, carbohydrate ingestion may blunt perceptual responses to exercise. *Research question*: Does carbohydrate ingestion alter predicted, momentary, or session RPE, pain and AR during exercise, making it more or less likely to evidence a mismatch between on-stimulus and off-stimulus responses?

16.5.8 Research Methodology Notes

1. How long will you require a subject to fast prior to exercise? According to a recent systematic review regarding carbohydrate mouth rinses, studies have required subjects to fast from 2 to 14 h (i.e., overnight) before exercise performance (e Silva et al. 2014). Carbohydrate ingestion studies have included a similar range of dietary restriction (O'Neal et al. 2013; Welsh et al. 2002).
2. When will the carbohydrate (ingested or mouth rinse) be administered during exercise? Research investigations have administered the carbohydrate immediately before exercise and at regular 10–15-min increments during exercise (Backhouse et al. 2005, 2007; Ball et al. 1995). Carbohydrate mouth rinses have been administered for 5–10 s per rinse from 4 to 12 times per session (e Silva et al. 2014).

3. How much carbohydrate will you administer to the subjects? Previous research involving the effect of carbohydrate ingestion on exercise performance, RPE and AR has used carbohydrate beverage solutions ranging from 6 to 6.4 % (Backhouse et al. 2005; O'Neal et al. 2013; Rollo et al. 2008; Winnick et al. 2005).

References

Backhouse SH, Bishop NC, Biddle SJ, Williams C. Effect of carbohydrate and prolonged exercise on affect and perceived exertion. Med Sci Sports Exerc. 2005;37:1768–73.

Backhouse SH, Ali A, Biddle SJ, Williams C. Carbohydrate ingestion during prolonged high-intensity intermittent exercise: impact on affect and perceived exertion. Scand J Med Sci Sports. 2007;17:605–10.

Ball TC, Headley SA, Vanderburgh PM, Smith JC. Periodic carbohydrate replacement during 50 min of high-intensity cycling improves subsequent sprint performance. Int J Sport Nutr. 1995;5:151–8.

Beleen M, Berghuis J, Bonaparte B, Ballak SB, Jeukendrup AE, Van Loon LJ. Carbohydrate mouth rinsing in the fed state: lack of enhancement of time-trial performance. Int J Sport Nutr Excerc Metab. 2009;19:400–9.

Burgess ML, Robertson RJ, Davis JM, Norris JM. RPE, blood glucose and carbohydrate oxidation during exercise: effects of glucose feedings. Med Sci Sports Exerc. 1991;23:353–9.

Carter JM, Jeukendrup AE, Jones DA. The effect of carbohydrate mouth rinse on 1-h cycle time trial performance. Med Sci Sports Exerc. 2004;36:2107–11.

Coyle EF, Coggan AR, Hemmert MK, Ivy JL. Muscle glycogen utilization during prolonged strenuous exercise when fed carbohydrate. J Appl Physiol. 1986;61:165–72.

Hargreaves M. Muscle glycogen and metabolic regulation. Proc Nutr Soc. 2004;63:217–20.

Hawley JA, Palmer GS, Noakes TD. Effects of 3 days of carbohydrate supplementation on muscle glycogen content and utilization during a 1-h cycling performance. Eur J Appl Physiol. 1997;74:407–12.

Jeukendrup AE, Brouns F, Wagenmakers AJM, Saris WHM. Carbohydrate-electrolyte feedings improve 1 h trial cycling performance. Int J Sports Med. 1997;18:125–9.

Jeunkendrup AE, Wagenmakers AJ, Stegen JH, Gijsen AP, Brouns F, Saris WH. Carbohydrate ingestion can completely suppress endogenous glucose production during exercise. Am J of Physiol. 1999;276:E672–83.

Jeukendrup A, Gleeson M. Sport nutrition: an introduction to energy production and performance. 2nd ed. Champaign, IL: Human Kinetics; 2010.

Kang J, Roberston RJ, Goss FL, DaSilva SG, Visisch P, Suminski RR, Utter AC, Denys BG. Effect of carbohydrate substrate availability on ratings of perceived exertion during prolonged exercise of moderate intensity. Percept Mot Skills. 1996;82:495–506.

O'Neal EK, Poulos SP, Wingo JE, Richardson MT, Bishop PA. Post-prandial carbohydrate ingestion during 1-h of moderate-intensity, intermittent cycling does not improve mood, perceived exertion, or subsequent power output in recreationally-active exercisers. J Int Soc Sports Nutr. 2013;10:1–9.

Pandolf KB. Differentiated ratings of perceived exertion during dynamic exercise. Med Sci Sports. 1982;14:397–405.

Pottier A, Bouckaert J, Gilis W, Roels T, Derave W. Mouth rinse but not ingestion of a carbohydrate solution improves 1-h cycle time trial performance. Scand J Med Sci Sports. 2010;20: 105–11.

Rollo I, Williams C, Gant N, Nute M. The influence of carbohydrate mouth rinse on self-selected speeds during a 30-min treadmill run. Int J Sport Nutr Exerc Metab. 2008;18:585–600.

e Silva T, de Souza ME, de Amorim JF, Stathis CG, Leandro CG, Lima-Silva AE. Can carbohydrate mouth rinse improve performance during exercise? A systematic review. Nutrients. 2014;6:1–10.
Timmons BW, Bar-Or O. RPE during prolonged cycling with and without carbohydrate ingestion in boys and men. Med Sci Sports Exerc. 2003;35:1901–7.
Tsintzas OK, Williams C, Boobis L, Greenhaff P. Carbohydrate ingestion and glycogen utilisation in different muscle fibre types in man. J Physiol. 1995;489:243–50.
Utter A, Kang J, Nieman D, Warren B. Effect of carbohydrate substrate availability on ratings of perceived exertion during prolonged running. Int J Sport Nutr. 1997;7:274–85.
Utter AC, Kang J, Nieman FW, Robertson RJ, Henson DA, Davis JM, Butterworth DE. Effect of carbohydrate ingestion and hormonal responses on ratings of perceived exertion during prolonged cycling and running. Eur J Appl Physiol. 1999;80:92–9.
Welsh RS, Davis JM, Burke JR, Williams HG. Carbohydrates and physical/mental performance during intermittent exercise fatigue. Med Sci Sports Exerc. 2002;34:723–31.
Whitham M, McKinney J. Effect of a carbohydrate mouthwash on running time performance. J Sports Sci. 2007;25:1385–92.
Wilber RL, Moffatt RJ. Influence of carbohydrate ingestion on blood glucose and performance in runners. Int J Sport Nutr. 1992;2:317–27.
Winnick JJ, Davis JM, Welsh RS, Carmichael MD, Murphy EA, Blackmon JA. Carbohydrate feedings during team sport exercise preserve physical and CNS function. Med Sci Sports Exerc. 2005;37:306–15.

Chapter 17
Effects of Music on Perceptual and Affective Responses to Exercise

Whether exercising at a fitness facility, a local park or greenway, or at home, many people listen to music during their workout. This has become much more common as media devices have become smaller and more portable. Gone are the days of the "boom box" radio in the gym; replaced by high quality sound systems often found in modern fitness facilities. However, now many participants have their own personal music player. As MP3 players and smartphones replaced cassette tape and CD players, people became much more likely to take their music along during outside exercise, weather permitting. Anecdotally, people discuss how they cannot exercise without accompanying music. They can exercise longer and harder because the music causes the exercise to feel easier. They will exercise more often because the music helps them feel more positive. The music can even help them forget about the exercise they are performing. These anecdotal reports have been tested through systematic experimentation. Research has studied the effects of varying types of music, including preferred versus nonpreferred and synchronous versus asynchronous, on perceived exertion, affect, and performance outcomes. This chapter presents recent evidence for the ergogenic effects of music during exercise as they pertain to perceptual and affective responses. In addition, selected laboratory applications of the perceptual methodologies presented in previous chapters are proposed regarding their use to study the ergogenic effects of listening to music during exercise.

17.1 Effect of Music on RPE During Constant Workload Exercise

Listening to music during exercise can reduce perceived exertion responses in comparison to exercise without music. Nethery (2002) studied the effect of music on the RPE response to 15 min of cycling at either 50 or 80 % VO_2peak in untrained males. For each workload, subjects performed four different conditions: control (no music

L. Haile et al., *Perceived Exertion Laboratory Manual*,
DOI 10.1007/978-1-4939-1917-8_17

or video but can still see and hear the surroundings), sensory deprived (cannot see or hear the surroundings), video, and music. Regardless of workload, RPE was significantly lower for the music condition compared to all others. In addition, RPE was significantly higher for the sensory deprived than all other conditions. HR was the same across conditions for each workload. Listening to music may have allowed the subjects to dissociate exertional perceptions from the effects of physiological mediators during exercise. In contrast, sensory deprivation may have caused the subjects to associate (i.e., focus on) the signals arising from active body regions, intensifying perceptions of exertion. Sensory dissociation has been indicated as a strategy that individuals use to distract themselves from physiological cues during exercise. Focusing attention on the external environment, which includes music, shifts the individual's conscious awareness away from bodily sensations such as increased V_E that would drive the RPE response upward (Baden et al. 2004; Tenenbaum et al. 2004). Interestingly, the control and video conditions produced similar RPE responses (Nethery 2002). This indicates that music may allow more dissociation than a video distraction.

The effect of music on perceived exertion was confirmed by Potteiger and colleagues (2000), who measured undifferentiated and differentiated RPE (legs, chest/breathing) during 20 min of cycling at 70 % VO_2peak in young physically active adults. The subjects performed the exercise under four different conditions: control (no music), fast upbeat music, classical music, and self-selected music. Listening to music resulted in significantly lower undifferentiated and differentiated RPE responses compared to the control condition, regardless of whether it was fast upbeat, classical or self-selected. HR was the same across conditions (Potteiger et al. 2000).

The effect of music on perceived exertion may not hold true for high intensity exercise. Tenenbaum and colleagues (2004) measured the RPE response to treadmill running at 90 % VO_2max, with performance continuing until volitional termination owing to fatigue. Runners performed the exercise under four conditions: control (i.e., no music) and listening to rock, dance and inspirational music. Exercise time until fatigue was similar across conditions (Tenenbaum et al. 2004). High intensity exercise may not allow an individual to dissociate from sensations associated with noxious stimuli arising from certain physiological mediators such as lactacidemia (Tenenbaum 2001). In a qualitative examination of the effect of music on exertional perceptions, 30 % of subjects reported that focusing on the music was beneficial near the beginning of exercise because they were able to dissociate from noxious elements of the exercise performance, which helped motivate them to continue exercise (Tenenbaum et al. 2004). Although this effect did not result in significant increases in exercise duration, it may hold important implications for exercise adherence. Individuals may be more likely to engage in exercise while listening to music because less attention is paid to unpleasant feelings of physical exertion.

17.2 Effect of Music on Optimal Pacing Strategy for Exercise Performance

Music may be a mediating factor of pacing strategy during time trial exercise performance, potentially due to its effect on perceived exertion. However, results in the current literature have not been consistent. Atkinson and colleagues (2004) found that listening to high tempo dance music during cycling significantly improved 10-km time trial performance in young adult males. This was primarily due to a significantly increased cycling speed during the first 3 km of the performance. The RPE response was higher in the music compared to the control condition, i.e., a response similar to that observed for exercise intensity. The investigators used the Brunel Music Rating Inventory to ask questions regarding the subjects' opinion of specific aspects of the music. The subjects reported that the tempo and rhythm of the music were more important than harmony and melody regarding their effect on motivation during exercise (Atkinson et al. 2004).

In contradiction to these results, the same research group found no effect of music on 10-km cycling time trial performance in a similar subject sample (Lim et al. 2009). Based on the results of the study by Atkinson and colleagues (2004) in which the music condition improved cycling speed during the initial segment of the time trial, Lim et al. (2009) used two separate music conditions. Rather than have subjects listen to music continuously during the entire exercise bout, one condition involved music played during the first 5 km only while the other condition involved music played during the second 5 km only. Cycling speed was faster during the first 5 km of the trial in which the music was not introduced until the second half of the time trial. This finding was contrary to that reported by Atkinson et al. (2004). The temporarily increased speed during the initial half of the trial was not sufficient to improve overall performance (Lim et al. 2009).

Lima-Silva and colleagues (2012) extended this line of research to a 5-km running time trial performed by male recreational runners. They compared a no music control condition to two music conditions: music played during the first 1.5 km or the last 1.5 km of the 5-km time trial. Running pace was significantly faster with music played during the first 1.5 km of the time trial, as evidenced by an improvement in 5-km running time compared to the control condition. However, total 5-km time was similar between music trials. In this study, the investigators assessed associative thoughts at 1 km intervals. The music resulted in a decrease in associative thoughts only when played during the first 1.5 km of the exercise trial (Lima-Silva et al. 2012).

During time trial performances such as those often used in research protocols, the intensity of exertional perception (i.e., RPE) increases throughout exercise until race completion. This is a common pacing strategy during time trial performances, the intent of which is that maximal RPE and the point of exhaustion occur simultaneously at the end of a race (Tucker 2009). In the study by Lima-Silva et al. (2012), regardless of changes in running speed, the linear increase in RPE as a function of elapsed exercise time was similar across both the control condition and those employing music during different segments of the race. Since music has been shown to reduce RPE at submaximal exercise intensities, the increased running speed that occurred when music was played during the first 1.5 km may have been to maintain the steady rate of increase in perceived exertion. However, at the high exercise intensities experienced at the end of the race, the dissociative effect of music is lessened and associative cognitive strategies linked to physiological cues become more dominant. Based on the data from constant workload exercise, the threshold at which the effect of music on RPE is no longer significant must be greater than 80 % VO_2peak during cycling exercise (Nethery 2002) but less than 90 % VO_2max during running exercise in most subjects (Tenenbaum et al. 2004).

17.3 Effect of Music on Performance in Trained Athletes

Previous research indicates that listening to music during exercise results in significant alterations in RPE and performance for untrained and recreationally active individuals. However, little research has investigated the effect of music on RPE-based pacing strategy in well-trained athletes. Hagen and colleagues (2013) studied male and female club level cyclists and triathletes during 10-km cycling time trials under control, sensory-deprived, and self-selected motivational music conditions. The subjects displayed similar RPE responses and performances across the three trials. These perceptual and performance response patterns were consistent with previous research in well-trained athletes that did not involve music (Foster et al. 2009; Joseph et al. 2008). Hagen et al. (2013) also reported comments from their subjects that the music caused a greater level of exercise enjoyment, allowed dissociation from the exercise performance, and that the exercise "felt easier" (Hagen et al. 2013). Such comments confirm that exercise while listening to music is more positive than that of exercise without music. Even though performance was not changed, a positive perceptual and affective memory of exercise could promote to future exercise participation. However, such a rèsult would be more significant for those who are not well-trained athletes.

17.4 Preferred Versus Nonpreferred Music

Some research has shown that while listening to music during an exercise performance can alter the RPE response, the specific type of music is not an important factor mitigating the change in RPE. For example, music resulted in significantly lower undifferentiated and differentiated RPE responses to cycling at 70 % VO_2peak regardless of whether it was fast and upbeat in tempo, classical or self-selected (Potteiger et al. 2000). It is possible that the subjects found each type of music to be enjoyable or at least tolerable to an extent that its presence did not change associative/dissociative cognitive strategy and have an effect on perceived exertion. Further research has employed experimental conditions involving preferred music and nonpreferred music to specifically compare external auditory stimuli that are pleasurable or unpleasurable. Preferred music deemed to be pleasurable may allow the individual to dissociate from the exercise performance and noxious physiological mediators of exertional perceptions. In contrast, nonpreferred music considered to be unpleasurable has the opposite effect. It interferes with the individual's cognitive ability to dissociate from feelings of fatigue, resulting in an increased perceived exertion response (Gfeller 1988; Tenenbaum et al. 2004).

Lin and Lu (2013) classified a number of popular songs based on induced feelings of enjoyment and motivational qualities, as rated by a pilot sample prior to their study. Songs were placed into one of four categories: high motivation and preference, high motivation and low preference, low motivation and preference, low motivation and high preference. The college-aged participants were randomly assigned to listen to music from one of the four categories during exercise or to a control condition (no music). They performed 12 min of running on a track at maximal pace. Music preference significantly affected performance, such that highly preferred music resulted in significantly greater performance distance compared to music of low preference. Motivational quality of the music did not affect running performance (Lin and Lu 2013). Preferred music may have allowed the individuals to dissociate from sensations arising from noxious physiological cues during exercise, blunting perceived exertion intensity and prolonging exercise duration. However, it is important to note that RPE were not measured during the investigation.

Nakamura et al. (2010) measured RPE during cycle ergometer exercise to exhaustion at critical power intensity in young male recreational cyclists. Subjects performed three trials: no music control, preferred music, and nonpreferred music. Each subject identified ten preferred songs and ten nonpreferred songs to be used during the cycling trials. Performance was significantly greater for the preferred music than the nonpreferred music. Exercise performance for the no music condition was similar to both the preferred and nonpreferred conditions. HR was similar across conditions, but RPE was significantly greater for the nonpreferred music condition as compared to both the preferred music and no music conditions. These results suggest that listening to preferred music is superior to listening to nonpreferred music during exercise performance. This conclusion was based on the blunted RPE response and increased time to exhaustion during high intensity cycling where subjects listened to music selections that they preferred (Nakamura et al. 2010). However, the preferred music condition did not significantly improve performance or decrease RPE in comparison to the no music condition. These findings were similar to previous research involving high intensity exercise (Tenenbaum et al. 2004). Therefore, as long as music is not seen as unpleasurable, it may not adversely affect exercise performance at high intensities. Listening to nonpreferred as compared to preferred music, however, may decrease exercise performance during a wide variety of conditions yet to be studied. As an example, such an unpleasant auditory stimulus may decrease AR during exercise. This could lead to a decreased motivation to initiate and/or continue participation in a PA program.

17.5 Asynchronous Versus Synchronous Music and AR

In the literature previously cited, investigations employed asynchronous music, such that no conscious effort was expected on the part of the subjects to synchronize motor patterns with the musical rhythm. A recent investigation has compared the effects of asynchronous music to synchronous music on neuromuscular responses during exercise (Lim et al. 2014). Subjects were instructed to consciously synchronize exercise movements with the tempo of the music (Karageorghis 2008). The physiological rationale for the effect of synchronous music goes beyond the fact that music in general can arouse the neuromuscular system (Rossignol and Melvill-Jones 1976). The synchronization of skeletal muscle contractions with a rhythmic auditory stimulus can decrease the metabolic cost of exercise (Bacon et al. 2012; Terry et al. 2012). This response may be due to functioning of the cerebellum in coordinating rhythmic sensory stimuli and motor responses leading to enhanced neuromuscular efficiency (Molinari et al. 2007).

In the study by Lim and colleagues (2014), recreationally active men performed four trials of cycle ergometer exercise at 90 % of VT intensity. The trials were: no music control, metronome guided, synchronous music, and asynchronous music. The metronome condition controlled for the effect of a synchronized auditory stimulus alone. Although the study found no effect on the metabolic cost of cycle exercise, significant between condition differences were found regarding FS ratings of AR. Both music conditions resulted in significantly higher FS ratings compared to the metronome and control conditions. Such a response has been found in numerous other investigations where the comparison conditions have generally been no music or even total sensory deprivation (Boutcher and Trenske 1990; Elliot et al. 2005; Karageorghis et al. 2009, 2010). One such study compared two types of synchronous music, motivational and oudeterous, i.e., neutral regarding its motivational qualities (Karageorghis et al. 2009). Subjects walked to exhaustion at an intensity initially eliciting 75 % of HR reserve for each condition. A no music control condition was also employed. Both music conditions improved time to exhaustion compared to the control condition, but motivational synchronous music had the greatest ergogenic effect. This ergogenic effect was accompanied by FS ratings significantly higher than the control condition. RPE was similar between conditions, most likely due to the high intensity of each exercise trial. The study also included the Exercise-Induced Feeling Inventory (EFI) to ask subjects about various affective states post-exercise. These states included the degree to which the exercise made them feel happy, refreshed, peaceful, or physically tired. Interestingly, although AR during exercise was significantly affected by the various music conditions, no differences were found between conditions in EFI responses (Karageorghis et al. 2009).

17.6 Selected Applications of Perceptual Methodology

The ergogenic effects of music during exercise may have public health implications. It is proposed that music may have the ability to decrease perceptions of effort and pain, and improve AR during acute exercise. Therefore, ergogenic properties of listening to music may extend those effects across multiple exercise sessions typically employed in an exercise and/or weight loss behavioral intervention. Public health programming could assist individuals with the cost of purchasing music media devices, teach them to exercise safely in their community while listening to music, and help them explore the type of music that is uniquely motivating to them. Research employing an ecological approach is needed to study the long-term effects of listening to music on regular PA participation.

The following are selected applications of the perceptual methodologies presented in previous chapters of this book that can be applied to the ergogenic effects of music during exercise. In addition, methodological notes are presented that should be taken into consideration when developing the research design.

17.6.1 Perceived Exertion Scale Validation

Rationale: Load-incremented endurance and resistance exercise protocols have been used to validate perceptual scales. The research paradigms employed concurrent measurement of perceptual, physiological and physical variables, each of which was expected to increase linearly with exercise intensity. Listening to music during exercise has been shown to blunt perceptual responses to moderate but not high exercise intensities and may increase the motivation for maximal exercise performance. *Research question*: Does listening to music during exercise alter validity coefficients for perceived exertion and affect scales?

17.6.2 Target RPE at the VT

Rationale: A target RPE corresponding to a specific physiological intensity, such as the VT, can be calculated following performance of a perceptual estimation protocol that includes both perceptual and metabolic measurements. This target RPE can then be used as an effective method of prescribing intensity of an exercise program that can be performed without the need to calculate a target HR. However, listening to music during exercise may alter the perceptual response to submaximal workloads. *Research questions*: Does listening to music during load-incremented exercise have an effect on the RPE at the VT? Is there an effect of preferred versus nonpreferred music?

17.6.3 Prediction of VO_2max Using RPE

Rationale: VO_2max/peak can be predicted from RPE assessed during submaximal exercise workloads. Music during exercise has been shown to blunt the RPE response to submaximal workloads during aerobic exercise. *Research question*: Does listening to music during exercise have an effect on the prediction of VO_2max/peak using RPE responses to submaximal exercise?

17.6.4 Estimation–Production Paradigms for Exercise Intensity Self-Regulation

Rationale: Estimation–production paradigms are used to test individuals' ability to perceptually self-regulate exercise intensity by comparing physiological variables corresponding to a priori determined target RPE's. The production protocol can use either a single target RPE when self-regulating continuous exercise or two target RPE's when self-regulating intensity of an interval exercise format. Listening to music during submaximal exercise has been shown to allow individuals to cognitively dissociate from noxious properties of physiological and performance-related exertional mediators and may have an impact upon the accuracy of exercise intensity self-regulation. *Research questions*: Does listening to music have an effect on exercise intensity self-regulation error when performing a continuous RPE production protocol or an interval RPE production protocol? Is there a differential effect of asynchronous versus synchronous music when self-regulating exercise intensity using either a continuous or interval protocol?

17.6.5 Teleoanticipation and the Perceived Exertion JND

Rationale: The effect of music on associative and dissociative cognitive strategies during exercise may influence teleoanticipation and the perceived exertion JND. An increase in dissociative thoughts during exercise caused by listening to music may affect teleoanticipation and in turn reduce perceptual acuity. *Research questions*: Does listening to music during exercise influence teleoanticipation? Does listening to music during exercise alter the perceived exertion JND for an exercise intensity corresponding to the OMNI RPE at the VT?

17.6.6 Self-Selected Versus Imposed Exercise Intensities

Rationale: Self-selected exercise intensity produces a comparatively more positive AR than imposed exercise intensity. In addition, many individuals self-select exercise intensity near their VT, often an appropriate intensity for exercise programming. Therefore, self-selected exercise intensity has been posited as a method to optimize AR to exercise and promote program adherence. In addition, listening to music during exercise has been shown to improve submaximal exercise performance and blunt perceived exertion responses to exercise. *Research questions*: Does listening to music alter perceptual and psychosocial responses to self-selected exercise? Does listening to music change the preferred intensity of exercise? Is there an effect of preferred versus nonpreferred music on perceptual and psychosocial responses during both self-selected and imposed exercise intensities?

17.6.7 Predicted, Momentary and Session RPE

Rationale: Listening to music may improve AR during exercise and may blunt perceived exertion responses to exercise. However, music also allows cognitive dissociation from the noxious influence of some physiological exertional mediators experienced during exercise. This dissociative strategy may affect the memory of perceive exertion responses experienced during previous exercise experiences. *Research question*: Does listening to music during exercise alter predicted, momentary, or session RPE, pain and AR, making it more or less likely to evidence a mismatch between on-stimulus and off-stimulus responses?

17.6.8 Research Methodology Notes

1. Control of the external environment is especially important for studies of external auditory stimuli. The space in which exercise is performed (i.e., laboratory or indoor track) should be quiet and free of distractions. Cycle ergometer studies have used sensory-deprivation (i.e., noise canceling headphones or ear plugs and eye patches or blindfolds) for no music control conditions to eliminate the negative influence of both auditory and visual external stimuli. Running studies have placed large blank screens in front of the treadmill to reduce visual interference when examining the effect of auditory stimuli.
2. The effect of preferred versus nonpreferred music must be considered in the experimental paradigm even when preference is not a primary independent variable. If music selected by investigators is not preferred by a subject, psychoperceptual and performance assessments can be significantly biased, leading to erroneous conclusions. Lim et al. (2014) chose songs that were unknown to their sample population (and subsequently confirmed this by querying subjects) to limit extraneous factors related to music preference and sociocultural background. Some investigations have used a pre-experimental assessment to determine individual subjects' preferred music from various genres (Karageorghis et al. 2009; Lin and Lu 2013).
3. When a study includes preferred and nonpreferred music conditions, subjects have been asked to provide song selections or select them from popular music of different genres. When subjects submit their own song selections, significant

time burden could be placed on investigators if the editing of tracks is necessary.

4. When a study compares asynchronous versus synchronous music conditions, tracks must be digitally altered to synchronize the tempo of the music with movement frequency patterns so the same songs can be used in both conditions to remove the effect of song preference (Lim et al. 2014).

References

Atkinson G, Wilson D, Eubank M. Effects of music on work-rate distribution during a cycling time trial. Int J Sports Med. 2004;25:611–5.

Bacon CJ, Myers TR, Karageorghis CI. Effect of music-movement synchrony on exercise oxygen consumption. J Sport Med Phys Fitness. 2012;52:359–65.

Baden DA, Warwick-Evans L, Lakomy J. Am I nearly there? The effect of anticipated running distance on perceived exertion and the attentional focus. J Sport Exerc Psychol. 2004;26: 215–31.

Boutcher SH, Trenske M. The effects of sensory deprivation and music on perceived exertion and affect during exercise. J Sport Exerc Psychol. 1990;12:167–76.

Elliot D, Carr S, Orme D. The effect of motivational music on submaximal exercise. Eur J Sport Sci. 2005;5:97–106.

Foster C, Hendrickson K, Peyer K, Reiner B, de Koning JJ, Lucia A, Battista RA, Hettinga FJ, Porcari JP, Wright G. Pattern of developing the performance template. Br J Sports Med. 2009; 43:765–9.

Gfeller GA. Musical components and style preferred by young adults for aerobic fitness activities. J Music Ther. 1988;25:28–43.

Hagen J, Foster C, Rodriguez-Marroyo J, de Koning JJ, Mikat RP, Hendrix CR, Porcari JP. The effect of music on 10-km time-trial performance. Int J Sports Physiol Perform. 2013;8:104–6.

Joseph T, Johnson B, Battista R, Wright G, Dodge C, Porcari JP, de Koning JJ, Foster C. Perception of fatigue during simulated competition. Med Sci Sports Exerc. 2008;40:381–6.

Karageorghis CI. The scientific application of music in sport and exercise. In: Lane AM, editor. Sport and exercise psychology. London, UK: Hodder Education; 2008. p. 109–37.

Karageorghis CI, Mouzourides DA, Priest D, Sasso TA, Morrish DJ, Walley CL. Psychophysical and ergogenic effects of synchronous music during treadmill walking. J Sport Exerc Psychol. 2009;31:18–36.

Karageorghis CI, Priest DL, Williams LS, Hirani RM, Lannon KM, Bates BJ. Ergogenic and psychological effects of synchronous music during circuit-type exercise. Psychol Sport Exerc. 2010;11:551–9.

Lim HB, Atkinson G, Karageorghis CI, Eubank MR. Effects of differentiated music on cycling time trial. Int J Sports Med. 2009;30:435–42.

Lim HBT, Karageorghis CI, Romer LM, Bishop DT. Psychophysiological effects of synchronous versus asynchronous music during cycling. Med Sci Sports Exerc. 2014;46:407–13.

Lima-Silva AE, Silva-Cavalcante MD, Pires FO, Bertuzzi R, Oliveira RSF, Bishop D. Listening to music in the first, but not the last 1.5 km of a 5-km running trial alters pacing strategy and improves performance. Int J Sports Med. 2012;33:813–8.

Lin J, Lu FJ. Interactive effects of visual and auditory intervention on physical performance and perceived effort. J Sports Sci Med. 2013;12:388–93.

Molinari M, Leggio MG, Thaut MH. The cerebellum and neural networks for rhythmic sensorimotor synchronization in the human brain. Cerebellum. 2007;6:18–23.

Nakamura PM, Pereira G, Papini CB, Nakamura FY, Kokobun E. Effects of preferred and nonpreferred music on continuous cycling exercise performance. Percept Mot Skills. 2010;110: 257–64.

Nethery VM. Competition between internal and external sources of information during exercise: influence on RPE and the impact of exercise load. J Sports Med Phys Fitness. 2002;42:172–8.

Potteiger JA, Schroeder JM, Goff KL. Influence of music on ratings of perceived exertion during 20 minutes of moderate intensity exercise. Percept Mot Skills. 2000;91:848–54.

Rossignol S, Melvill-Jones G. Audiospinal influences in man studied by H-reflex and its possible role in rhythmic movement synchronized to sound. Electroencephalogr Clin Neurophysiol. 1976;41:83–92.

Tenenbaum G. A social-cognitive perspective of perceived exertion and exercise tolerance. In: Singer RN, Hausenblas HA, Janelle C, editors. Handbook of sport psychology. New York, NY: Wiley and Sons; 2001. p. 810–20.

Tenenbaum G, Lidor R, Lavyan N, Morrow K, Tonnel S, Gershgoren A, Meis J, Johnsonet M. The effect of music type on running perseverance and coping with effort sensations. Psychol Sport Exerc. 2004;5:89–109.

Terry PC, Karageorghis CI, Saha AM, D'Auria S. Effects of synchronous music on treadmill running among elite athletes. J Med Sci Sport. 2012;15:52–7.

Tucker R. The anticipatory regulation of performance: the physiological basis for pacing strategies and the development of a perception-based model for exercise performance. Br J Sports Med. 2009;43:392–400.

Appendix A
Selected OMNI Scales

Fig. A.1 Children's OMNI-walk/run RPE scale (Robertson 2004)

L. Haile et al., *Perceived Exertion Laboratory Manual*,
DOI 10.1007/978-1-4939-1917-8

Fig. A.2 Adult OMNI-walk/run RPE scale (Robertson 2004)

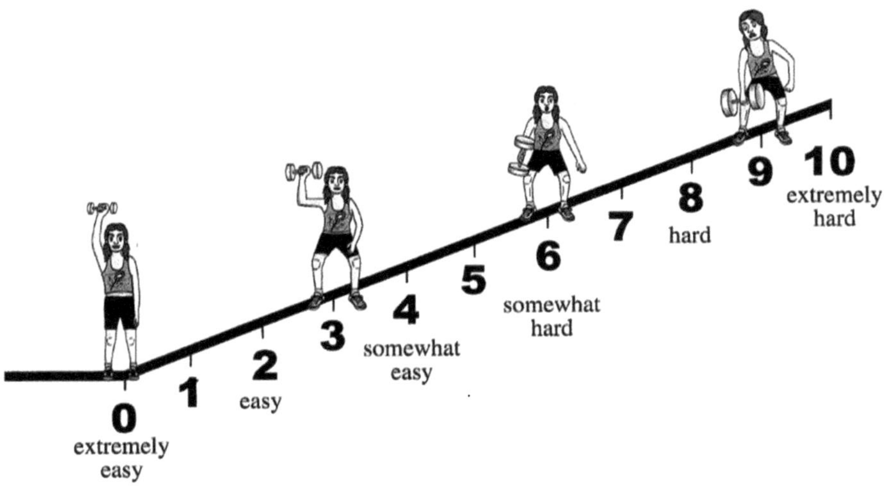

Fig. A.3 Children's OMNI-resistance exercise RPE scale, female (Robertson 2004)

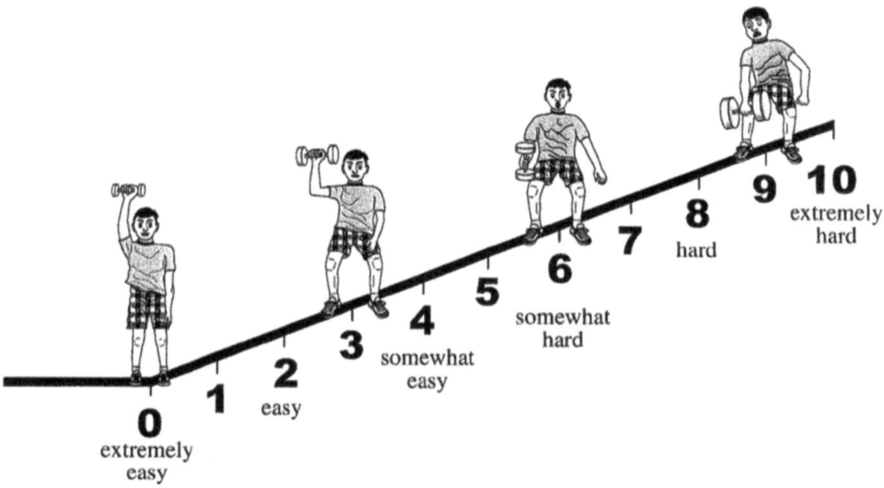

Fig. A.4 Children's OMNI-resistance exercise RPE scale, male (Robertson 2004)

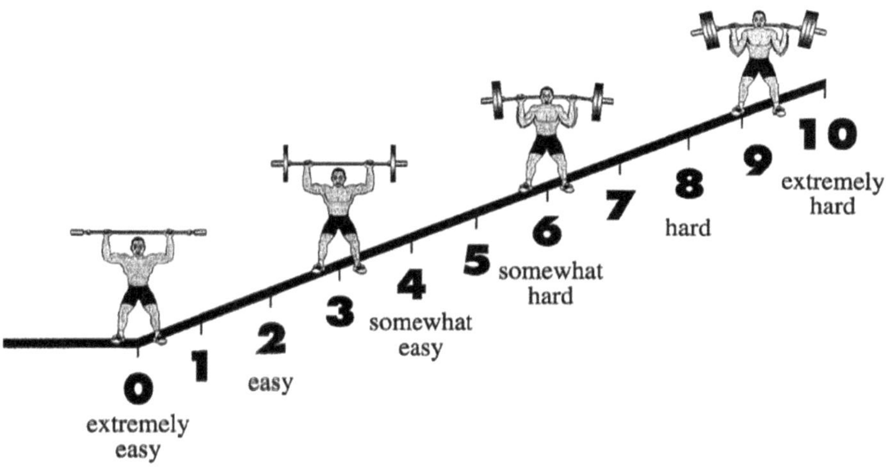

Fig. A.5 Adult OMNI-resistance exercise RPE scale, male (Robertson 2004)

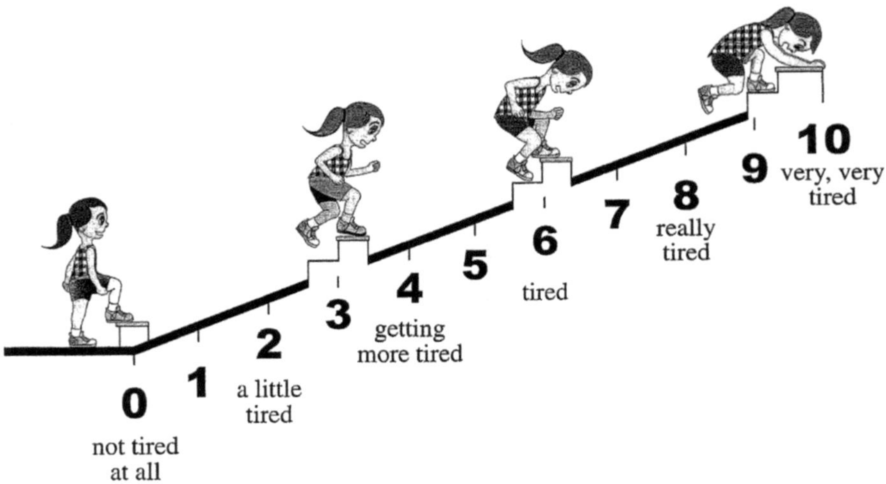

Fig. A.6 Children's OMNI-step RPE scale, female (Robertson 2004)

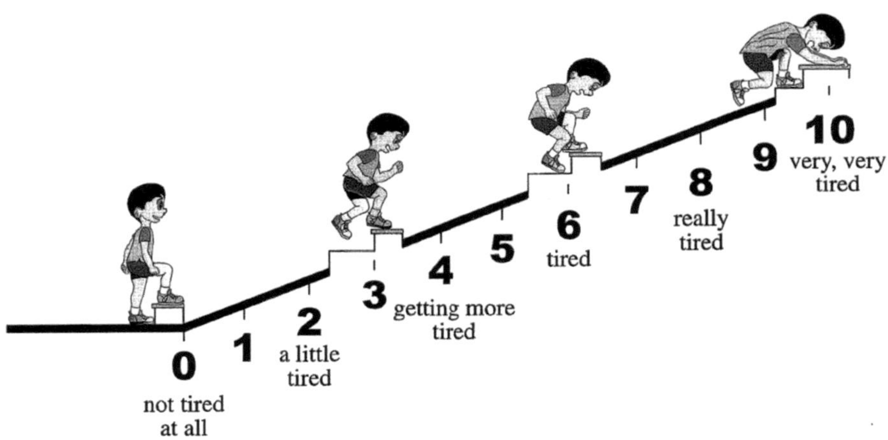

Fig. A.7 Children's OMNI-step RPE scale, male (Robertson 2004)

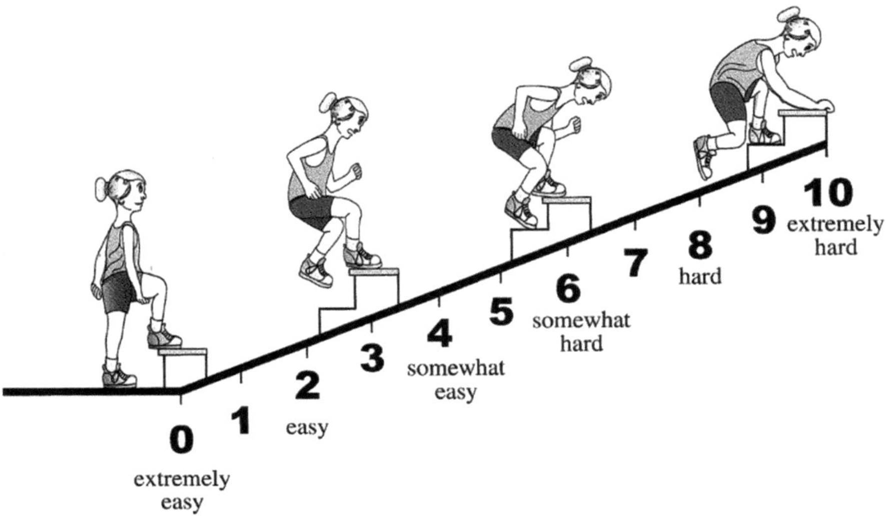

Fig. A.8 Adult OMNI-step RPE scale, female (Robertson 2004)

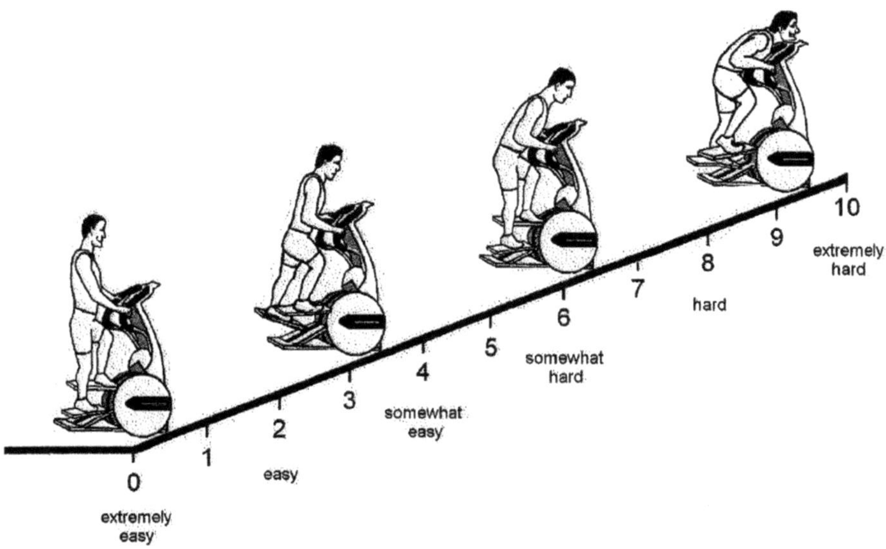

Fig. A.9 Adult OMNI-elliptical RPE scale (Mays et al. 2010)

Appendix B
RPE Scale Instructions

Adult OMNI-Walk/Run RPE Scale Instructions for RPE-O Only

This perceived exertion scale includes numerical categories from 0 to 10. You will use it to assess your perceptions of exertion while you exercise. The numbers on the scale represent a range of exertion levels from 0, "extremely easy," to 10, "extremely hard." To help you select a number that represents your level of exertion, consider the following. When the exercise exertion you are experiencing is "extremely easy," respond with a 0. Think about a time when you exercised and the level of exertion was "extremely easy" and most likely equivalent to a rating of 0. As an example, you should respond with a 0 when you are walking very slowly on the treadmill. When the exertion you are experiencing is "extremely hard," respond with a 10. Think about a time when you exercised and the perception of exertion was "extremely hard," likely attained at your maximal performance level. Most likely the exertional level would be equivalent to a rating of 10. As an example, you should respond with a 10 when you are running up a steep incline on the treadmill and you may not be able to exercise much longer owing to fatigue. Please rate your level of exertion for your overall body, taking into consideration the exertion experienced in your legs and your chest/breathing. When asked, use both the pictures and words to help you select one rating number that represents the level of exertion your body is experiencing. Each number response is called a rating of perceived exertion, or RPE. Please point to the number that best represents your RPE at the moment you are asked.

L. Haile et al., *Perceived Exertion Laboratory Manual*,
DOI 10.1007/978-1-4939-1917-8

Adult OMNI-Walk/Run RPE Scale Instructions for Undifferentiated and Differentiated RPE

This perceived exertion scale includes numerical categories from 0 to 10. You will use it to assess your perceptions of exertion while you exercise. The numbers on the scale represent a range of exertion levels from 0, 'extremely easy,' to 10, 'extremely hard.' To help you select a number that represents your level of exertion, consider the following. When the exercise exertion you are experiencing is 'extremely easy,' respond with a 0. Think about a time when you exercised and the level of exertion was 'extremely easy' and most likely equivalent to a rating of 0. As an example, you should respond with a 0 when you are walking very slowly on the treadmill. When the exertion you are experiencing is 'extremely hard,' respond with a 10. Think about a time when you exercised and the perception of exertion was 'extremely hard,' likely attained at your maximal performance level. Most likely the exertional level would be equivalent to a rating of 10. As an example, you should respond with a 10 when you are running up a steep incline on the treadmill and you may not be able to exercise much longer owing to fatigue. You will be asked to rate your level of exertion for your overall body, your legs and your chest/breathing. When asked, use both the pictures and words to help you select one rating number that represents the level of exertion your overall body, legs, or chest/breathing are experiencing. Each number response is called a rating of perceived exertion, or RPE. Please point to the number that best represents your RPE at the moment you are asked.

Borg (6–20) Scale Instructions for RPE-O Only During Treadmill Exercise

This perceived exertion scale includes numerical categories from 6 to 20. You will use it to assess your perceptions of exertion while you exercise. The numbers on the scale represent a range of exertion levels from 6, "no exertion at all," to 20, "maximal exertion." To help you select a number that represents your level of exertion, consider the following. When the exercise exertion you are experiencing is "no exertion at all," respond with a 6. Think about a time when you exercised and the level of exertion was "no exertion at all" and most likely equivalent to a rating of 6. As an example, you should respond with a 6 when you are walking very slowly on the treadmill. When the exertion you are experiencing is "maximal exertion," respond with a 20. Think about a time when you exercised and the perception of exertion was "maximal exertion," likely attained at your maximal performance level. Most likely the exertional level would be equivalent to a rating of 20. As an example, you should respond with a 20 when you are running up a steep incline on the treadmill and you may not be able to exercise much longer owing to fatigue. Please rate your level of exertion for your overall body, taking into consideration the exertion experienced in your legs and your chest/breathing. When asked, use the words to help you select one rating number that represents the level of exertion your body is experiencing. Each number response is called a rating of perceived exertion, or RPE. Please point to the number that best represents your RPE at the moment you are asked.

Adult OMNI-Cycle RPE Scale Instructions for RPE-L Only

This perceived exertion scale includes numerical categories from 0 to 10. You will use it to assess your perceptions of exertion while you exercise. The numbers on the scale represent a range of exertion levels from 0, "extremely easy," to 10, "extremely hard." To help you select a number that represents your level of exertion, consider the following. When the exercise exertion you are experiencing is "extremely easy," respond with a 0. Think about a time when you exercised and the level of exertion was "extremely easy" and most likely equivalent to a rating of 0. As an example, you should respond with a 0 when you are pedaling against no resistance on the cycle. When the exertion you are experiencing is "extremely hard," respond with a 10. Think about a time when you exercised and the perception of exertion was "extremely hard," likely attained at your maximal performance level. Most likely the exertional level would be equivalent to a rating of 10. As an example, you should respond with a 10 when you are pedaling against a very heavy resistance on the cycle and may not be able to exercise any longer owing to fatigue. You will be asked to rate the level of exertion of your legs only, not for your chest/breathing or your overall body. When asked, use both the pictures and words to help you select one rating number that represents the level of exertion your body is experiencing. Each number response is called a rating of perceived exertion, or RPE. Please point to the number that best represents your RPE at the moment you are asked.

Adult OMNI-Cycle RPE Scale Instructions for Undifferentiated and Differentiated RPE

This perceived exertion scale includes numerical categories from 0 to 10. You will use it to assess your perceptions of exertion while you exercise. The numbers on the scale represent a range of exertion levels from 0, "extremely easy," to 10, "extremely hard." To help you select a number that represents your level of exertion, consider the following. When the exercise exertion you are experiencing is "extremely easy," respond with a 0. Think about a time when you exercised and the level of exertion was "extremely easy" and most likely equivalent to a rating of 0. As an example, you should respond with a 0 when you are pedaling against no resistance on the cycle. When the exertion you are experiencing is "extremely hard," respond with a 10. Think about a time when you exercised and the perception of exertion was "extremely hard," likely attained at your maximal performance level.

Most likely the exertional level would be equivalent to a rating of 10. As an example, you should respond with a 10 when you are pedaling against a very heavy resistance on the cycle and may not be able to exercise any longer owing to fatigue. You will be asked to rate your level of exertion for your overall body, your legs and your chest/breathing. When asked, use both the pictures and words to help you select one rating number that represents the level of exertion your overall body, legs, or chest/breathing are experiencing. Each number response is called a rating of perceived exertion, or RPE. Please point to the number that best represents your RPE at the moment you are asked.

Borg (6–20) Scale Instructions for RPE-L Only During Cycle Exercise

This perceived exertion scale includes numerical categories from 6 to 20. You will use it to assess your perceptions of exertion while you exercise. The numbers on the scale represent a range of exertion levels from 6, "no exertion at all," to 20, "maximal exertion." To help you select a number that represents your level of exertion, consider the following. When the exercise exertion you are experiencing is "no exertion at all," respond with a 6. Think about a time when you exercised and the level of exertion was "no exertion at all" and most likely equivalent to a rating of 6. As an example, you should respond with a 6 when you are pedaling against no resistance on the cycle. When the exertion you are experiencing is "maximal exertion," respond with a 20. Think about a time when you exercised and the perception of exertion was "maximal exertion," likely attained at your maximal performance level. Most likely the exertional level would be equivalent to a rating of 20. As an example, you should respond with a 20 when you are pedaling against a very heavy resistance on the cycle and may not be able to exercise any longer owing to fatigue. Please rate your level of exertion for your legs only, not for your chest/breathing or your overall body. When asked, use the words to help you select one rating number that represents the level of exertion your legs are experiencing. Each number response is called a rating of perceived exertion, or RPE. Please point to the number that best represents your RPE at the moment you are asked.

Adult OMNI-Resistance Exercise RPE Scale for RPE-AM Only

This perceived exertion scale includes numerical categories from 0 to 10. You will use it to assess your perceptions of exertion while you perform resistance exercise. The numbers on the scale represent a range of exertion levels from 0, "extremely easy," to 10, "extremely hard." To help you select a number that represents your level of exertion, consider the following. When the resistance exercise exertion you are experiencing is "extremely easy," respond with a 0. Think about a time when you exercised and the level of exertion was "extremely easy" and most likely equivalent to a rating of 0. As an example, you should respond with a 0 when you are lifting a very light weight that is extremely easy to lift. When the exertion you are experiencing is "extremely hard," respond with a 10. Think about a time when you performed resistance exercise and the perception of exertion was "extremely hard," likely attained at your maximal performance level. Most likely the exertional level would be equivalent to a rating of 10. As an example, you should respond with a 10 when you are lifting the heaviest weight you can lift and may not be able to lift for one more repetition owing to fatigue. You will be asked to rate the level of exertion of your active muscles only, not for your chest/breathing or your overall body. When asked, use both the pictures and words to help you select one rating number that represents the level of exertion your active muscles are experiencing. Each number response is called a rating of perceived exertion, or RPE. Please point to the number that best represents your RPE at the moment you are asked.

Adult OMNI-Resistance Exercise RPE Scale for Undifferentiated and Differentiated RPE

This perceived exertion scale includes numerical categories from 0 to 10. You will use it to assess your perceptions of exertion while you perform resistance exercise. The numbers on the scale represent a range of exertion levels from 0, "extremely easy," to 10, "extremely hard." To help you select a number that represents your level of exertion, consider the following. When the resistance exercise exertion you are experiencing is "extremely easy," respond with a 0. Think about a time when you exercised and the level of exertion was "extremely easy" and most likely equivalent to a rating of 0. As an example, you should respond with a 0 when you are lifting a very light weight that is extremely easy to lift. When the exertion you are experiencing is "extremely hard," respond with a 10. Think about a time when you performed resistance exercise and the perception of exertion was "extremely hard," likely attained at your maximal performance level. Most likely the exertional level would be equivalent to a rating of 10. As an example, you should respond with a 10 when you are lifting the heaviest weight you can lift and may not be able to lift for one more repetition owing to fatigue. You will be asked to rate your level of exertion for your overall body, your active muscles and your chest/breathing. When asked, use both the pictures and words to help you select one rating number that represents the level of exertion your overall body, active muscles, or chest/breathing are experiencing. Each number response is called a rating of perceived exertion, or RPE. Please point to the number that best represents your RPE at the moment you are asked.

Borg (6–20) Scale Instructions for RPE-AM Only During Resistance Exercise

This perceived exertion scale includes numerical categories from 6 to 20. You will use it to assess your perceptions of exertion while you perform resistance exercise. The numbers on the scale represent a range of exertion levels from 6, "no exertion at all," to 20, "maximal exertion." To help you select a number that represents your level of exertion, consider the following. When the resistance exercise exertion you are experiencing is "no exertion at all," respond with a 6. Think about a time when you exercised and the level of exertion was "no exertion at all" and most likely equivalent to a rating of 6. As an example, you should respond with a 6 when you are lifting a very light weight that is extremely easy to lift. When the exertion you are experiencing is "maximal exertion," respond with a 20. Think about a time when you performed resistance exercise and the perception of exertion was "maximal exertion," likely attained at your maximal performance level. Most likely the exertional level would be equivalent to a rating of 20. As an example, you should respond with a 20 when you are lifting the heaviest weight you can lift and may not be able to lift for one more repetition owing to fatigue. You will be asked to rate the level of exertion of your active muscles only, not for your chest/breathing or your overall body. When asked, use the words to help you select one rating number that represents the level of exertion your active muscles are experiencing. Each number response is called a rating of perceived exertion, or RPE. Please point to the number that best represents your RPE at the moment you are asked.

Appendix C
Determination of Validity Coefficients: An Example Using Cycle Ergometry Graded Exercise Test Results

1. In a Microsoft Excel spreadsheet, label columns of data as shown in Fig. A.10.
   (a) For resistance exercise, Exercise Stage may be replaced with %1RM and Weight Lifted may take the place of physiological variables such as VO_2 and HR.
   (b) For treadmill exercise, VO_2 may be expressed in $ml \cdot kg \cdot min^{-1}$.
   (c) RPE-O and other differentiated RPE's such as RPE-C or RPE-AM may be used in your experiment.

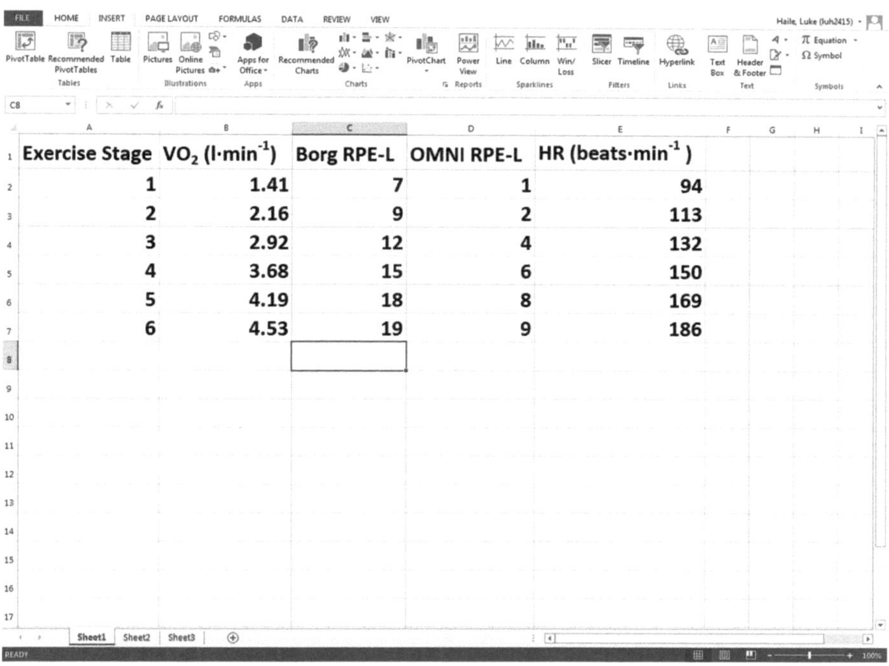

Exercise Stage	VO_2 ($l \cdot min^{-1}$)	Borg RPE-L	OMNI RPE-L	HR ($beats \cdot min^{-1}$)
1	1.41	7	1	94
2	2.16	9	2	113
3	2.92	12	4	132
4	3.68	15	6	150
5	4.19	18	8	169
6	4.53	19	9	186

Fig. A.10 ∎

L. Haile et al., *Perceived Exertion Laboratory Manual*,
DOI 10.1007/978-1-4939-1917-8

2. Plot of VO_2 and Borg RPE-L for determination of concurrent validity:

    (a) Click on the **INSERT** tab and in the **CHARTS** section click on **SCATTER**. Select the first available chart option. A blank or example scatter plot will appear on your screen (Fig. A.11).

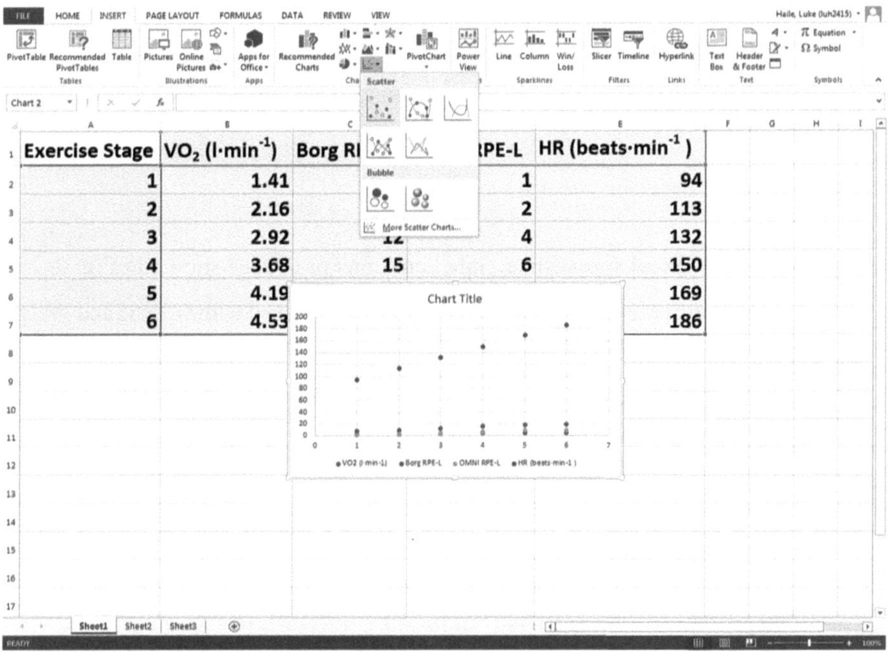

Fig. A.11 ∎

    (b) Click on the **SELECT DATA** tab (Fig. A.12).

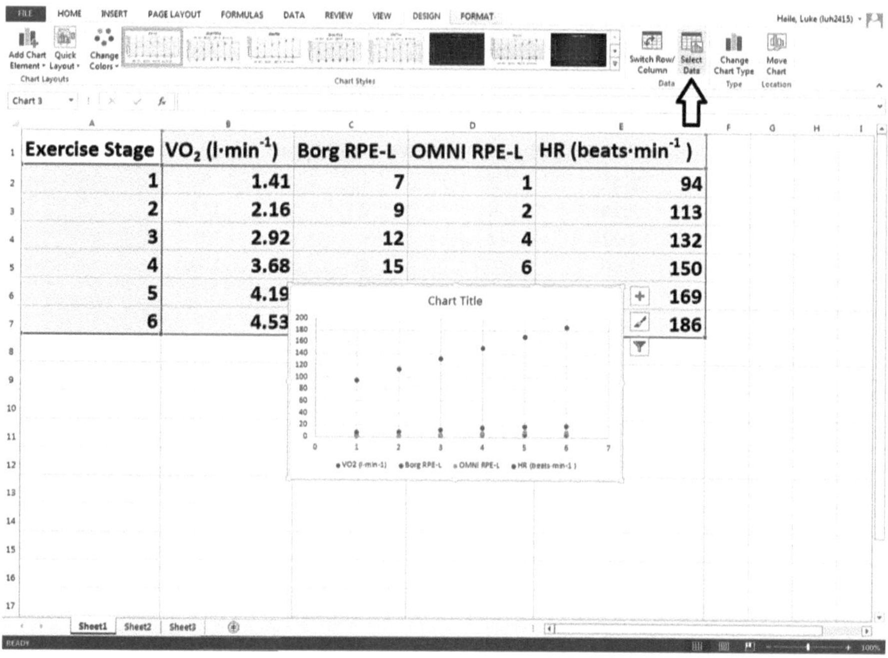

Fig. A.12 ∎

(c) Remove any entries found in the **LEGEND ENTRIES** text box then click **ADD** (Figs. A.13 and A.14).

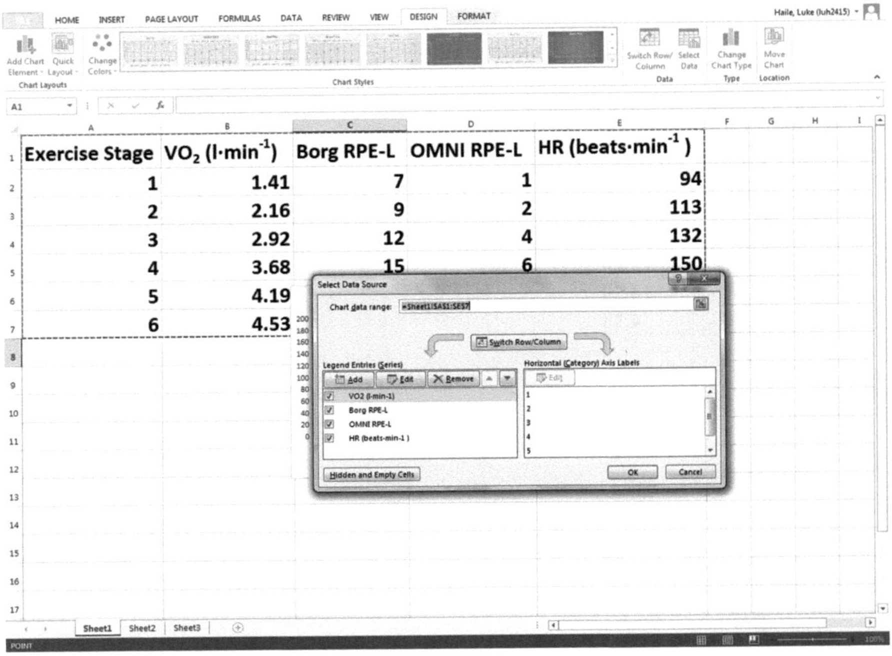

Fig. A.13 ∎

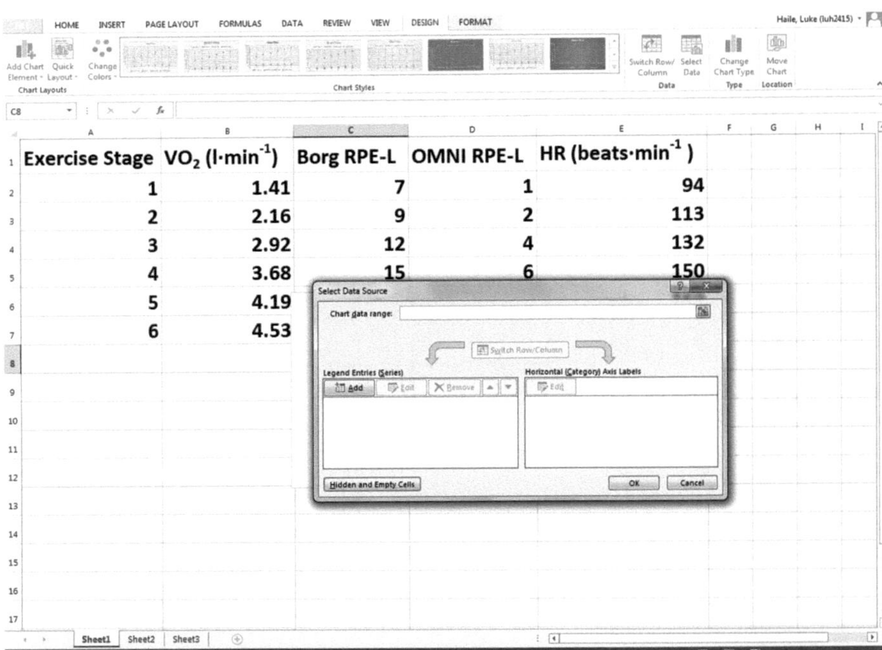

Fig. A.14 ∎

(d) Under **<u>SERIES NAME</u>**, enter VO_2 and Borg RPE-L. Then click on the icon to the right of the **<u>SERIES X VALUES</u>** text box and highlight the VO_2 values. After the values are highlighted click the icon on the box that appeared (Fig. A.15).

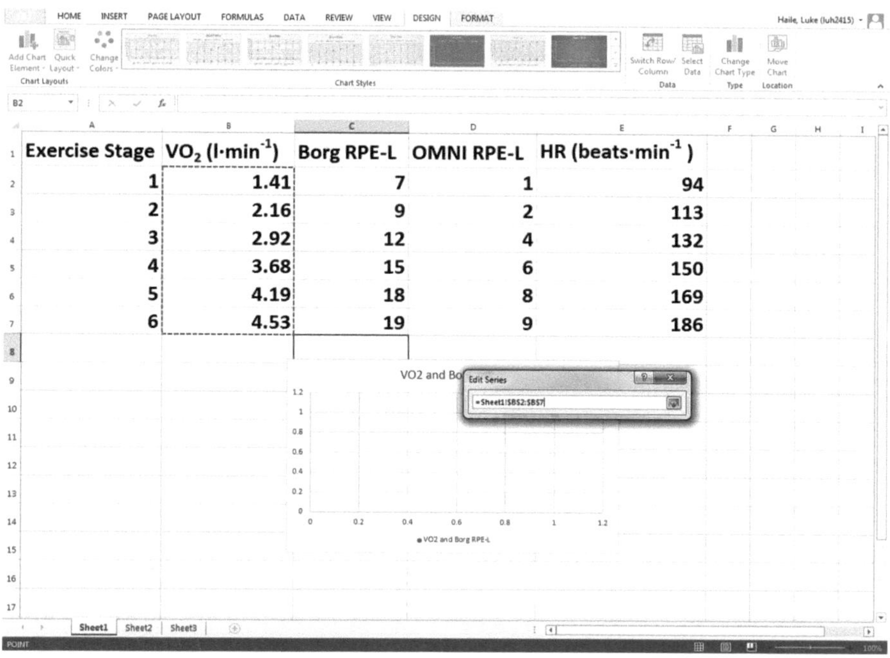

Fig. A.15 ∎

(e) Then click on the icon to the right of the **SERIES Y VALUES** text box and highlight the Borg RPE-L values. After the values are highlighted click the icon on the box that appeared (Fig. A.16).

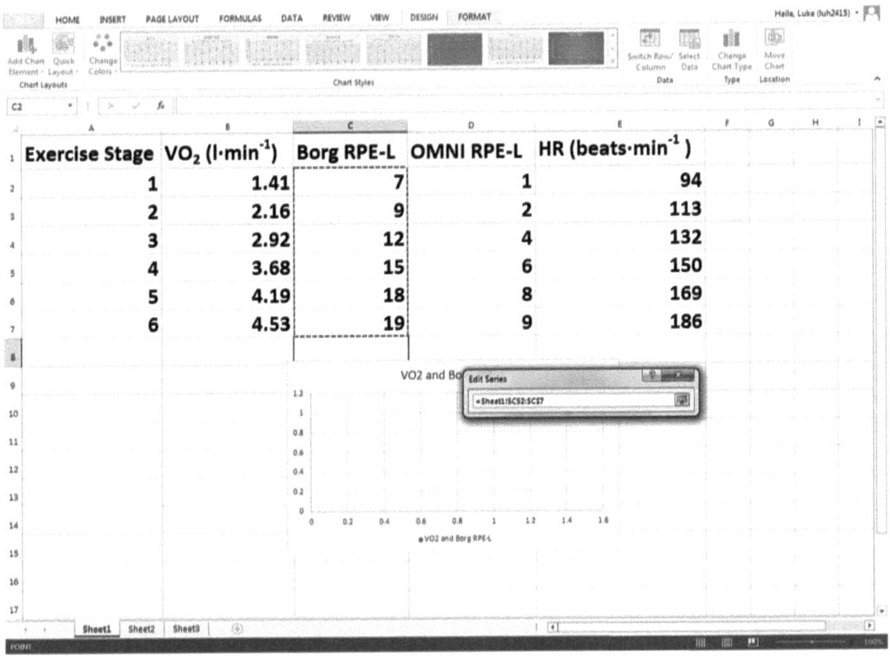

Fig. A.16 ∎

(f) Click **OK** on the next two screens. You should now have a scatter plot with Borg RPE-L on the *y*-axis and VO_2 on the *x*-axis (Fig. A.17).

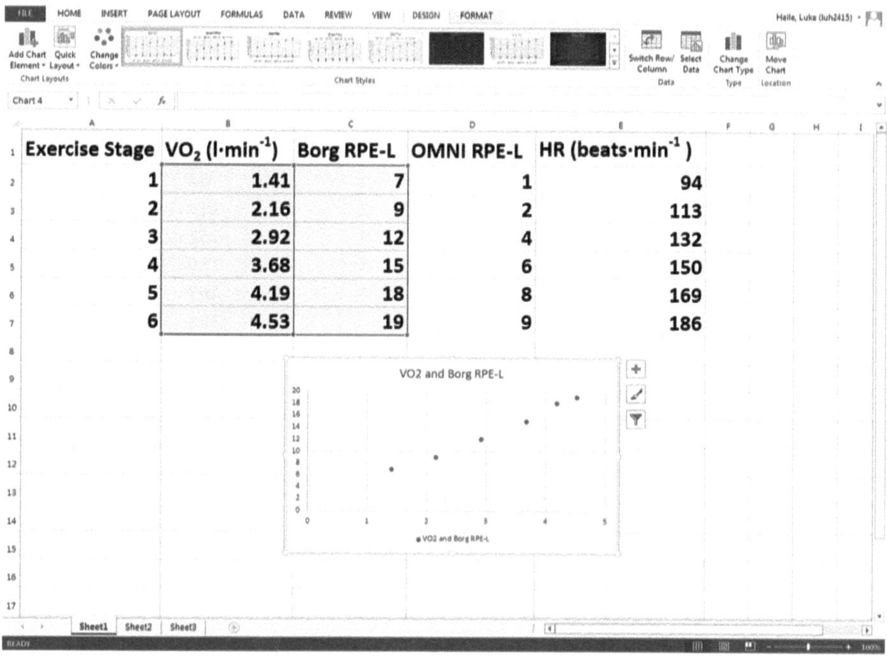

Fig. A.17 ∎

(g) Create a title for the plot and enter the appropriate axis labels and units of measure (Fig. A.18).

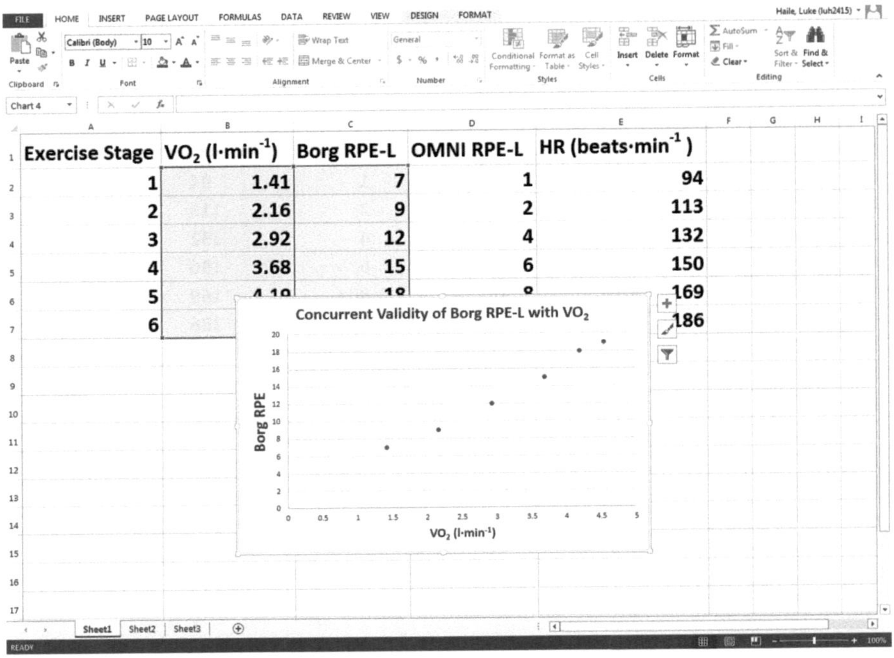

Fig. A.18 ∎

(h) To determine the validity coefficient, click on one of the data points to highlight the entire data series. Right click on one of the data points and a menu will appear. Click **ADD TRENDLINE** and the **FORMAT TRENDLINE** menu will appear. Select **LINEAR**, **DISPLAY EQUATION ON CHART**, and **DISPLAY R-SQUARED VALUE ON CHART** then click **CLOSE**. The trendline and equation will be displayed on the chart. Take the square root of the R^2 value to determine the Pearson correlation coefficient (Fig. A.19).

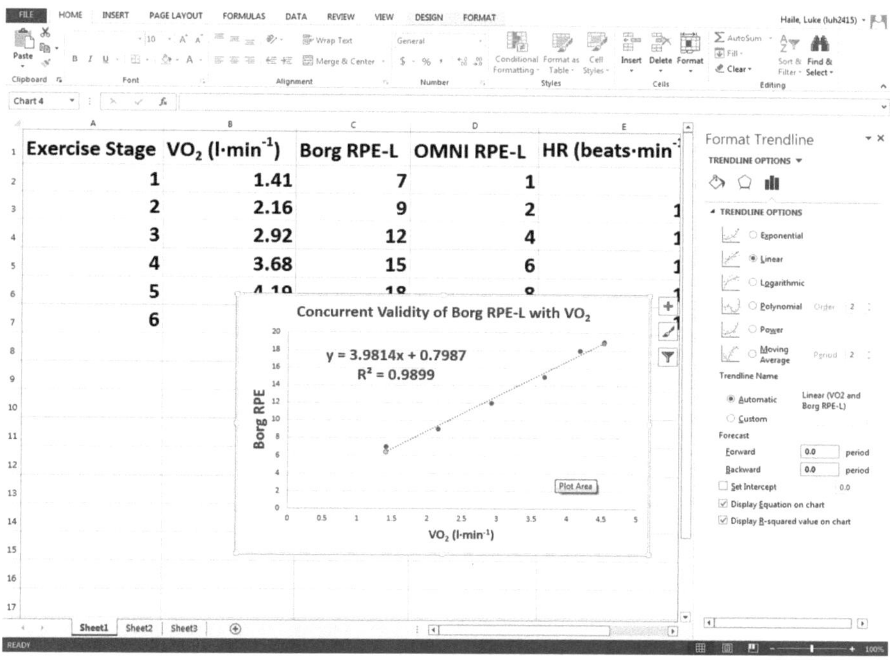

Fig. A.19 ∎

Appendix D
Determination of VT and RPE-VT

Calculation of $V_E \cdot VO_2^{-1}$ and $V_E \cdot VCO_2^{-1}$

1. Obtain a printout containing the 15-s exercise VO_2, VCO_2, and V_E values in $l \cdot min^{-1}$ from the respiratory-metabolic measurement system. In a Microsoft Excel spreadsheet, label columns for VO_2, VCO_2, V_E (each in $l \cdot min^{-1}$), $V_E \cdot VO_2^{-1}$ and $V_E \cdot VCO_2^{-1}$.
2. In the columns for VO_2, VCO_2, and V_E, enter each 15-s value measured during exercise as listed in the printout (Fig. A.20).

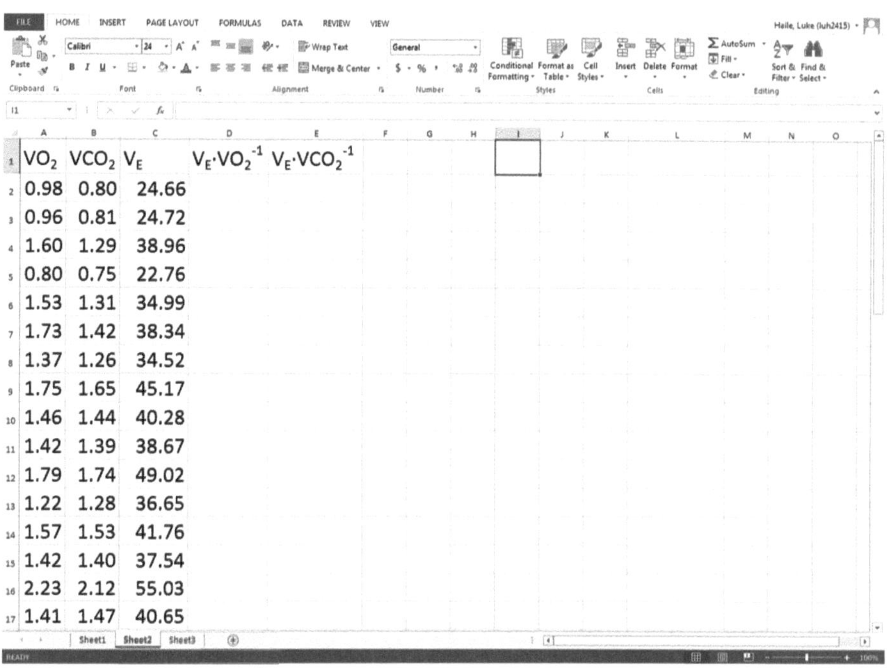

VO_2	VCO_2	V_E	$V_E \cdot VO_2^{-1}$	$V_E \cdot VCO_2^{-1}$
0.98	0.80	24.66		
0.96	0.81	24.72		
1.60	1.29	38.96		
0.80	0.75	22.76		
1.53	1.31	34.99		
1.73	1.42	38.34		
1.37	1.26	34.52		
1.75	1.65	45.17		
1.46	1.44	40.28		
1.42	1.39	38.67		
1.79	1.74	49.02		
1.22	1.28	36.65		
1.57	1.53	41.76		
1.42	1.40	37.54		
2.23	2.12	55.03		
1.41	1.47	40.65		

Fig. A.20 ∎

L. Haile et al., *Perceived Exertion Laboratory Manual*,
DOI 10.1007/978-1-4939-1917-8

3. Calculate $V_E \cdot VO_2^{-1}$ by dividing V_E ($l \cdot min^{-1}$) by VO_2 ($l \cdot min^{-1}$) for each row and enter the value in the appropriate cell. This can be completed by typing the equation seen in Fig. A.21 in the first cell available under $V_E \cdot VO_2^{-1}$ then hitting the **ENTER** key on the keyboard. This equation can be copied into the remaining cells below $V_E \cdot VO_2^{-1}$ to complete the calculation for each 15-s interval.

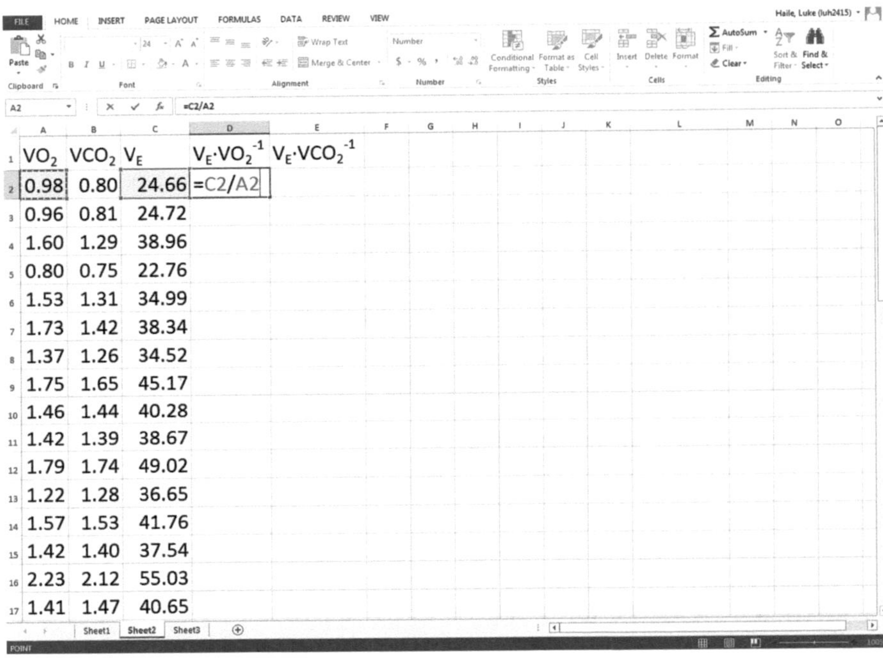

	A	B	C	D	E
1	VO_2	VCO_2	V_E	$V_E \cdot VO_2^{-1}$	$V_E \cdot VCO_2^{-1}$
2	0.98	0.80	24.66	=C2/A2	
3	0.96	0.81	24.72		
4	1.60	1.29	38.96		
5	0.80	0.75	22.76		
6	1.53	1.31	34.99		
7	1.73	1.42	38.34		
8	1.37	1.26	34.52		
9	1.75	1.65	45.17		
10	1.46	1.44	40.28		
11	1.42	1.39	38.67		
12	1.79	1.74	49.02		
13	1.22	1.28	36.65		
14	1.57	1.53	41.76		
15	1.42	1.40	37.54		
16	2.23	2.12	55.03		
17	1.41	1.47	40.65		

Fig. A.21 ∎

4. Calculate $V_E \cdot VCO_2^{-1}$ by dividing V_E ($l \cdot min^{-1}$) by VCO_2 ($l \cdot min^{-1}$) for each row and enter the value in the appropriate cell. This can be completed by typing the equation seen in Fig. A.22 in the first cell available under $V_E \cdot VO_2^{-1}$ then hitting the **ENTER** key on the keyboard. This equation can be copied into the remaining cells below $V_E \cdot VCO_2^{-1}$ to complete the calculation for each 15-s interval.

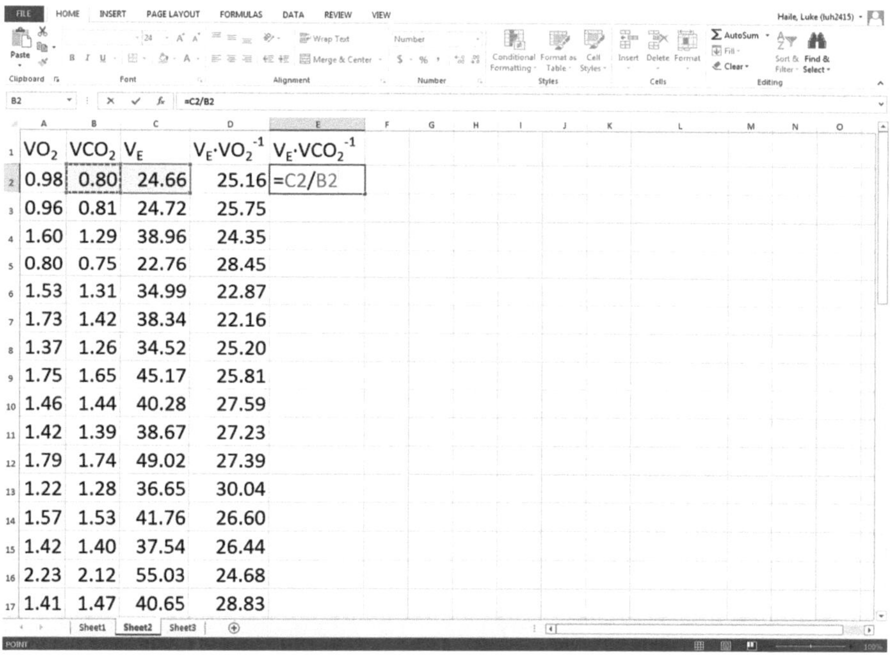

VO2	VCO2	VE	VE·VO2-1	VE·VCO2-1
0.98	0.80	24.66	25.16	=C2/B2
0.96	0.81	24.72	25.75	
1.60	1.29	38.96	24.35	
0.80	0.75	22.76	28.45	
1.53	1.31	34.99	22.87	
1.73	1.42	38.34	22.16	
1.37	1.26	34.52	25.20	
1.75	1.65	45.17	25.81	
1.46	1.44	40.28	27.59	
1.42	1.39	38.67	27.23	
1.79	1.74	49.02	27.39	
1.22	1.28	36.65	30.04	
1.57	1.53	41.76	26.60	
1.42	1.40	37.54	26.44	
2.23	2.12	55.03	24.68	
1.41	1.47	40.65	28.83	

Fig. A.22 ∎

Plot of $V_E \cdot VO_2^{-1}$ and $V_E \cdot VCO_2^{-1}$ for Determination of VT

1. Obtain a printout from the respiratory-metabolic measurement system containing the 15-s exercise values of VO_2 ($l \cdot min^{-1}$) and the ventilatory equivalents ($V_E \cdot VO_2^{-1}$ and $V_E \cdot VCO_2^{-1}$). These variables have no units of measure because they are a ratio between two variables with the same units of measure.
2. In a Microsoft Excel spreadsheet, label columns for VO_2 ($l \cdot min^{-1}$), $V_E \cdot VO_2^{-1}$ and $V_E \cdot VCO_2^{-1}$. Enter each 15-s value measured during exercise as listed in the printout. If $V_E \cdot VO_2^{-1}$ and $V_E \cdot VCO_2^{-1}$ were calculated using a Microsoft Excel spreadsheet as described above, you may continue to use that spreadsheet (Fig. A.23).

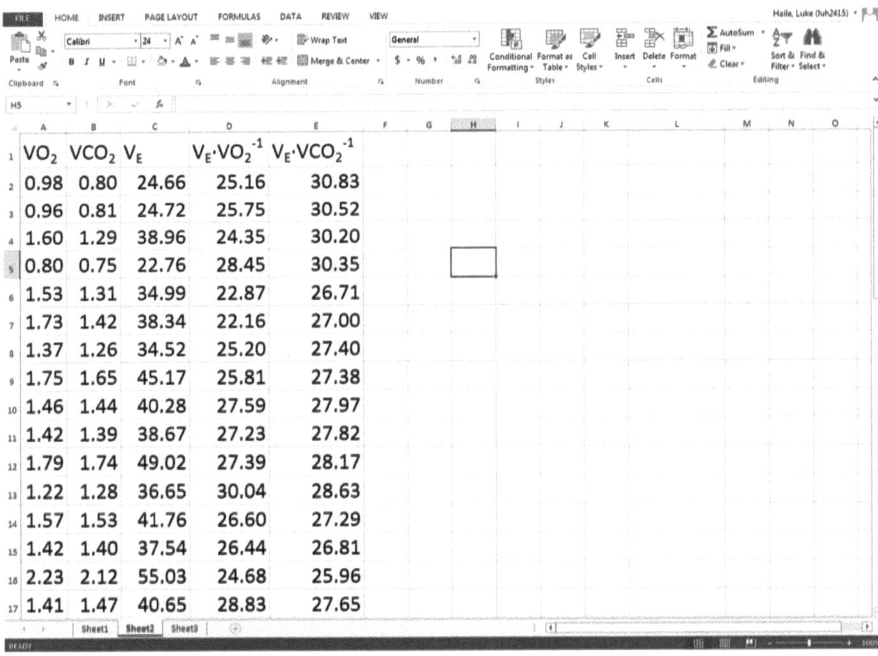

VO_2	VCO_2	V_E	$V_E \cdot VO_2^{-1}$	$V_E \cdot VCO_2^{-1}$
0.98	0.80	24.66	25.16	30.83
0.96	0.81	24.72	25.75	30.52
1.60	1.29	38.96	24.35	30.20
0.80	0.75	22.76	28.45	30.35
1.53	1.31	34.99	22.87	26.71
1.73	1.42	38.34	22.16	27.00
1.37	1.26	34.52	25.20	27.40
1.75	1.65	45.17	25.81	27.38
1.46	1.44	40.28	27.59	27.97
1.42	1.39	38.67	27.23	27.82
1.79	1.74	49.02	27.39	28.17
1.22	1.28	36.65	30.04	28.63
1.57	1.53	41.76	26.60	27.29
1.42	1.40	37.54	26.44	26.81
2.23	2.12	55.03	24.68	25.96
1.41	1.47	40.65	28.83	27.65

Fig. A.23 ■

3. Create a line graph with $V_E \cdot VO_2^{-1}$ and $V_E \cdot VCO_2^{-1}$ on the *y*-axis and VO_2 $(l \cdot min^{-1})$ on the *x*-axis.

   (a) Click on the **INSERT** tab then click on the **INSERT LINE CHART** icon. Select the first option for a basic 2D line chart (Fig. A.24).

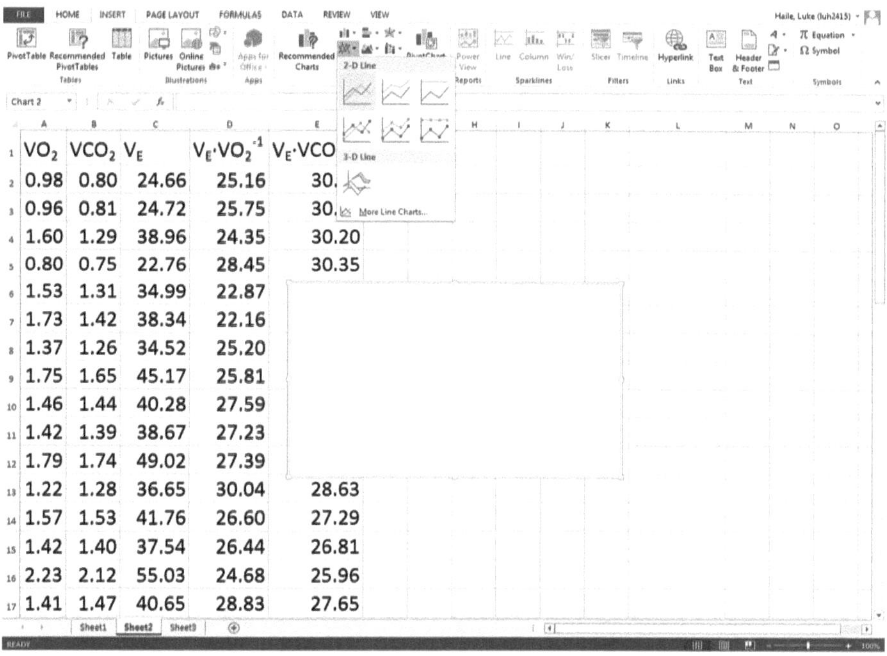

Fig. A.24

(b) A blank chart will appear on the spreadsheet. Right click on the chart and click **SELECT DATA** (Fig. A.25).

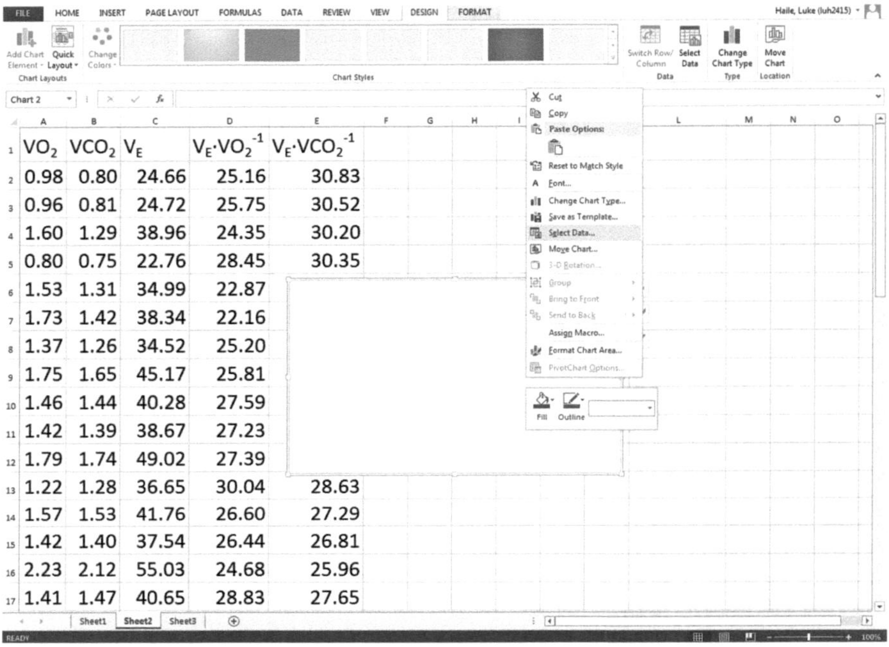

VO2	VCO2	VE	VE·VO2^-1	VE·VCO2^-1
0.98	0.80	24.66	25.16	30.83
0.96	0.81	24.72	25.75	30.52
1.60	1.29	38.96	24.35	30.20
0.80	0.75	22.76	28.45	30.35
1.53	1.31	34.99	22.87	
1.73	1.42	38.34	22.16	
1.37	1.26	34.52	25.20	
1.75	1.65	45.17	25.81	
1.46	1.44	40.28	27.59	
1.42	1.39	38.67	27.23	
1.79	1.74	49.02	27.39	
1.22	1.28	36.65	30.04	28.63
1.57	1.53	41.76	26.60	27.29
1.42	1.40	37.54	26.44	26.81
2.23	2.12	55.03	24.68	25.96
1.41	1.47	40.65	28.83	27.65

Fig. A.25 ∎

(c) The **SELECT DATA SOURCE** box will appear (Fig. A.26).

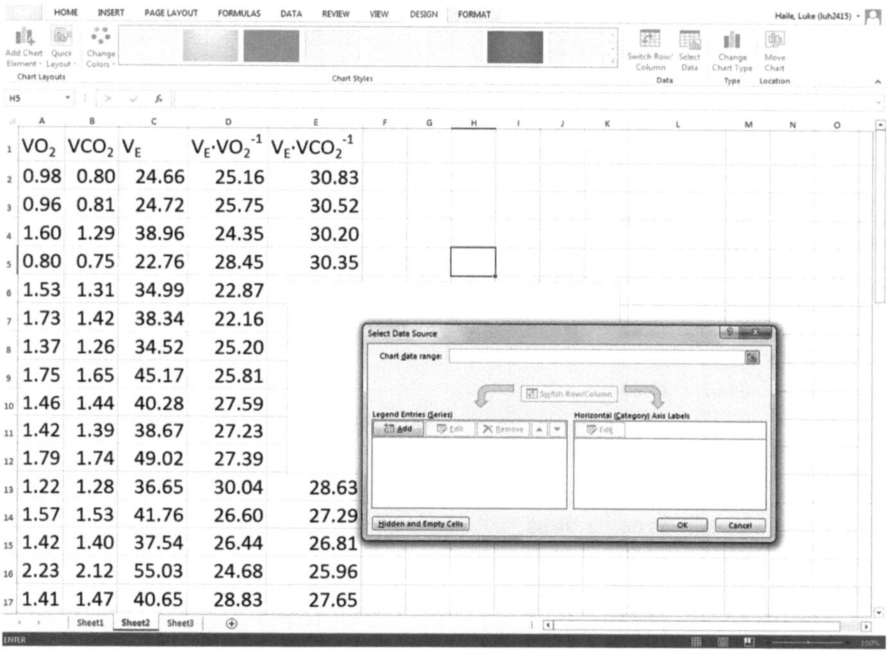

Fig. A.26 ∎

(d) Click **ADD** under **LEGEND ENTRIES** and the **EDIT SERIES** box will appear (Fig. A.27).

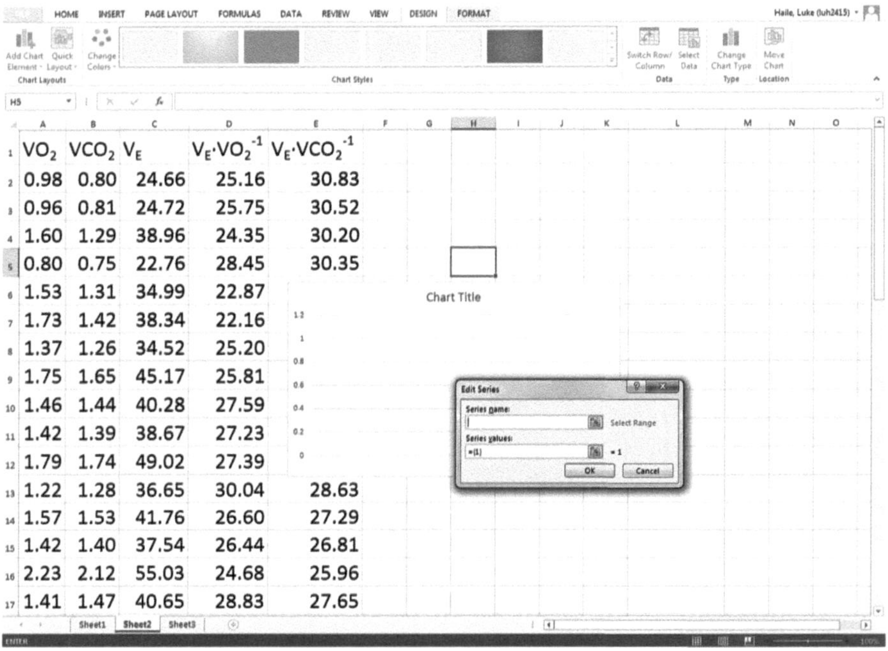

Fig. A.27 ∎

(e) Click on the icon to the right of the **SERIES NAME** text box then click on the cell in the spreadsheet containing $V_E \cdot VO_2^{-1}$. Click the icon on the right side of the box that appeared to return to the **EDIT SERIES** box (Fig. A.28).

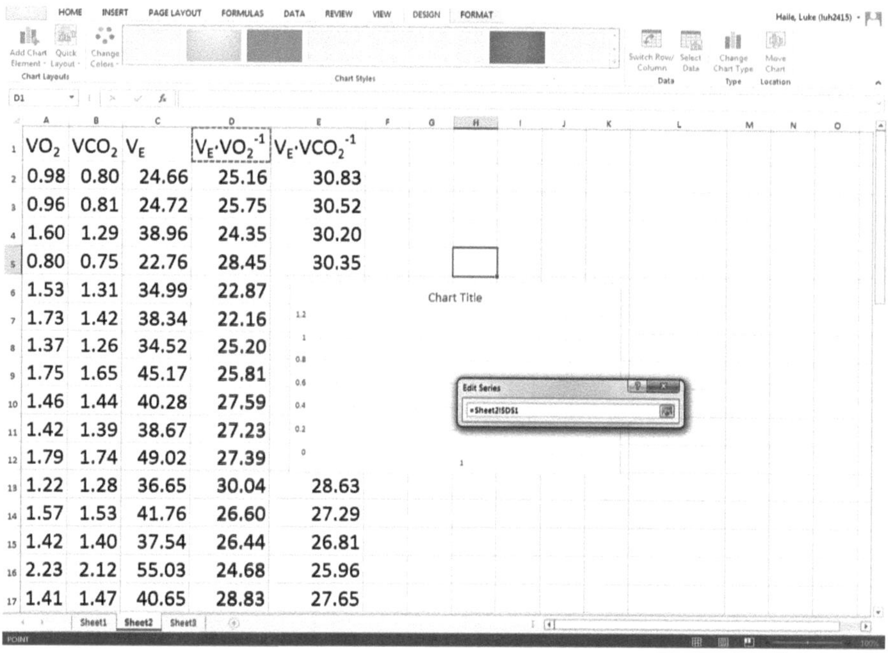

Fig. A.28 ∎

(f) Click on the icon to the right of the **SERIES VALUES** text box then highlight all the cells in the spreadsheet containing data under $V_E \cdot VO_2^{-1}$. Click the icon on the right side of the box that appeared to return to the **EDIT SERIES** box. Click **OK** to return to the **SELECT DATA SOURCE BOX** (Fig. A.29).

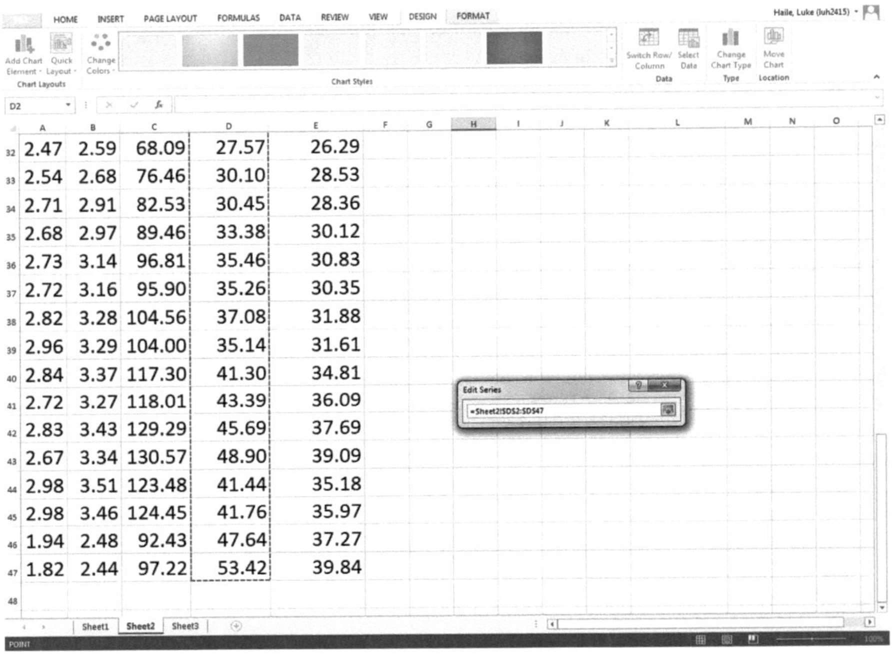

Fig. A.29 ∎

(g) Click **ADD** under **LEGEND ENTRIES** and the **EDIT SERIES** box will appear.

(h) Click on the icon to the right of the **SERIES NAME** text box then click on the cell in the spreadsheet containing $V_E \cdot VCO_2^{-1}$. Click the icon on the right side of the box that appeared to return to the **EDIT SERIES** box.

(i) Click on the icon to the right of the **SERIES VALUES** text box then highlight all the cells in the spreadsheet containing data under $V_E \cdot VCO_2^{-1}$. Click the icon on the right side of the box that appeared to return to the **EDIT SERIES** box. Click **OK** to return to the **SELECT DATA SOURCE BOX**.

(j) Click **EDIT** under **HORIZONTAL (CATEGORY) AXIS LABELS** (Fig. A.30).

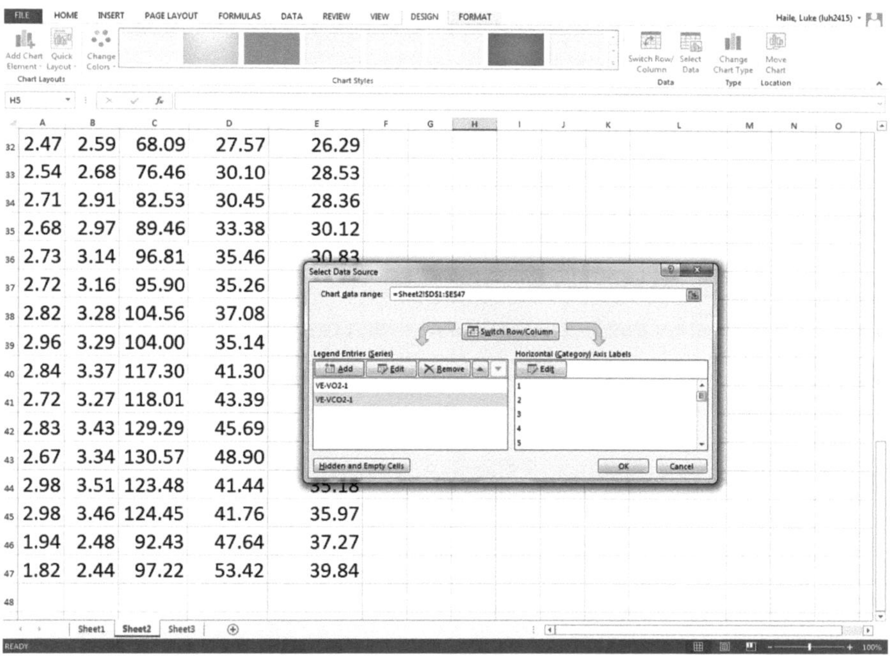

Fig. A.30 ∎

(k) Highlight all the cells containing data under VO_2 then click **OK** (Fig. A.31).

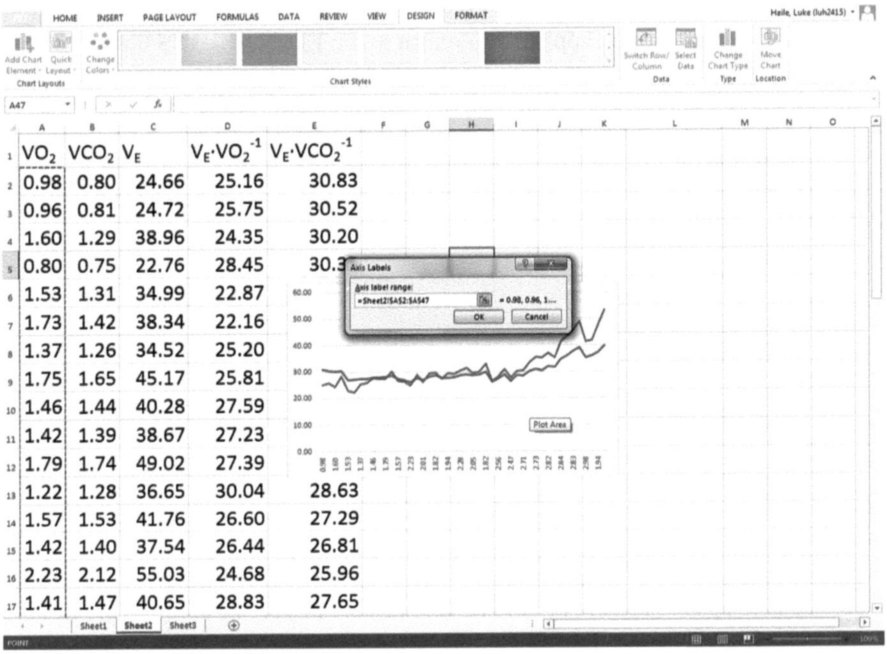

Fig. A.31 ∎

(l) Click **OK** on the **SELECT DATA SOURCE** box and the plot will appear. It may be beneficial to enlarge the plot so the labels on the *x*-axis are easily viewable. Locate the point on the graph where $V_E \cdot VO_2^{-1}$ begins to increase without an increase in $V_E \cdot VCO_2^{-1}$. Draw a vertical line from that point down to the *x*-axis and identify the VO_2 equivalent to this divergent point. Convert the units of this VO_2 value from $l \cdot min^{-1}$ to $\%VO_2max/peak$ (Fig. A.32).

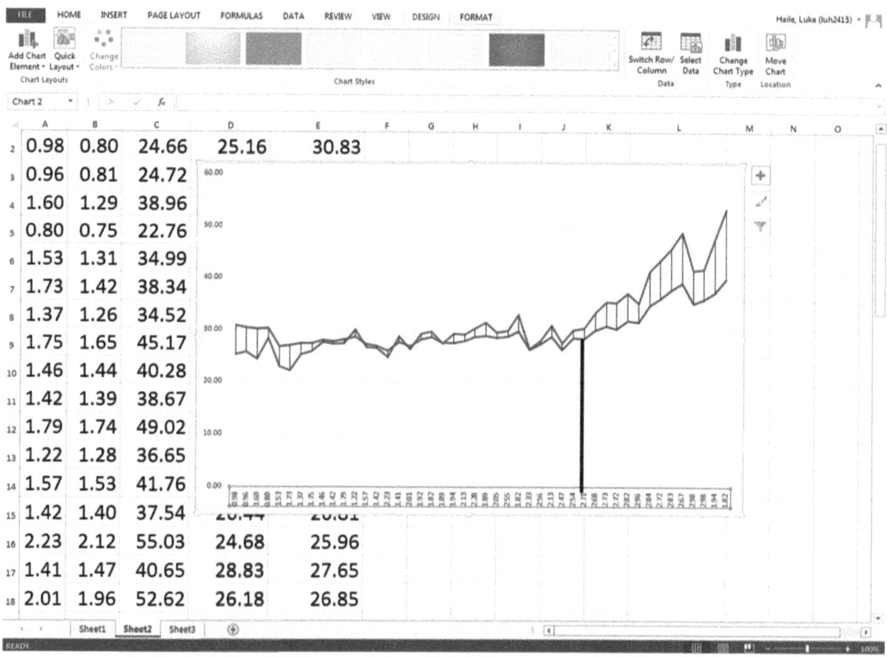

Fig. A.32 ∎

Adjustment of Automatic VT Calculation in a Respiratory Metabolic Measurement System

1. Focus your attention on the figure showing $V_E \cdot VO_2^{-1}$ and $V_E \cdot VCO_2^{-1}$ on the *y*-axis and Time on the *x*-axis. There will be a vertical line indicating the position of the VT on this figure.
2. If the vertical line is not located over the point where $V_E \cdot VO_2^{-1}$ begins to increase without an increase in $V_E \cdot VCO_2^{-1}$, adjust the vertical line until it is located directly over this point.
3. If the vertical line is located over the point where $V_E \cdot VO_2^{-1}$ begins to increase without an increase in $V_E \cdot VCO_2^{-1}$, do not adjust it.
4. The computer program will automatically provide the VO_2 value ($l \cdot min^{-1}$) and $\%VO_2max$/peak associated with the VT.

Determination of RPE-VT: An Example Using Cycle Ergometry Graded Exercise Test Results

1. In a Microsoft Excel spreadsheet, label columns of data as shown in Fig. A.33.

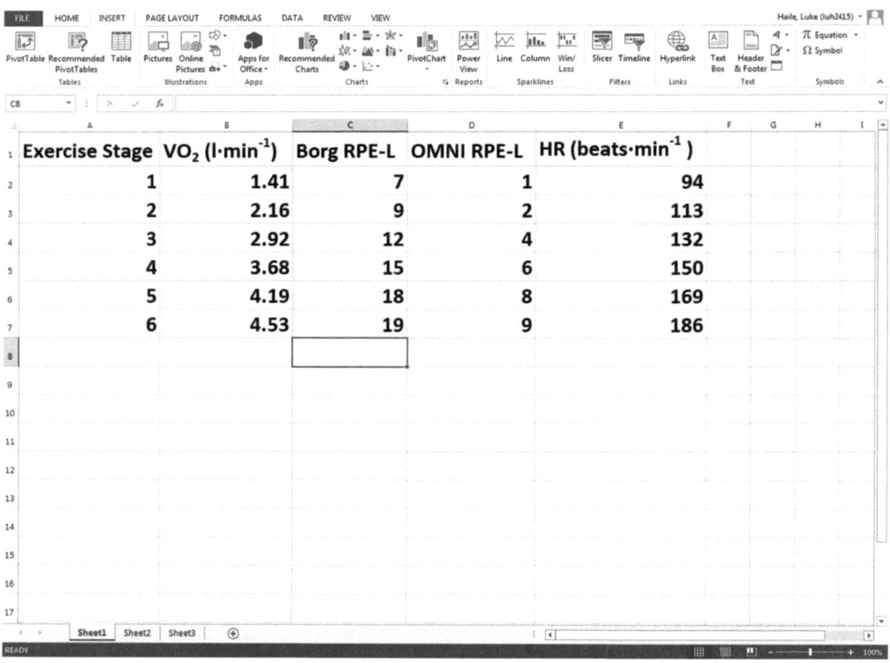

Exercise Stage	VO_2 ($l \cdot min^{-1}$)	Borg RPE-L	OMNI RPE-L	HR ($beats \cdot min^{-1}$)
1	1.41	7	1	94
2	2.16	9	2	113
3	2.92	12	4	132
4	3.68	15	6	150
5	4.19	18	8	169
6	4.53	19	9	186

Fig. A.33 ∎

(a) For treadmill exercise, VO_2 may be expressed in ml · kg · min^{-1}.
(b) RPE-O and differentiated RPE such as RPE-C may be used in your experiment.

2. Plot of VO_2 as a function of Borg RPE-L for determination of RPE-VT.

(a) Click on the **INSERT** tab and in the **CHARTS** section click on **SCATTER**. Select the first available chart option. A blank or example scatter plot will appear on your screen (Fig. A.34).

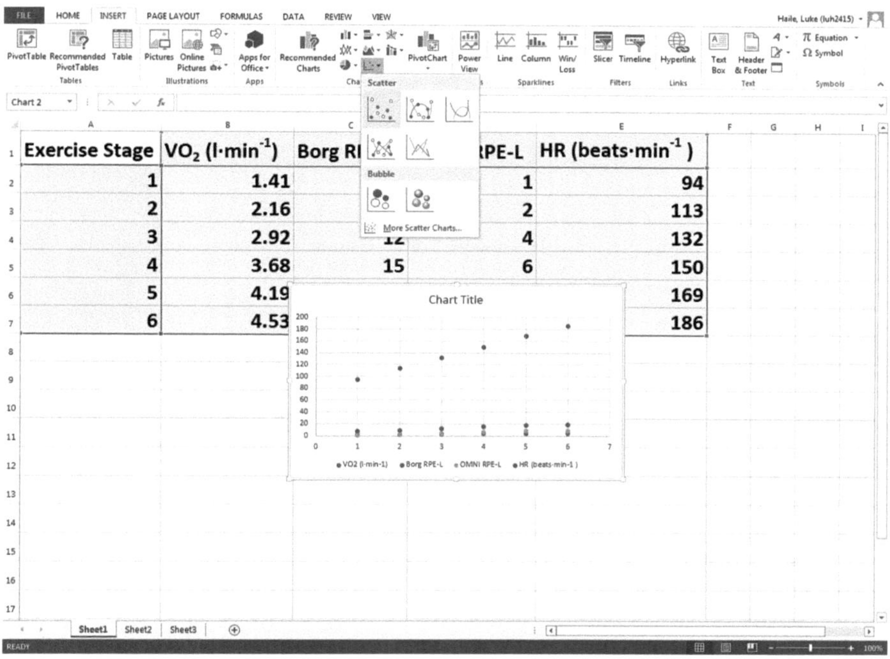

Fig. A.34 ∎

(b) Click on the **SELECT DATA** tab (Fig. A.35).

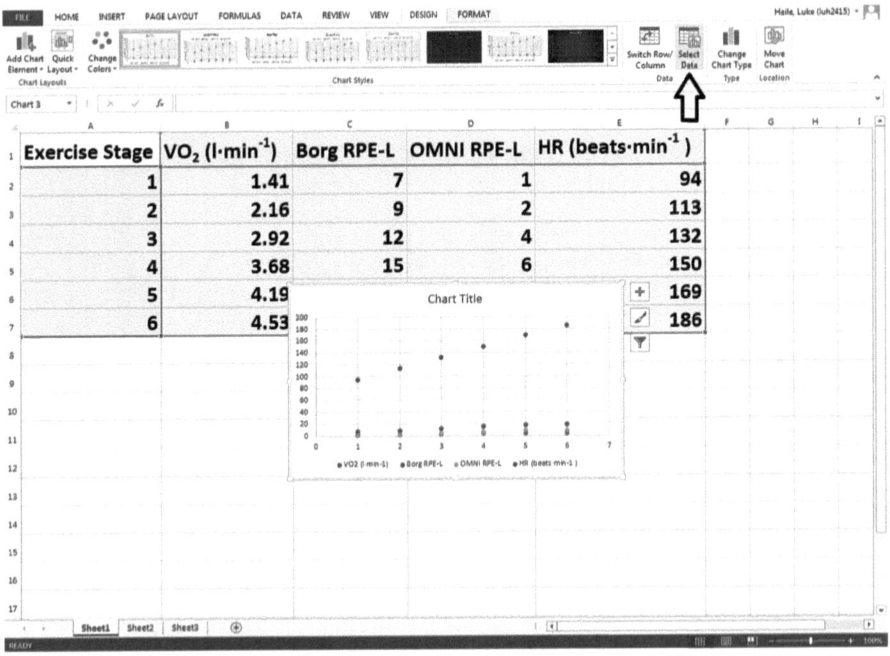

Fig. A.35 ∎

(c) Remove any entries found in the **LEGEND ENTRIES** text box then click **ADD** (Fig. A.36).

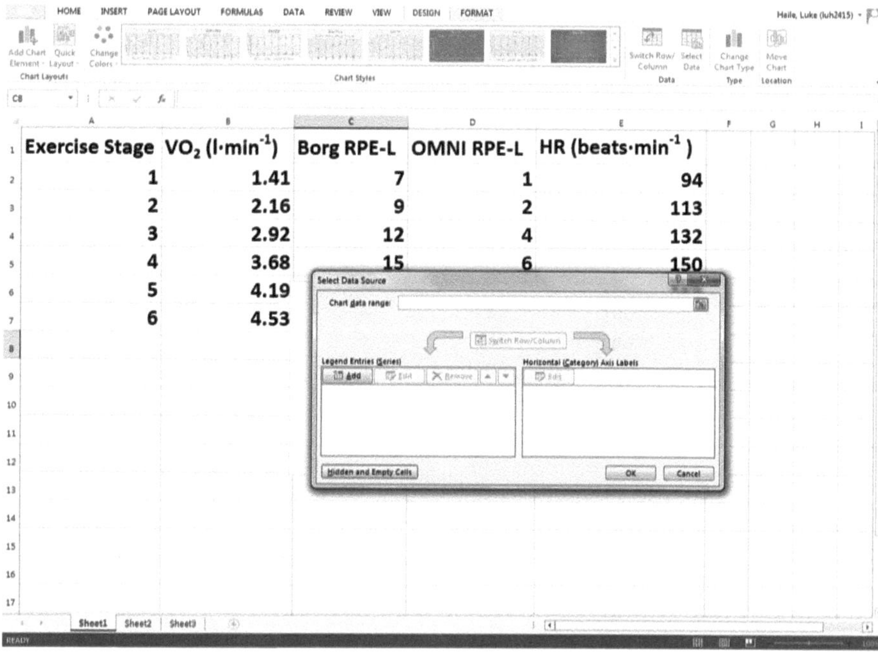

Fig. A.36 ∎

(d) Under **SERIES NAME**, enter VO_2 and Borg RPE-L. Then click on the icon to the right of the **SERIES X VALUES** text box and highlight the VO_2 values. After the values are highlighted click the icon on the box that appeared (Fig. A.37).

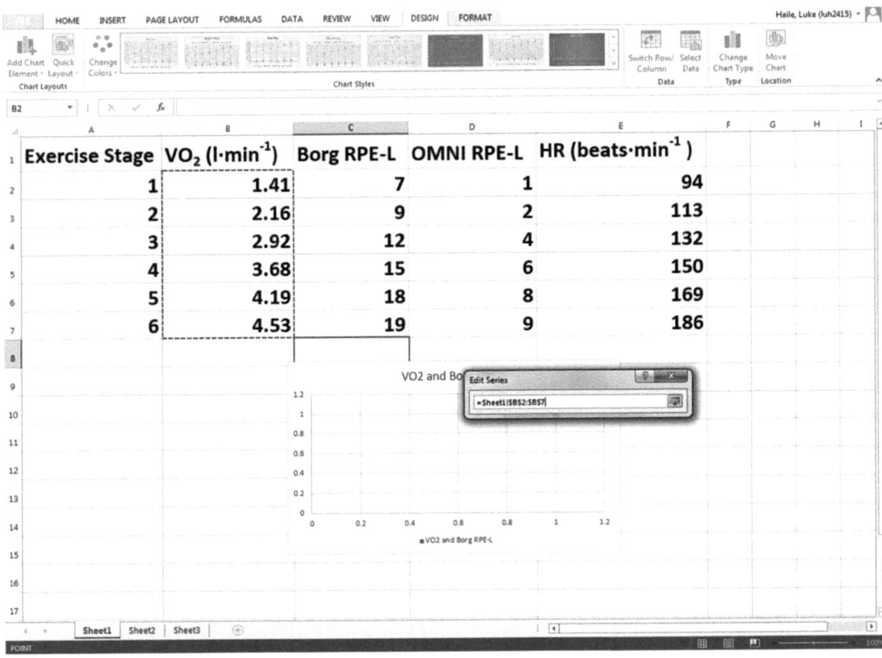

Fig. A.37

(e) Then click on the icon to the right of the **SERIES Y VALUES** text box and highlight the Borg RPE-L values. After the values are highlighted click the icon on the box that appeared (Fig. A.38).

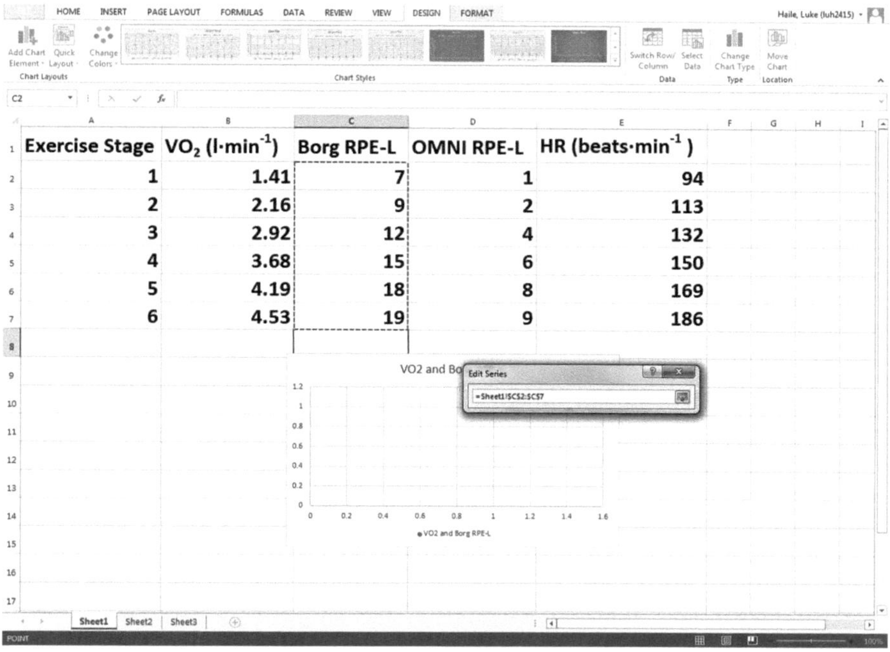

Fig. A.38 ∎

(f) Click **OK** on the next two screens. You should now have a scatter plot with Borg RPE-L on the *y*-axis and VO_2 on the *x*-axis (Fig. A.39).

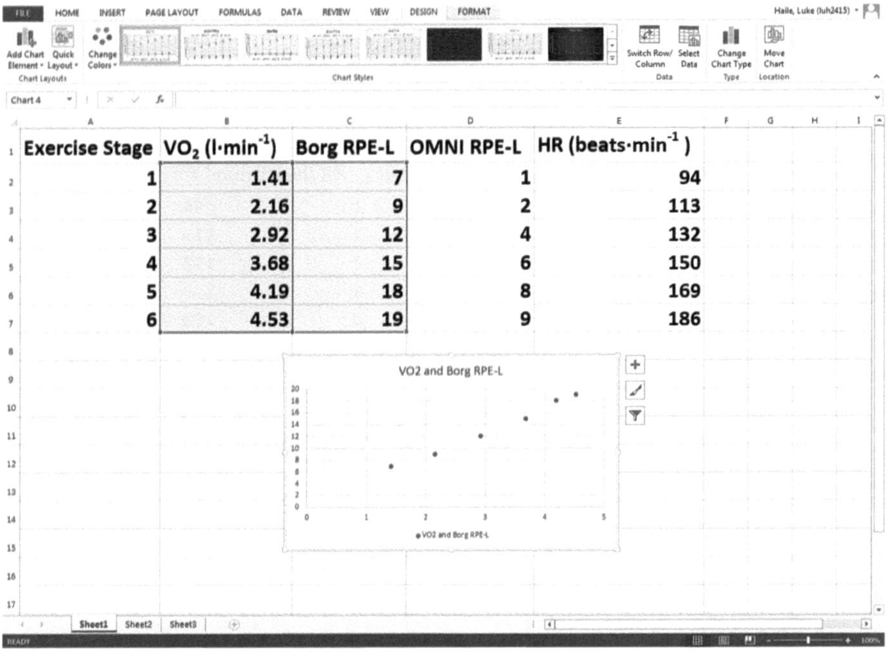

Fig. A.39 ∎

(g) To determine the Borg RPE-VT, click on one of the data points to highlight the entire data series. Right click on one of the data points and a menu will appear. Click **ADD TRENDLINE** and the **FORMAT TRENDLINE** menu will appear. Select **LINEAR** and **DISPLAY EQUATION ON CHART** then click **CLOSE**. The trendline and its linear equation will be displayed on the chart. Use this linear equation to calculate RPE-VT. Use VO_2 ($l \cdot min^{-1}$) corresponding to the VT as the "x" value in the equation and solve for "y." The calculated "y" value is the Borg RPE-VT (Fig. A.40).

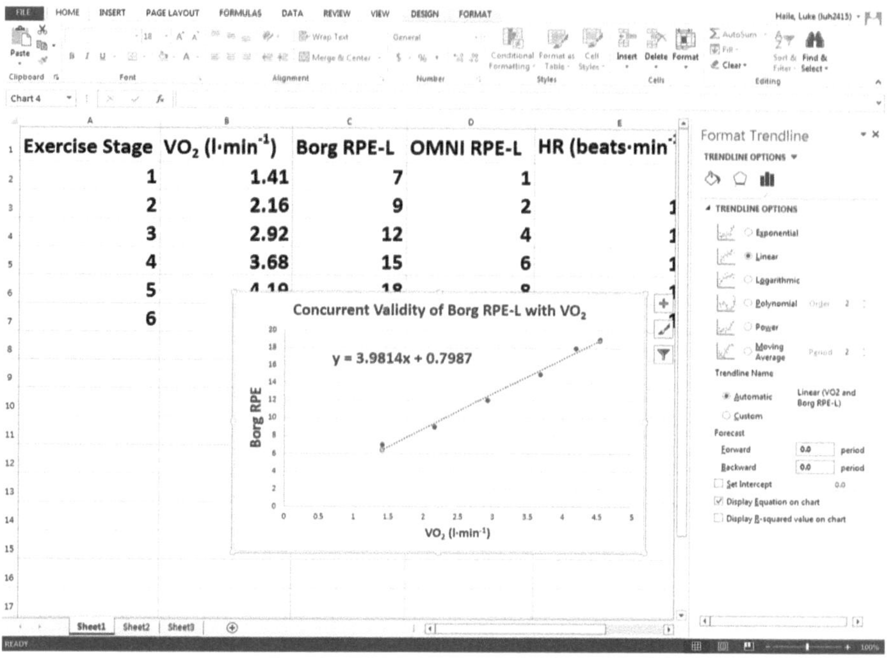

Fig. A.40 ∎

Appendix E
Prediction of VO_2peak: An Example Using Cycle Ergometry Graded Exercise Test Results

1. In a Microsoft Excel spreadsheet, label columns of data as shown in Fig. A.41.
   (a) For treadmill exercise, VO_2 may be expressed in $ml \cdot kg \cdot min^{-1}$.
   (b) RPE-O and differentiated RPE such as RPE-C may be used in your experiment.

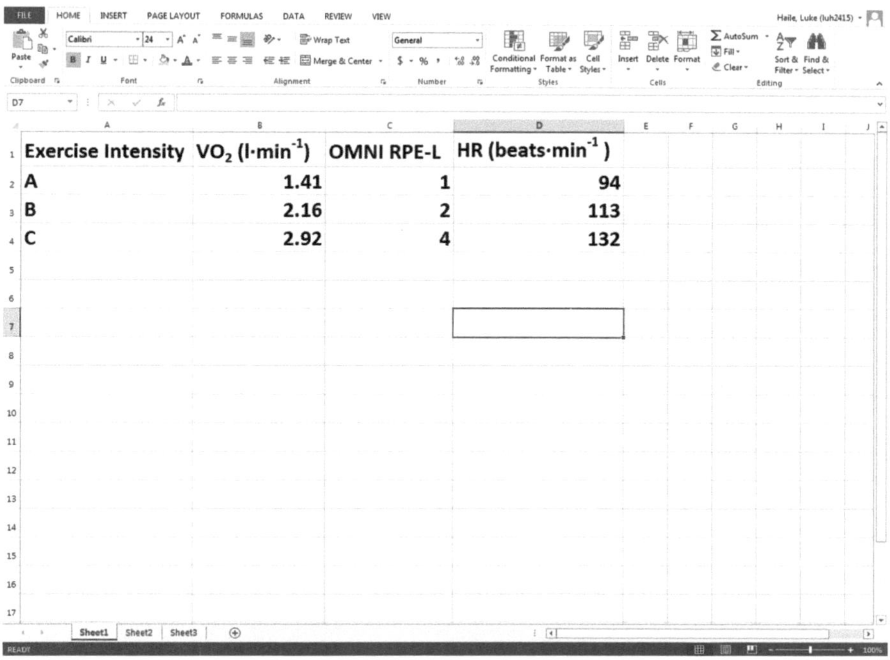

Exercise Intensity	VO_2 ($l \cdot min^{-1}$)	OMNI RPE-L	HR ($beats \cdot min^{-1}$)
A	1.41	1	94
B	2.16	2	113
C	2.92	4	132

Fig. A.41 ■

L. Haile et al., *Perceived Exertion Laboratory Manual*,
DOI 10.1007/978-1-4939-1917-8

2. Plot of VO_2 and OMNI RPE-L for prediction of VO_2peak.

   (a) Click on the **INSERT** tab and in the **CHARTS** section click on **SCATTER**. Select the first available chart option. A blank or example scatter plot will appear on your screen (Fig. A.42).

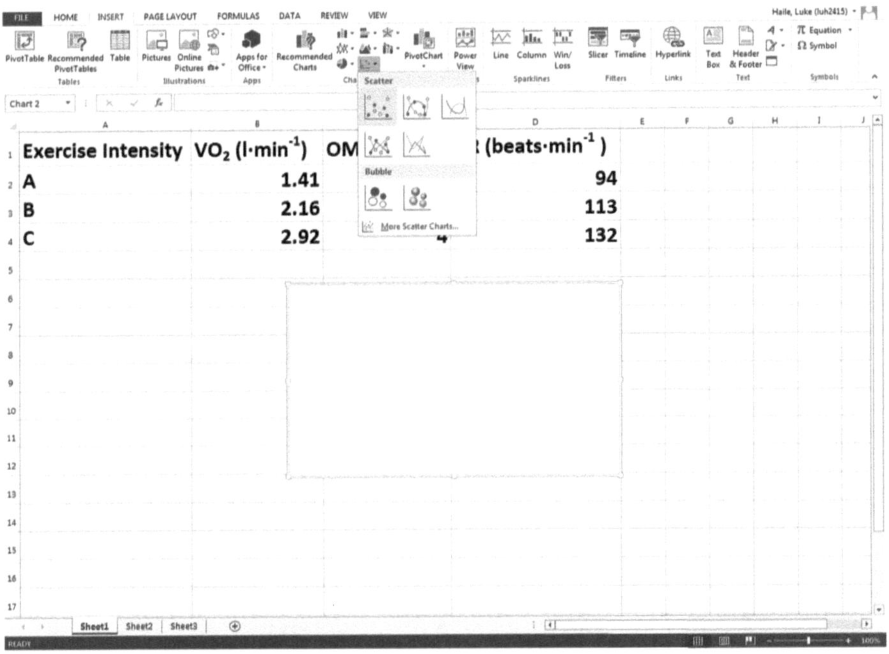

Fig. A.42

(b) Click on the **SELECT DATA** tab (Fig. A.43).

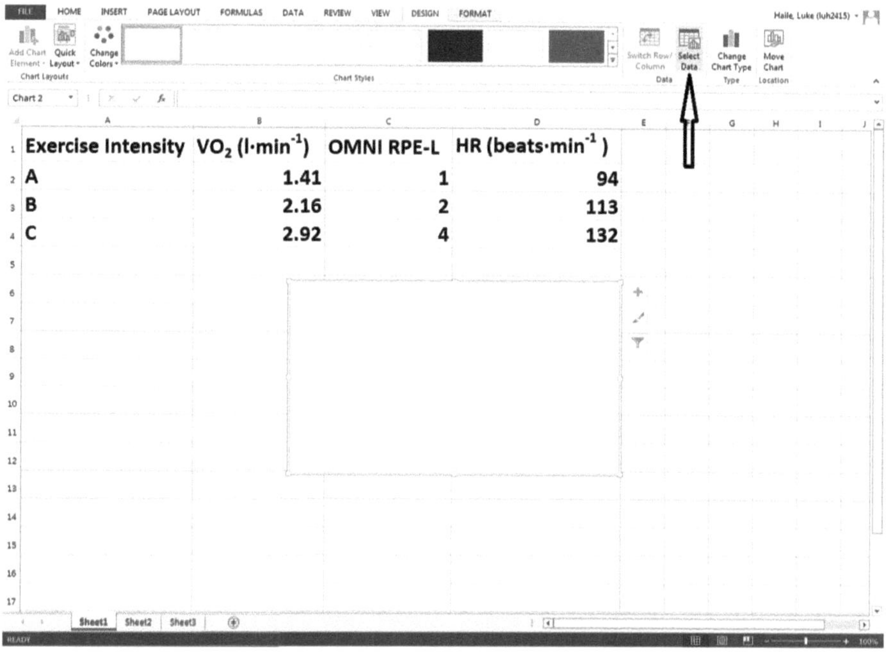

Fig. A.43 ∎

(c) Remove any entries found in the **LEGEND ENTRIES** text box then click **ADD** (Fig. A.44).

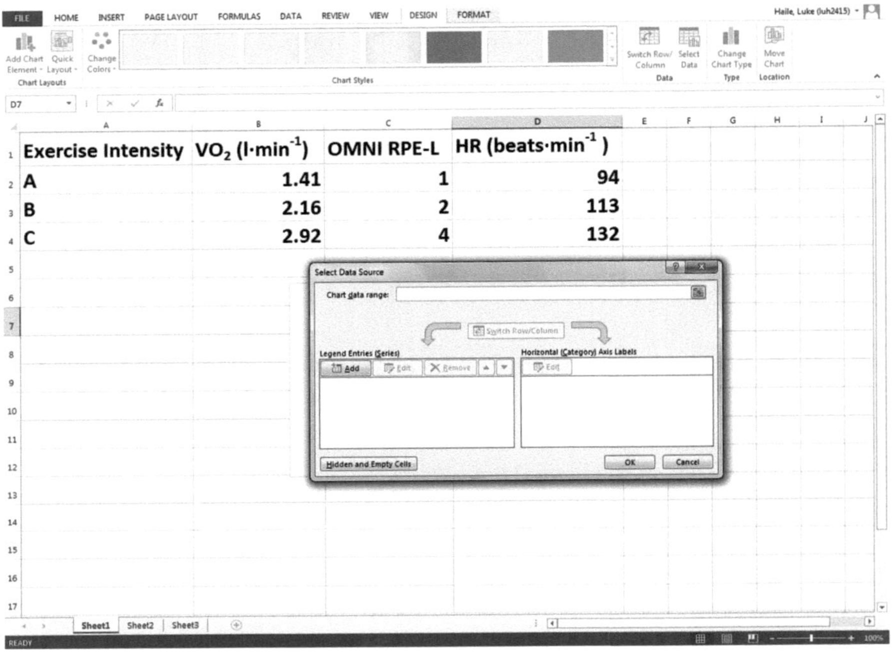

Fig. A.44 ∎

(d) Under **SERIES NAME**, enter VO_2 and OMNI RPE-L. Then click on the icon to the right of the **SERIES X VALUES** text box and highlight the OMNI RPE-L values. After the values are highlighted click the icon on the box that appeared (Fig. A.45).

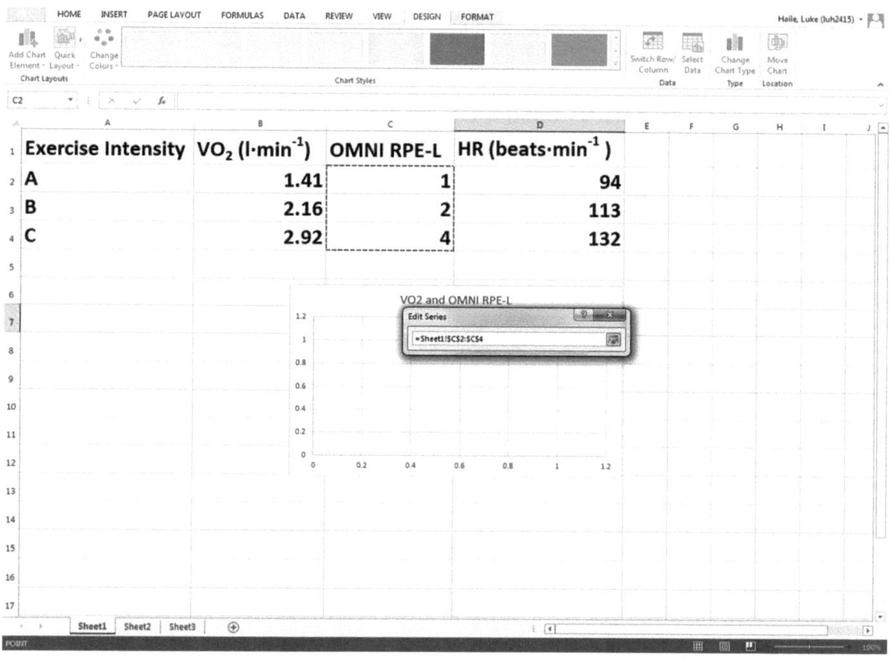

Fig. A.45 ∎

(e) Then click on the icon to the right of the **SERIES Y VALUES** text box and highlight the VO_2 values. After the values are highlighted click the icon on the box that appeared (Fig. A.46).

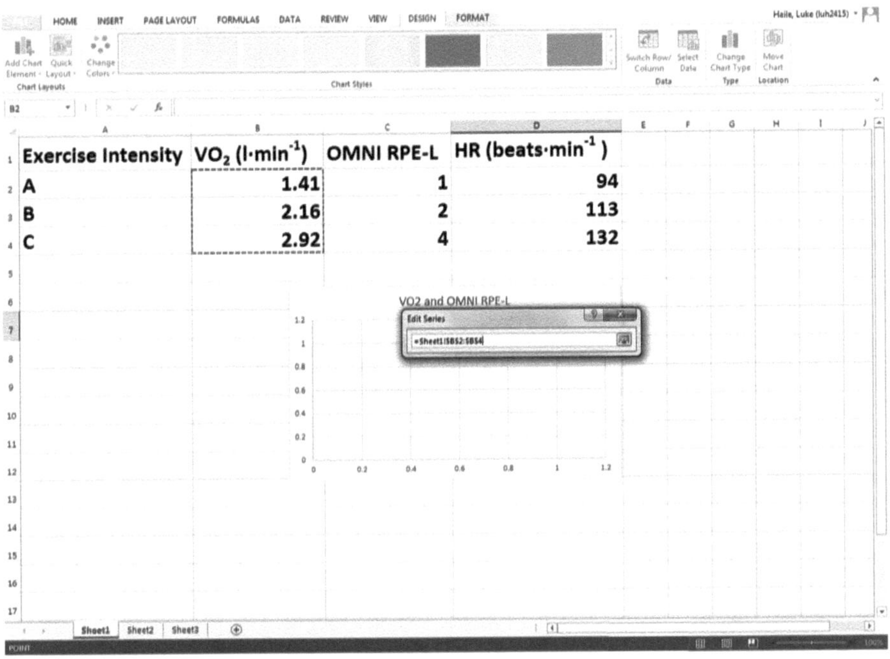

Fig. A.46

(f) Click **OK** on the next two screens. You should now have a scatter plot with VO_2 on the *y*-axis and OMNI RPE-L on the *x*-axis (Fig. A.47).

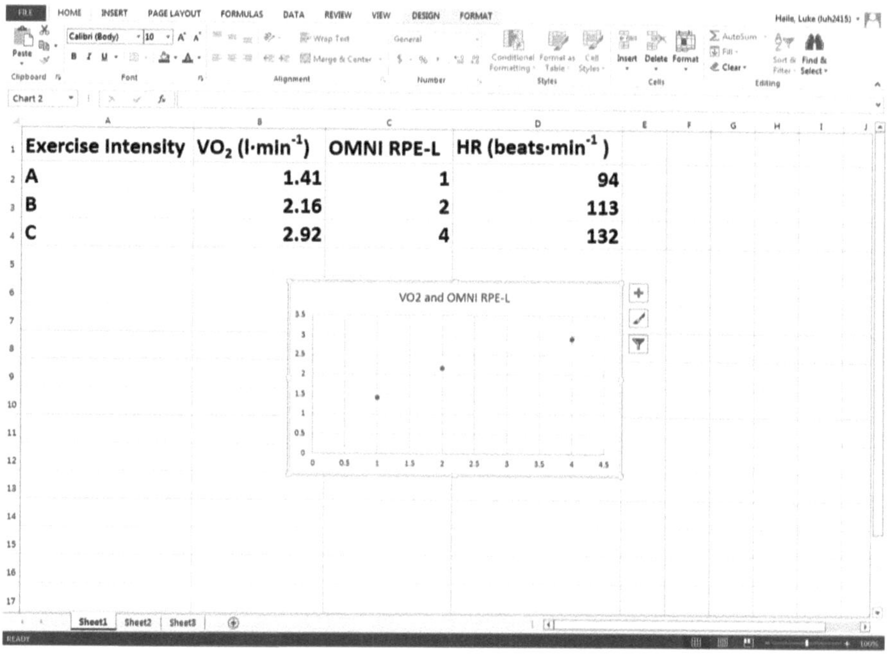

Fig. A.47 ∎

(g) To determine the equation from which VO_2peak will be predicted, click on one of the data points to highlight the entire data series. Right click on one of the data points and a menu will appear. Click **ADD TRENDLINE** and the **FORMAT TRENDLINE** menu will appear. Select **LINEAR** and **DISPLAY EQUATION ON CHART** then click **CLOSE**. The trendline and its linear equation will be displayed on the chart. Use this linear equation to calculate RPE-VT. Use VO_2 ($l \cdot min^{-1}$) corresponding to the VT as the "x" value in the equation and solve for "y." The calculated "y" value is the Borg RPE-VT (Fig. A.48).

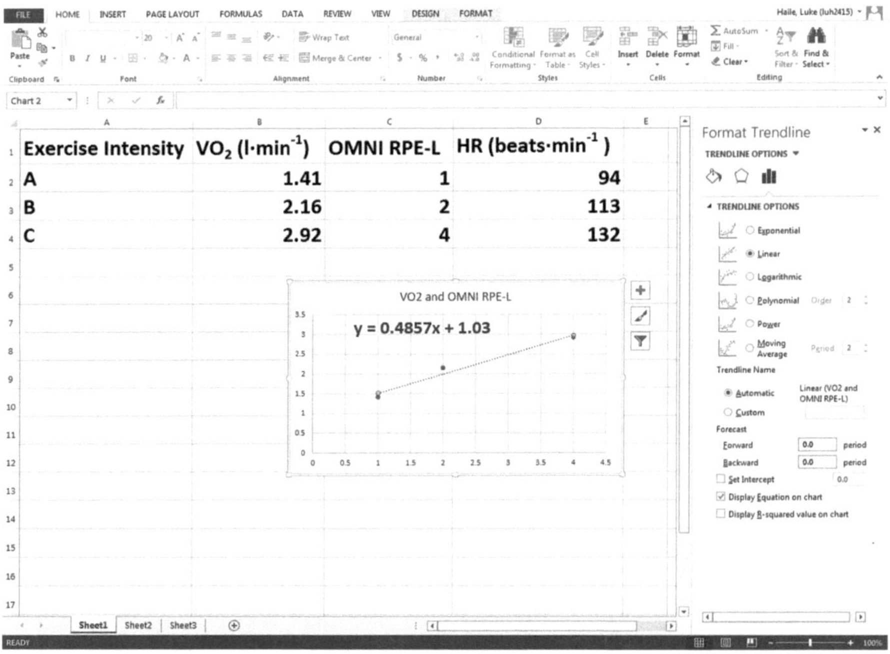

Fig. A.48 ∎

Appendix F
Advanced Perceived Exertion Scaling Procedure for Use Prior to an RPE-Based, Interval Exercise Program

For cycle exercise, the low intensity bout begins with unloaded pedaling and subsequently increases in power output every 15 or 30 s. The client provides RPE just prior to each power output increase until moderate intensity is reached, identified by a rating of 5–6 on the OMNI Scale or 11–13 on the Borg Scale. If the client shows physiological signs of moderate to high intensity exercise, such as increased rate of breathing or heart rate, yet continues to rate exertion levels as low, further discussion about the link between RPE and physiological intensity may be necessary. Then, to further test the subject's understanding, he/she is asked to produce a specific level of perceived exertion on the cycle. The test administrator decides on a specific target RPE, from two to four on the OMNI Scale, and then instructs the client to adjust the power output until the intensity of cycling produces an exertion level equal to that specific RPE. The subject should be allowed to adjust the intensity for 1 or 2 min before the power output is checked for accuracy. If the intensity chosen does not match the intensity at which the client previously rated that particular RPE, additional practice and feedback may be necessary. However, if the intensity chosen matches the intensity at which the client previously reported that particular RPE, provide him/her with positive reinforcement and continue with the next phase.

For the moderate intensity phase, begin the load-incremented bout at the low intensity that the client previously produced and increase power output every 15 or 30 s. The client provides RPE just prior to each power output increase until high intensity is reached, identified by a rating of 8 on the OMNI Scale or 17–18 on the Borg Scale. If the client terminates exercise due to fatigue yet rates exertion levels as moderate, further discussion about RPE and maximal physiological intensity may be necessary. For some clients, especially those who may not have performed maximal exercise previously, the experience of maximal exercise facilitates their ability to rate exertion levels at moderate intensities of exercise. For the production bout at moderate intensity, RPE's from 5 to 7 on the OMNI Scale are appropriate. This bout is important to determine if additional practice and feedback may be necessary to rate exertion accurately at moderate exercise intensity.

L. Haile et al., *Perceived Exertion Laboratory Manual*,
DOI 10.1007/978-1-4939-1917-8

For the high/maximal intensity phase, begin the load-incremented bout at the intensity the client previously produced in the moderate phase and subsequently increase power output every 15 or 30 s until volitional termination owing to fatigue. Similar to the standard exercise anchoring procedure, instruct the subject to assign a maximal RPE value (10 on the OMNI Scale, 20 on the Borg Scale) to that intensity. Minimal rest (approximately 2 min) is necessary between phases and between load-incremented and production bouts within phases. However, ample rest sufficient for complete or near complete recovery is advised between advanced exercise anchoring and subsequent administration of aerobic fitness testing. For unfit and/or sedentary individuals, advanced exercise anchoring and aerobic fitness testing may need to be performed on separate days.

Glossary

Cardiorespiratory fitness The ability to perform dynamic exercise of a moderate to vigorous intensity using large muscle groups for a prolonged period; this fitness measurement is dependent upon the functional capacity of the cardiovascular and respiratory systems and the oxidative capacity of skeletal muscle.

Differentiated RPE RPE used to estimate the level of exertion for a specific anatomical region of the body, such as the chest/breathing (RPE-C), arms (RPE-A), legs (RPE-L), or active muscle mass (RPE-AM).

Estimation–production paradigm A perceptually based exercise prescription procedure whereby both the estimation and production test protocols are used to prescribe and self-regulate exercise intensity according to a target RPE or RPE zone.

Estimation protocol A research procedure used in perceived exertion scale validation studies involving a graded exercise test during which RPE and physiological responses are measured for each progressive exercise test stage, with intensity ranging from very low through maximal.

Exercise anchoring A procedure whereby the individual links the perception of exertion experienced while actually performing a very low exercise intensity and when performing a very high exercise intensity with the low and high anchor points on the perceived exertion scale, respectively.

Exercise intensity self-regulation error When an individual is not accurately self-regulating exercise intensity at a target RPE such that the physiological responses (VO_2, HR) during the production trial are different from those that were observed at the same target RPE during the estimation trial.

Exertional recall An estimate of the RPE for a bout of exercise or physical activity performed at least 1 week prior; may be included as part of a physical activity questionnaire.

Group-normalized perceptual response A range of RPE's that corresponds to a target physiological outcome during exercise and that is common to a specified group of individuals.

L. Haile et al., *Perceived Exertion Laboratory Manual*,
DOI 10.1007/978-1-4939-1917-8

Imposed exercise intensity When an individual performs a prescribed exercise intensity based on physical units (W), ergometer settings (speed/grade, intensity settings) or physiological measures (HR, VO_2) which has been determined by the health-fitness professional or exercise test administrator.

Intensity discrimination The ability to perceptually differentiate between separate target RPE's such that physiological responses are different between different self-regulated conditions.

Just noticeable difference in perceived exertion (perceived exertion JND) The smallest amount of change in a stimulus (exercise intensity), expressed in physiological (VO_2) or physical (PO) units, necessary to elicit a change in sensation (perception of physical exertion).

Maximal Aerobic Power (VO_2max/peak) The maximum amount of oxygen that can be consumed while breathing ambient air during load-incremented aerobic exercise at sea level; the terms maximal aerobic power and maximal or peak oxygen uptake (VO_2max/peak) can be used interchangeably.

Memory anchoring A procedure used to acquaint the user with the level of exertion perceived at the low and high anchor points of a category RPE scale.

Momentary RPE The acute level of perceived exertion rated at the moment the individual is asked during exercise or PA; also referred to as in-task or on-stimulus RPE.

Muscular strength, dynamic The ability of a muscle or muscle group to exert force using concentric or eccentric muscular action resulting in the movement of a resistance.

One-repetition maximum (1RM) The maximal amount of force that can be produced during a single contraction of a muscle or muscle group through the full range of motion.

Pacing Strategy The self-selected exercise pace or tactic that an athlete adopts, usually at the beginning of an event or competition, to ensure optimal metabolic requirements and performance outcomes.

Perceived exertion The subjective intensity of effort, strain, discomfort, and/or fatigue that is experienced during exercise and physical activity.

Perceptual augmenter A perceptual outlier who provides RPE values greater than what is appropriate based on a given physiological and/or physical marker of exercise intensity and may assign a maximal or near-maximal RPE to submaximal exercise intensity.

Perceptual outlier An individual who provides inappropriate RPE values that do not conform to the predictions of Borg's Range Model.

Perceptual reducer A perceptual outlier who provides RPE values less than what is appropriate based on a given physiological and/or physical marker of exercise intensity and may assign a submaximal RPE to maximal exercise intensity.

Prescription congruence When physiological responses (VO_2, HR) corresponding to a specific target RPE are similar between the estimation trial and production trial at a given submaximal intensity.

Predicted RPE A global estimate of the expected RPE for an entire bout of exercise or PA rated prior to performance of that activity.

Production protocol An exercise bout during which an individual self-regulates exercise intensity to produce a specific target RPE.

RPE warning zone A range of RPE's that indicate impending graded exercise test termination and, as such, the initiation of preliminary procedures to safely end the exercise test.

Segmented session RPE A global estimate of the average RPE experienced for a specific segment (time-period) of a bout of exercise or PA but rated after performance of that activity.

Self-selected exercise intensity When an individual performs exercise at a preferred intensity during which self-adjustment of ergometer settings are allowed.

Session RPE A global estimate of the average RPE experienced for an entire bout of exercise or PA but rated after performance of that activity.

Target RPE or target RPE range One RPE or a range of RPE's that indicate the level(s) of exertion to be achieved by self-regulating exercise intensity using a production perceptual protocol.

Teleoanticipation A sensory nervous system comprised of both feed-forward and feedback perceptual-cognitive information regarding muscle fiber recruitment and firing frequency during exercise performance; in this system the magnitude and frequency of efferent (i.e., descending) motor signals associated with previous exercise performance are stored in the sensory cortex; this information is further augmented by afferent signals reflecting the metabolic and biomechanical limits of muscular performance; subsequently, the stored perceptual-cognitive information is recalled to shape the upper limits of exercise performance as set by peak tolerable perceptual limits of heavy muscular exercise.

Undifferentiated RPE RPE used to estimate the level of exertion for the overall body, often referred to as RPE-O.

Validity The degree to which a test or test item measures the construct it is intended to measure.

Validity, concurrent (general definition) The extent to which test scores are associated with those of other accepted tests when both measures are obtained along a common stimulus range.

Validity, concurrent, of a perceived exertion scale The extent to which RPE are associated with accepted physical and physiological markers of exercise intensity across an individual's full physiological range.

Validity, construct (general definition) The ability of a test to represent the underlying construct.

Validity, construct, of a perceived exertion scale The extent to which RPE from a newly developed perceived exertion scale are associated with RPE derived from a perceived exertion scale for which concurrent validity has been previously established.

Ventilatory threshold (VT) Also known as the ventilatory breakpoint, can be defined as the point during exercise of increasing intensity when pulmonary ventilation begins to increase at a rate disproportionately faster than that of oxygen consumption; the respiratory analog to the lactate threshold (both commonly called the anaerobic threshold).

Index

L. Haile et al., *Perceived Exertion Laboratory Manual*,
DOI 10.1007/978-1-4939-1917-8

R

MIX
Papier aus verantwortungsvollen Quellen
Paper from responsible sources
FSC® C105338
FSC
www.fsc.org

If you have any concerns about our products,
you can contact us on
ProductSafety@springernature.com

In case Publisher is established outside the EU,
the EU authorized representative is:
Springer Nature Customer Service Center GmbH
Europaplatz 3, 69115 Heidelberg, Germany

Printed by Libri Plureos GmbH
in Hamburg, Germany